Fixed Point Theory in Modular Function Spaces

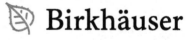
Birkhäuser

Mohamed A. Khamsi • Wojciech M. Kozlowski

Fixed Point Theory in Modular Function Spaces

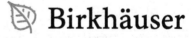 Birkhäuser

Mohamed A. Khamsi
Department of Mathematical Sciences
The University of Texas at El Paso
El Paso, Texas
USA

Department of Mathematics & Statistics
King Fahd University of Petroleum & Minerals
Dhahran
Saudi Arabia

Wojciech M. Kozlowski
School of Mathematics and Statistics
University of New South Wales
Sydney, New South Wales
Australia

ISBN 978-3-319-34635-9   ISBN 978-3-319-14051-3 (eBook)
DOI 10.1007/978-3-319-14051-3

Mathematics Subject Classification (2010): 47H10, 47H09, 46E30, 46B20, 47H20, 47H30

Springer Cham Heidelberg New York Dordrecht London
© Springer International Publishing Switzerland 2015
Softcover reprint of the hardcover 1st edition 2015

Printed on acid-free paper

Springer is part of Springer Science+Business Media (www.springer.com)

Foreword

Because of its many diverse applications, fixed point theory has been a flourishing area of mathematical research for decades. S. Banach's formulation of the contraction mapping principle in the early twentieth century signaled the advent of an intense interest in "metric" related aspects of the theory. The metric theory, often cast in a Banach space framework, has usually involved an intertwining of geometric and topological conditions. Fixed point theory in modular function spaces is closely related to the metric theory, in that it provides modular equivalents of norm and metric concepts. Modular spaces are extensions of the classical Lebesgue and Orlicz spaces, and in many instances conditions cast in this framework are more natural and more easily verified than their metric analogs.

This book is devoted to a comprehensive treatment of what is currently known about fixed point theory in modular function spaces. A unified treatment of this subject was initiated in 1988 with the appearance of W. Kozlowski's Marcel Dekker monograph *Modular Function Spaces*. Since the appearance of Kozlowski's monograph, there have been numerous further developments, both in the theory of modular function spaces and in metric fixed point theory. This book takes full account of these developments. It covers the foundations of the theory, taking as a point of departure a review of the central themes of metric fixed point theory and the basic structures of modular function spaces. Anyone interested in these topics would want to have this book readily at hand. The book is essentially self-contained and should be easily accessible to students. Also it provides a valuable resource for those already involved in this avenue of research.

Iowa City, May, 2014   *W. A. Kirk*

Preface

The single valued fixed point problem, which forms the basis of the fixed point theory, may be stated as :

Let X be a set, A and B two nonempty subsets of X such that $A \cap B \neq \emptyset$, and $f : A \to B$ be a map. When does a point $x \in A$ such that $f(x) = x$, also called a fixed point of f, exist? And if yes, how many such points exist and how to find them?

From the perspective of different settings, methods and applications, the fixed point theory is typically divided into the three major areas:

- Topological fixed point theory
- Metric fixed point theory
- Discrete fixed point theory

Historically, the boundary lines between the three areas were defined by the discovery of three major theorems:

- Brouwer's fixed point theorem [31]
- Banach's fixed point theorem [14]
- Tarski's fixed point theorem [198]

In this book, we will focus mainly on the second area, although from time to time we may touch the other areas as well. Conceptually, the metric fixed point theory deals with situations where the set X is equipped with some kind of a method allowing to assign to every two elements of X a numeric value that measures the level of difference between them; in other words—a distance between them. We need to keep in mind that this idea of a "distance" can take many forms and does have not to be a distance in a popular meaning of the word, or a metric in the strict mathematical sense. Such a situation is very common when dealing with any quantifiable events in science and technology. Many problems like solving equations or finding stationary points of a time dependent system can be actually reduced to finding fixed points of a suitably defined mapping acting within a suitably selected set equipped with a suitably selected "distance."

The authors would like to invite the reader to join them in a journey taking them from a well-known base of classical fixed point theory in Banach and metric spaces (and yes, with Banach's fixed point theorem as the starting point) to the world of the theory of fixed points of mappings defined in a class of spaces of measurable functions, known in the literature as modular function spaces.

The results and methods of fixed point theory, applied to spaces of measurable functions, have been used extensively in the field of integral and differential equations. Since the 1930s many prominent mathematicians like Orlicz and Birnbaum recognized that using the methods of L^p-norms alone created many complications and in some cases did not allow to solve some nonpower type integral equations. They considered spaces of functions with some growth properties different from the power type growth control provided by the L^p-norms. The possibility of introducing the structure of a linear metric as well as the interesting properties of these spaces, later named Orlicz spaces, and many applications to differential and integral equations with kernels of nonpower types were among the reasons for the development of the theory of Orlicz spaces, their applications and generalizations. Using the apparatus of the modular function spaces we can go much further: the operator itself is used for the construction of a function modular and hence of a space in which this operator has required properties. These techniques together with relevant modular function space fixed point theorems can be efficiently applied to solving many mathematical problems. The aim of this book is to familiarize the readers with the main concepts and results of the fixed point theory in modular function spaces, as well as to encourage them to use these results in the course of their research activities.

We would like to thank the University of Texas El Paso and the University of New South Wales for their continuing support during the preparation of this manuscript. We would like to acknowledge also the role of the King Fahd University of Petroleum and Minerals in extending a strong support from the beginning to the end, facilitating every means during the preparation of this book. We also owe a measure of gratitude to the Deanship of Scientific Research, for providing excellent opportunity of academic research activities.

Dhahran, El Paso, Sydney   *M. A. Khamsi*
November 2014   *W. M. Kozlowski*

Contents

Chapter 1
Introduction

Before we introduce the definition and the fundamental properties of modular function spaces, let us discuss a historical background, which will also provide a motivation for the introduction of the methods of modular function spaces into the fixed point theory.

The results and methods of fixed point theory, applied to spaces of measurable functions, have been used extensively in the field of integral equations and integral inequalities. Since the 1930s, many prominent mathematicians like Orlicz and Birnbaum recognized that using the methods of L^p-spaces alone created many complications and in some cases did not allow to solve some nonpower type integral equations, see [29]. They considered spaces of functions with some growth properties different from the power type growth control provided by the L^p-norms. Orlicz and Birnbaum considered for instance, function spaces defined as follows:

$$L^\varphi = \{f : \mathbb{R} \to \mathbb{R}; \ \exists \lambda > 0 : \int_\mathbb{R} \varphi(\lambda |f(x)|) \, dm(x) < \infty\},$$

where $\varphi : [0, \infty) \to [0, \infty)$ was assumed to be a convex function increasing to infinity, that is, the function which to some extent behaves similarly to power functions $\varphi(t) = t^p$. Let us mention two typical examples of such functions: $\varphi_1(t) = e^t - t - 1$ or $\varphi_2(t) = e^{t^2} - 1$. The possibility of introducing the structure of a linear metric in L^φ as well as the interesting properties of these spaces, later named Orlicz spaces, and many applications to differential and integral equations with kernels of nonpower types were among the reasons for the development of the theory of Orlicz spaces, their applications, and generalizations. Consider, for example, the following Hammerstein nonlinear integral equation, which plays an important role in the elasticity theory:

$$f(x) = \int_0^1 k(x, y) \varphi(f(y)) dy,$$

where $\varphi(u)$ is a function which increases more rapidly than an arbitrary power function. Krasnosel'skii and Rutickii, [147], showed that the Hammerstein opera-

© Springer International Publishing Switzerland 2015

M. A. Khamsi, W. M. Kozlowski, *Fixed Point Theory in Modular Function Spaces*,
DOI 10.1007/978-3-319-14051-3_1

tor defined by the right member of this integral equation does not operate in any of the L^p spaces. And yet, they showed how to find an Orlicz space where the Hammerstein operator is well defined and posses properties allowing to use some fixed point theorems for solving the corresponding integral equation.

Using the apparatus of the modular function spaces we can go much further: the operator itself is used for the construction of a function modular and hence of a space in which this operator has required properties. This technique together with a relevant modular function space fixed point theorem can be applied to solving integral equations like Urysohn integral equation.

An additional importance for applications of modular function spaces consists in the richness of structure of modular function spaces that, besides being Banach spaces (or F-spaces in a more general settings), are equipped with modular equivalents of norm or metric notions, and also are equipped with almost convergence everywhere and convergence in submeasure. In many situations in integral equations, approximation and fixed point theory, modular type conditions are much more natural, and modular type assumptions can be more easily verified than their metric or norm counterparts. There are also important results that can be proved only using the apparatus of modular function spaces. From this perspective, the fixed point theory in modular function spaces should be considered as complementary to the fixed point theory in normed and metric spaces.

The theory of contractions and nonexpansive mappings defined on convex subsets of Banach spaces has been well-developed since the 1960s (see, for example, [32, 58, 79, 82, 85, 120, 122]), and generalized to metric spaces (see, for example, [13, 84, 111]), and modular function spaces (see, for example, [109, 116, 117]). The corresponding fixed point results were then extended to larger classes of mappings like asymptotic mappings [112, 125], pointwise contractions [121] and asymptotic pointwise contractions, and nonexpansive mappings [96, 114, 115, 127, 128].

While Banach contraction principle and its generalizations usually provide a fixed point construction method as limits of orbits, the proofs of the fixed point existence results for nonexpansive mappings and their generalizations are of the existential nature, and do not describe any algorithms for constructing fixed points of an asymptotic pointwise nonexpansive mapping. It is well known that the fixed point construction iteration processes for generalized nonexpansive mappings have been successfully used to develop efficient and powerful numerical methods for solving various nonlinear equations and variational problems, often of great importance for applications in various areas of pure and applied science. In [133, 138], the author proved convergence to fixed points of some iterative algorithms applied to asymptotic pointwise nonexpansive mappings in Banach spaces and the existence of common fixed points of semigroups of pointwise Lipschitzian mappings in Banach spaces [134]. Recently, the weak and strong convergence of such processes to common fixed points of semigroups of mappings in Banach spaces was demonstrated in

[137, 144]. The pioneering work on the convergence of Mann and Ishikawa algorithms in modular function spaces can be found in [54]. Our approach is based on this paper. We would like to emphasize that all convergence theorems presented in this book, define constructive algorithms that can be actually implemented. When dealing with specific applications of these theorems, one should take into consideration how additional properties of the mappings, sets, and modulars involved, can influence the actual implementation of the algorithms defined in this book.

The existence of common fixed points for families of contractions and nonexpansive mappings in Banach spaces have been investigated since the early 1960s, see, for example, [19, 20, 32, 35, 57, 153]. The asymptotic approach for finding common fixed points of semigroups of Lipschitzian (but not pointwise Lipschitzian) mappings has been also investigated for some time, see, for example, [199]. It is worthwhile mentioning the recent studies on the special case, when the parameter set for the semigroup is equal to $\{0, 1, 2, 3, ...\}$ and $T_n = T^n$, the n-th iterate of an asymptotic pointwise nonexpansive mapping, that is, such a $T : C \to C$ that there exists a sequence of functions $\alpha_n : C \to [0, \infty)$ with $\|T^n(x) - T^n(y)\| \leq \alpha_n(x)\|x - y\|$. In [128], the authors proved the existence of fixed points for asymptotic pointwise contractions and asymptotic pointwise nonexpansive mappings in Banach spaces, while the paper [96] extended this result to metric spaces, and the papers [114, 115] to the case of modular function spaces. The existence of common fixed points for general semigroups of mappings acting in modular function spaces was proved in [135].

The book is structured as follows:

Chapter 2 provides an introduction to the general metric fixed point theory. The aim of this chapter is to use the classical fixed point results for contractions and nonexpansive mappings acting in metric or normed spaces, to set the scene for developing the fixed point theory in modular function spaces. The reader will be able to see many similarities but at the same time will be able to appreciate significant differences. Readers familiar with the classical theory may skip reading Chap. 2 and jump straight to Chap. 3, while the beginners can actually use Chap. 2 as a "cheat sheet" summarizing key results of the fixed point theory in metric spaces.

Chapter 3 introduces the foundation of the theory of modular function spaces defined by convex regular function modulars. The results discussed in this chapter are used throughout the rest of the book. This chapter also presents the reader with an exhaustive list of examples which frequently reoccur in later parts of the book. Section 3.4 "Generalizations and Special Cases" summarizes a more general approach which allows nonconvex function modulars and extends the setting to spaces of function with values in Banach spaces. While these results are not necessary for understanding the contents of the remaining chapters, the general framework introduced in this section may help to appreciate the full potential of the theory of modular function spaces. The readers who are interested in an active research in this area, can use this chapter for possible generalizations of the results presented in the book and for setting new research directions.

In Chap. 4 we discuss some fundamental notions of the geometry of modular functions spaces. Hence, we introduce several types of modular uniform convexity, as well as modular versions of Opial and Kadec—Klee properties. We will direct readers' attention to the fact that even though these properties formally resemble their Banach space counterparts, they are different and frequently they are satisfied, while their norm equivalents do not. Similarly, as in the fixed point theory in Banach spaces, these properties will be extensively used in proofs of our fixed point theorems.

Chapter 5 is probably the most important part of this book as it presents a long list of fixed point existence theorems for nonlinear mappings acting in modular functions spaces. We cover a range of different types of mappings including ρ-contractions and their pointwise asymptotic versions and ρ-nonexpansive mappings and pointwise asymptotic ρ-nonexpansive mappings, under various assumptions on the function modular ρ and on the sets these mappings are defined in. In addition some examples and applications to the theory of differential equations are provided. In this chapter, like in the whole book, we deal only with the fixed point theory for single-valued mappings acting in modular function spaces. At the time of this publication, very little was known about the set-valued case, which definitely creates an interesting direction for further research.

Chapter 6 is devoted to the fixed point construction processes and their convergence. These results form an important sequel to the existential results of the previous chapter since proving existence without any means to approximate fixed points would be of limited use for applications. This area of research is in its early stages and there are many untouched topics like implicit iterative methods or possible generalizations of the Halpern type method to the case of modular function spaces.

Chapter 7 discusses at length, the question of existence and construction of common fixed points of semigroups of nonlinear mappings in modular functions spaces. This chapter also demonstrates how solutions to some initial value problems, defined by nonlinear differential equations, form semigroups of ρ-nonexpansive mappings in modular functions spaces and interpret the meaning of common fixed points in terms of stationary points of corresponding processes. The structure of the set of common fixed points is also investigated in this chapter.

The final chapter, Chap. 8, provides a sneak preview into the fixed point theory for nonlinear mappings in modular metric spaces, which in essence is an attempt to generalize the modular function space theory to the nonlinear case, that is to remove an assumption that the base space is a vector space. These results are quite recent and they open a wide field for further research.

Chapter 2
Fixed Point Theory in Metric Spaces: An Introduction

This chapter is primarily intended to serve as an introduction to metric fixed point theory. It will set the foundation for the coming chapters. In terms of content this chapter overlaps in places with the following popular books on fixed point theory by A. Aksoy and M. A. Khamsi [5], by K. Goebel and W. A. Kirk [81], by J. Dugundji and A. Granas, [67], by M. A. Khamsi and W. A. Kirk [111], by I.A. Rus, A. Petrusel, and G. Petrusel [186], and by E. Zeidler [203]. Material on the general theory of Banach space geometry is drawn from many sources but the following books by B. Beauzamy [16], and by J. Diestel [59] are worth a special mention.

2.1 Banach Contraction Principle

In 1922, Banach [14] published his fixed point theorem, also known as *the Banach Contraction Principle*, which uses the concept of Lipschitz mappings .

Definition 2.1. Let (M,d) be a metric space. The map $T : M \to M$ is said to be *Lipschitzian* if there exists a constant $k > 0$ (called *Lipschitz constant*) such that

$$d\Big(T(x), T(y)\Big) \leq k\, d(x,y)$$

for all $x, y \in M$. A Lipschitzian mapping with a Lipschitz constant $k < 1$, is called *contraction* . Moreover if the Lipschitz constant $k = 1$, the Lipschitzian mapping is called *nonexpansive* . Observe that all contractions and nonexpansive mappings are uniformly continuous.

Theorem 2.1 (Banach Contraction Principle). *Let (M,d) be a complete metric space and let $T : M \to M$ be a contraction mapping, with Lipschitz constant $k < 1$. Then T has a unique fixed point ω in M, and for each $x \in M$, we have*

$$\lim_{n \to \infty} T^n(x) = \omega.$$

© Springer International Publishing Switzerland 2015
M. A. Khamsi, W. M. Kozlowski, *Fixed Point Theory in Modular Function Spaces*,
DOI 10.1007/978-3-319-14051-3_2

Moreover, for each $x \in M$, we have

$$d(T^n(x), \omega) \leq \frac{k^n}{1-k} d(T(x), x).$$

Proof. Since T is a contraction mapping we know that for each $x \in M$ and $n \geq 1$,

$$d(T^n(x), T^{n+1}(x)) \leq k d(T^{n-1}(x), T^n(x)) \leq k^n d(x, T(x)).$$

Let $n, m \in \mathbb{N}$, with $n \geq 1$, we get

$$d(T^n(x), T^{n+m}(x)) \leq \sum_{i=n}^{n+m} d(T^i(x), T^{i+1}(x)) \leq \sum_{i=n}^{n+m} k^i d(x, T(x)),$$

for any $x \in M$. Since $k < 1$, we get

$$\sum_{i=n}^{n+m} k^i = k^n \frac{1 - k^{m+1}}{1-k} \leq \frac{k^n}{1-k}.$$

Hence

$$d(T^n(x), T^{n+m}(x)) \leq \frac{k^n}{1-k} d(x, T(x)). \tag{2.1}$$

This forces $\{T^n(x)\}$ to be a Cauchy sequence. Since M is complete, then $\lim_{n \to \infty} T^n(x) = \omega_x$ exists in M. Note that a priori the limit point depends on the starting point x. Since T is continuous

$$\omega_x = \lim_{n \to \infty} T^n(x) = \lim_{n \to \infty} T^{n+1}(x) = T(\omega_x).$$

Thus ω_x is a fixed point of T, for any $x \in M$. Next we show that ω_x is independent of the starting point x. Let $y \in M$ and ω_y be the fixed point of T associated to y. Then

$$d(\omega_x, \omega_y) = d(T(\omega_x), T(\omega_y)) \leq k d(\omega_x, \omega_y).$$

Since $k < 1$, we get $d(\omega_x, \omega_y) = 0$, i.e., $\omega_x = \omega_y$. This shows that T has one fixed point ω and for any $x \in M$, we have $\lim_{n \to \infty} T^n(x) = \omega$. If we let $m \to \infty$ in the inequality (2.1), we get

$$d(T^n(x), \omega) \leq \frac{k^n}{1-k} d(x, T(x)),$$

for any $x \in M$ and $n \in \mathbb{N}$. This completes the proof of Theorem 2.1.
□

An easy implication of the Banach Contraction Principle is the following theorem.

Theorem 2.2. *Let (M, d) be a complete metric space. Suppose $T : M \to M$ is a mapping for which T^N is a contraction mapping, for some positive integer $N \geq 1$. Then T has a unique fixed point.*

Proof. Since T^N is a contraction, Theorem 2.1 implies the existence of a unique fixed point x_0 of T^N. Since

$$T^N(T(x_0)) = T^{N+1}(x_0) = T(T^N(x_0)) = T(x_0),$$

then $T(x_0)$ is also a fixed point of T^N. Since T^N has only one fixed point, we conclude that $T(x_0) = x_0$.

$\square$

It is not clear in general whether T has a fixed point whenever T^N has a fixed point. Note that fixed points of T^N are also known as periodic points of T.

Remark 2.1. We all have learned that the origins of the metric contraction principle and, ergo, metric fixed point theory itself, rest in the method of successive approximations for proving existence and uniqueness of solutions of differential equations. This method is associated with the names of such celebrated nineteenth century mathematicians as Cauchy, Liouville, Lipschitz, Peano, and specially Picard. In fact the iterative process used in the proof of the Banach Contraction Theorem bears the name of Picard iterates. It is quite interesting to know that in 1429, Al-Kashani [23] published a book entitled: "The Calculator's Key", where he used Picard iterates. In fact, Al-Kashani set the stage for the so-called numerical techniques some 600 years ago. He was keen to develop ideas with practical matters, like the approximate values of $\sin(1^o)$ which enabled scientists after him to come up with very good approximations to the circumference of the Earth.

2.2 Pointwise Lipschitzian Mappings

The notion of pointwise Lipschitzian mappings was introduced in [125, 127, 128]. The main motivation behind this generalization is to find a larger class of mappings for which the conclusion of Banach's celebrated contraction mapping theorem is still valid. Pointwise Lipschitzian mappings are defined as follows.

Definition 2.2. Let (M,d) be a metric space. A mapping $T : M \to M$ is called a pointwise Lipschitzian mapping if there exists a mapping $\alpha : M \to [0, +\infty)$ such that

$$d(T(x), T(y)) \le \alpha(x)\, d(x, y),$$

for any $x, y \in M$. The map T is called pointwise contraction if $\alpha(x) < 1$, for any $x \in M$.

Note that if T is a pointwise contraction, then it is continuous although not necessarily uniformly continuous. Moreover, if $\alpha(x) = 0$ for some $x \in M$, then T is a constant map. It is easy to prove that a pointwise contraction $T : M \to M$ has at most one fixed point, and if x_0 is its fixed point, then the orbit $\{T^n(x)\}$ converges to x_0, for each $x \in M$. Indeed, we have

$$d(x_0, T^n(x)) \le \alpha(x_0)^n d(x_0, x)$$

for any $x \in M$. The above conclusion follows because $\alpha(x_0) < 1$. It is not clear how to prove the existence of the fixed point from the convergence of the orbits which is the case in the classical proof given to the Banach Contraction Principle, Theorem 2.1.

In order to extend the main conclusion of [128] to metric spaces, the authors in [96] needed the following definition.

Definition 2.3. Let (M,d) be a metric space. A subset C of M is *admissible* if it is a nonempty intersection of closed balls. The family of all admissible subsets of M is denoted by $\mathscr{A}(M)$. We will say that $\mathscr{A}(M)$ is compact if any family $(A_\alpha)_{\alpha \in \Gamma}$ of elements of $\mathscr{A}(M)$, has a nonempty intersection provided $\bigcap_{\alpha \in F} A_\alpha \neq \emptyset$ for any finite subset $F \subset \Gamma$. Moreover for any nonempty subset A of M, the cover of A is defined by

$$\mathrm{cov}(A) = \bigcap \{B : B \text{ is a closed ball and } A \subset B\}.$$

The admissible subsets were introduced in metric fixed point theory because of their similarities with convex subsets in linear vector spaces.

Banach Contraction Principle for pointwise contraction mappings may be stated as:

Theorem 2.3. *Let M be a bounded metric space. Assume that the convexity structure $\mathscr{A}(M)$ is compact. Let $T : M \to M$ be a pointwise contraction. Then T has a unique fixed point x_0. Moreover the orbit $\{T^n(x)\}$ converges to x_0, for each $x \in M$.*

Proof. Since $\mathscr{A}(M)$ is compact, there exists a minimal nonempty $K \in \mathscr{A}(M)$ such that $T(K) \subset K$. It is easy to check that $\mathrm{cov}(T(K)) = K$. Let $a \in K$, then we have $K \subset B(a, r_a(K))$, where

$$r_a(K) = \sup\{d(a,x) : x \in K\}.$$

Since T is a pointwise contraction, there exists a mapping $\alpha : M \to [0,1)$ such that

$$d(T(x),T(y)) \leq \alpha(x)d(x,y) \text{ for any } x,y \in M.$$

In particular, we have then $T(K) \subset B(T(a), \alpha(a)r_a(K))$, which implies

$$K = \mathrm{cov}(T(K)) \subset B(T(a), \alpha(a)r_a(K)).$$

So $r_{T(a)}(K) \leq \alpha(a)r_a(K)$. This will force $\mathrm{diam}(K) = 0$. Indeed let $a \in K$ and define

$$K_a = \{x \in K : r_x(K) \leq r_a(K)\}.$$

Clearly K_a is not empty. Moreover, we have

$$K_a = \bigcap_{x \in K} B(x, r_a(K)) \cap K \in \mathscr{A}(M).$$

And since $r_{T(x)}(K) \leq \alpha(x)r_x(K)$, for any $x \in K$, we get $T(K_a) \subset K_a$. The minimality behavior of K implies $K_a = K$. In particular we have $r_x(K) = r_a(K)$ for any $x \in K$. Hence $\mathrm{diam}(K) = r_a(K)$, for any $a \in K$, i.e., a is a diametral point of K. Hence $\mathrm{diam}(K) \leq \alpha(a)\mathrm{diam}(K)$. And since $\alpha(a) < 1$, we get $\mathrm{diam}(K) = 0$, i.e., K is reduced to one point which is fixed by T. The remaining conclusion of the theorem follows from the general properties of pointwise contractions.
□

As we did in the case of Banach's Contraction Principle, we extend Theorem 2.3 to the case when iterates are pointwise contractions. First we need the following definition.

Definition 2.4. Let (M,d) be a metric space. A mapping $T : M \to M$ is called an *asymptotic pointwise mapping* if there exists a sequence of mappings $\alpha_n : M \to [0,\infty)$ such that

$$d(T^n(x), T^n(y)) \leq \alpha_n(x)d(x,y), \quad \text{for any } y \in M.$$

(i) If $\{\alpha_n\}$ converges pointwise to $\alpha : M \to [0,1)$, then T is called an *asymptotic pointwise contraction* .

(ii) If $\limsup\limits_{n\to\infty} \alpha_n(x) \leq 1$, then T is called *asymptotic pointwise nonexpansive* .

(ii) If $\limsup\limits_{n\to\infty} \alpha_n(x) \leq k$, with $0 < k < 1$, then T is called a *strongly asymptotic pointwise contraction* .

Let (M,d) be a metric space. We will say that a function $\Phi : M \to [0,\infty)$ is *convex* if $\{x : \Phi(x) \leq r\} \in \mathscr{A}(M)$, for any $r \geq 0$. Also we define a *type* to be a function $\Phi : M \to [0,\infty)$ defined as

$$\Phi(u) = \limsup_{n\to\infty} d(x_n, u)$$

where $\{x_n\}$ is a bounded sequence in M. Types are very useful in the study of the geometry of Banach spaces and the existence of fixed point of mappings. We will say that $\mathscr{A}(M)$ is *T-stable* if types are convex. We have the following lemma.

Lemma 2.1. *Let M be a metric space. Assume that $\mathscr{A}(M)$ is compact and T-stable. Then for any type Φ, there exists $x_0 \in M$ such that*

$$\Phi(x_0) = \inf\{\Phi(x) : x \in M\}.$$

Proof. Let Φ be a type. Then there exists a bounded sequence $\{x_n\}$ in M such that $\Phi(x) = \limsup\limits_{n\to\infty} d(x_n,x)$, for any $x \in M$. Set $\Phi_0 = \inf\limits_{x\in M} \Phi(x)$. Then the set

$$M_n = \left\{x \in M : \ \Phi(x) \leq \Phi_0 + \frac{1}{n}\right\},$$

is a nonempty admissible subset of M because $\mathscr{A}(M)$ is convex. Since $\{M_n\}$ is a decreasing sequence and $\mathscr{A}(M)$ is compact, then $M_\infty = \bigcap_{n \geq 1} M_n$ is not empty. Note that for any $x \in M_\infty$ we have $\Phi(x) = \Phi_0$.

$\square$

We have the following fixed point theorem for strongly asymptotic pointwise contractions in metric spaces.

Theorem 2.4. *Let M be a fixed bounded metric space. Assume that $\mathscr{A}(M)$ is compact. Let $T : M \to M$ be a strongly asymptotic pointwise contraction. Then T has a unique fixed point x_0. Moreover any orbit $\{T^n(x)\}$ converges to x_0.*

Proof. First note that T has, at most, one fixed point. Indeed, let $a, b \in M$ be two fixed points of T. Then we have

$$d(a,b) = d(T^n(a), T^n(b)) \leq \alpha_n(a) \, d(a,b).$$

If we let n go to infinity, we get $d(a,b) \leq k \, d(a,b)$ for some $k \in (0,1)$. This will force $d(a,b) = 0$. Let us fix $x \in M$ and define the type

$$\Phi(u) = \limsup_{n \to \infty} d(T^n(x), u), \quad \text{for } u \in M.$$

Since $\mathscr{A}(M)$ is compact, then

$$\Omega(x) = \bigcap_{n \geq 1} \operatorname{cov}\Big(\{T^k(x) : k \geq n\}\Big) \neq \emptyset.$$

Let $\omega \in \Omega(x)$. Then for every $n, m \in \mathbb{N}$ we have

$$d(T^{m+n+h}(x), T^{m+h}(x)) \leq \alpha_h(T^m(x)) d(T^n(x), T^m(x)).$$

If we let n go to infinity, we get

$$\Phi(T^{m+h}(x)) \leq \alpha_h(T^m(x)) \Phi(T^m(x)).$$

Next we let h go to infinity to get

$$\limsup_{n \to \infty} \Phi(T^n(x)) \leq k \, \Phi(T^n(x))$$

for some $k \in (0,1)$, which easily implies that $\limsup_{n \to \infty} \Phi(T^n(x)) = 0$. Notice that

$$\Phi(\omega) \leq \limsup_{n \to \infty} \Phi(T^n(x)) = 0.$$

Indeed let $u \in M$, then for any $\varepsilon > 0$, then there exists $n_0 \geq 1$ such that for any $n \geq n_0$

$$d(T^n(x), u) \leq \Phi(u) + \varepsilon.$$

In particular we have $T^n(x) \in B(u, \Phi(u) + \varepsilon)$, for any $n \geq n_0$. So

$$\Omega(x) \subset \mathrm{cov}\left(\{T^n(x) : n \geq n_0\}\right) \subset B(u, \Phi(u) + \varepsilon),$$

which implies $\omega \in B(u, \Phi(u) + \varepsilon)$. This is true for any $\varepsilon > 0$. Hence for any $u \in M$ we have $d(\omega, u) \leq \Phi(u)$, which implies that

$$\Phi(\omega) = \limsup_{n \to \infty} d(T^n(x), \omega) \leq \limsup_{n \to \infty} \Phi(T^n(x)).$$

Therefore we have $\Phi(\omega) = 0$ which implies that $\{T^n(x)\}$ converges to ω. Since T is continuous, then ω is a fixed point of T. Since T has at most one fixed point, then the limit of the orbit $\{T^n(x)\}$ is independent of $x \in M$. Therefore T has a unique fixed point x_0 and any orbit converges to x_0.
□

Let us relax now the strong behavior of T but assume that types are convex to obtain the following result.

Theorem 2.5. *Let M be a bounded metric space. Assume that $\mathscr{A}(M)$ is compact and T-stable. Let $T : M \to M$ be an asymptotic pointwise contraction. Then T has a unique fixed point x_0. Moreover the orbit $\{T^n(x)\}$ converges to x_0, for each $x \in M$.*

Proof. As we did in the beginning of the proof of Theorem 2.4, we can easily show that T has at most one fixed point. Let $x \in M$ and define the type

$$\Phi(u) = \limsup_{n \to \infty} d(T^n(x), u), \quad \text{for each } u \in M.$$

Since $\mathscr{A}(M)$ is compact and T-stable, then by Lemma 2.1 there exists $x_0 \in M$ such that

$$\Phi(x_0) = \inf \{\Phi(u) : u \in M\}.$$

Let us show that $\Phi(x_0) = 0$. Indeed we have

$$d(T^{n+m}(x), T^m(x_0)) \leq \alpha_m(x_0)\, d(T^n(x), x_0),$$

for any $n, m \geq 1$. If we let n go to infinity, we get

$$\Phi(T^m(x_0)) \leq \alpha_m(x_0)\, \Phi(x_0),$$

which implies

$$\Phi(x_0) = \inf \{\Phi(u) : u \in M\} \leq \Phi(T^m(x_0)) \leq \alpha_m(x_0)\, \Phi(x_0).$$

If we let m go to infinity, we get $\Phi(x_0) \leq \alpha(x_0)\Phi(x_0)$. Since $\alpha(x_0) < 1$, we get $\Phi(x_0) = 0$, which implies that $\{T^n(x)\}$ converges to x_0. This will force x_0 to be a fixed point of T. Since we already noticed that T has at most one fixed point, then T has a fixed point x_0 and any orbit converges to x_0.
□

2.3 Caristi-Ekeland Extension

This is one of the most interesting extensions of the Banach Contraction Principle. In order to understand its context, let us first go over the proof of Theorem 2.1 given by Caristi [40, 41]. Let $T : M \to M$ be a contraction with Lipschitz constant $k < 1$. Then we have

$$d(T(x), T^2(x)) \le k\, d(x, T(x)),$$

for any $x \in M$. Adding $d(x, T(x))$ to both sides of the above inequality yields

$$d(x, T(x)) + d(T(x), T^2(x)) \le d(x, T(x)) + k\, d(x, T(x))$$

which can be rewritten

$$d(x, T(x)) - k\, d(x, T(x)) \le d(x, T(x)) - d(T(x), T^2(x)).$$

This in turn is equivalent to

$$d(x, T(x)) \le \frac{1}{1-k}\Big[d(x, T(x)) - d(T(x), T^2(x))\Big],$$

for any $x \in M$. Now define the function $\varphi : M \to \mathbb{R}^+$ by setting

$$\varphi(x) = \frac{1}{1-k} d(x, T(x)).$$

This gives us the basic inequality

$$d(x, T(x)) \le \varphi(x) - \varphi(T(x)),$$

for any $x \in M$. As a generalization to contraction mappings, Caristi [41] and Ekeland [73] considered mappings $T : M \to M$ which satisfy the following property

$$d(x, T(x)) \le \varphi(x) - \varphi(T(x)), \quad x \in M,$$

where $\varphi : M \to \mathbb{R}^+ = [0, +\infty)$. Both Caristi and Ekeland investigated this new class of mappings to find out when a fixed point exists. Recall that the function φ is said to be *lower semicontinuous(l.s.c.)*, if for any sequence $\{x_n\} \subset M$, if $\lim\limits_{n \to \infty} x_n = x$ and $\lim\limits_{n \to \infty} \varphi(x_n) = r$, then $\varphi(x) \le r$.

Theorem 2.6 (Ekeland Variational Principle, 1974). *Let (M, d) be a complete metric space and $\varphi : M \to \mathbb{R}^+$ l.s.c. Define the Brønsted partial order :*

$$x \preceq y \Leftrightarrow d(x, y) \le \varphi(x) - \varphi(y), \; x, y \in M.$$

Then $(M, \preceq)$ has a maximal element.

Theorem 2.7 (Caristi Fixed Point Theorem, 1975). *Let (M,d) be a complete metric space and $\varphi : M \to \mathbb{R}^+$ l.s.c. Suppose $T : M \to M$ satisfies:*

$$d(x, T(x)) \leq \varphi(x) - \varphi(T(x)), \; x \in M.$$

Then T has a fixed point.

Although both theorems have different settings, in fact they are equivalent, [171]. The proof of Caristi-Ekeland's theorems is based on the discrete technique: Zorn's lemma and Axiom of Choice. There are some attempts to find a pure metric proof of Caristi's fixed point theorem, without success so far.

2.4 Some Applications

There are many known examples of the Banach Contraction Principle. Here, we will discuss only two of them.

2.4.1 ODE and Integral Equations

Consider the integral equation

$$f(x) = \lambda \int_a^x K(x,t) f(t) dt + \phi(x)$$

for a fixed real number λ, where $K(x,t)$ is continuous on $[a,b] \times [a,b]$. Consider the metric space $\mathscr{C}[a,b]$ of continuous real-valued functions defined on $[a,b]$. Consider the map $T : \mathscr{C}[a,b] \to \mathscr{C}[a,b]$ defined by

$$T(f)(x) = \lambda \int_a^x K(x,t) f(t) dt + \phi(x).$$

For $f_1, f_2 \in \mathscr{C}[a,b]$, we have

$$d\left(T^n(f_1), T^n(f_2)\right) \leq |\lambda|^n M^n \frac{(b-a)^n}{n!} d(f_1, f_2)$$

where

$$d(f_1, f_2) = \max\{|f_1(x) - f_2(x)| : x \in [a,b]\},$$

and

$$M = \max\{|K(x,t)| : (x,t) \in [a,b] \times [a,b]\}.$$

Clearly there exists $n \geq 1$ such that T^n is a contraction, which implies that the above integral equation has a unique solution $f(x)$. In general, the map T may not be a contraction on $[a,b]$. Bielecki [26] discovered another way to find a remedy to this problem. Indeed, for $\lambda > 0$, set

$$\|f\|_\lambda = \max_{a \leq x \leq b} \left\{ e^{-\lambda (x-a)} |f(x)| \right\},$$

it is now possible to prove that for any $f_1, f_2 \in \mathscr{C}[a,b]$, we have

$$d_\lambda \left(T(f_1), T(f_2) \right) = \|T(f_1) - T(f_2)\|_\lambda \leq \frac{M}{\lambda} \|f_1 - f_2\|_\lambda = \frac{M}{\lambda} d_\lambda(f_1, f_2),$$

where $M = \max\limits_{a \leq x,y \leq b} |K(x,y)|$ is as before. It is then clear that for λ sufficiently large T is a contraction for the new distance d_λ.

2.4.2 Cantor and Fractal sets

Let (M,d) be a complete metric space and let $\mathscr{M}$ denote the family of all nonempty bounded closed subsets of M, and let $\mathscr{C}$ denote the subfamily of $\mathscr{M}$ consisting of all compact sets. For $A \in \mathscr{M}$ and $\varepsilon > 0$ define the ε-neighborhood of A to be the set

$$N_\varepsilon(A) = \{x \in M : dist(x,A) < \varepsilon\}$$

where $dist(x,A) = \inf\limits_{y \in A} d(x,y)$. Now for $A, B \in \mathscr{M}$, set

$$H(A,B) = \inf\{\varepsilon > 0 : A \subseteq N_\varepsilon(B) \text{ and } B \subseteq N_\varepsilon(A)\}.$$

Then $(\mathscr{M}, H)$ is a metric space, and H is called the *Hausdorff metric* on $\mathscr{M}$. Notice that if (M,d) is complete, then $(\mathscr{M}, H)$ is also complete. Let $T_i : M \to M, i = 1, \ldots, n$, be a family of contractions. Define the map $T^* : \mathscr{C} \to \mathscr{C}$ by

$$T^*(X) = \bigcup_{i=1}^n T_i(X).$$

Then T^* is a contraction and its Lipschitz constant is smaller than the maximum of all Lipschitz constants of the mappings T_i, $i = 1, \ldots, n$. Then Banach Contraction Principle implies the existence of a unique nonempty compact subset X of M such that

$$X = \bigcup_{i=1}^n T_i(X).$$

As an application of this, consider the real interval $[0,1]$ and the two contractions

$$T_1(x) = \frac{1}{3}x \text{ and } T_2(x) = \frac{1}{3}x + \frac{2}{3}.$$

Then the compact X which satisfies $X = T_1(X) \cup T_2(X)$ is the well-known Cantor set. Moreover this fixed point set is the limit of the iterate of $[0, 1]$.

2.5 Metric Fixed Point Theory in Banach Spaces

The formal definition of Banach spaces is due to Banach himself. But examples like the finite dimensional vector space $\mathbb{R}^n$ was studied prior to Banach's formal definition. In 1912, Brouwer [31] proved the following:

Theorem 2.8 (Brouwer Fixed Point Theorem). *Let B be a closed ball in $\mathbb{R}^n$. Then any continuous mapping $T : B \to B$ has at least one fixed point.*

This theorem has a long history. The ideas used in its proof were known to Poincare as early as 1886. In 1909, Brouwer proved the theorem for $n = 3$, and in 1910, Hadamard gave the first proof for arbitrary n. Brouwer gave another proof in 1912. All of these are older results than the Banach Contraction Principle. Though the two theorems are different in nature, they bear some similarities. A combination of the two led to the so-called metric fixed point theorem in Banach spaces. Indeed, in Brouwer's theorem, the convexity, compactness, and the continuity of T are crucial, while the Lipschitz behavior of the contraction and completeness are crucial in Banach's fixed point theorem. In infinite normed linear vector spaces, we lose the compactness of the bounded closed convex sets (like closed balls). So if we assume completeness, we get Banach spaces. On these, we have another natural topology (other than the norm topology), that is the weak topology. A weakly compact convex set does not need to be compact for the norm. The best example here is the Hilbert space, where any bounded closed convex set is weakly compact. The Lipschitz condition considered is when the Lipschitz constant is equal to 1, i.e., nonexpansive mappings.

The metric fixed point problem in Banach spaces becomes:

> *Let X be a Banach space, and C a nonempty bounded closed convex subset of X. When does any nonexpansive mapping $T : C \to C$ have a fixed point?*

Other interesting problems closely related to this one are:

- The structure of the fixed points sets [34]
- The approximation of fixed points.

Recognition of fixed point theory for nonexpansive mappings as a noteworthy avenue of research almost surely dates from the 1965 publication of

1. *Kirk's Theorem [120].* Let K be a weakly-compact convex subset of a Banach space X. Assume that K has the normal structure property (see Definition 2.5 in Section 2.5.2), then any nonexpansive mapping $T : K \to K$ has a fixed point.

2. *Browder-Göhde's Theorem [32, 85].* If K is a bounded closed convex subset of a uniformly convex Banach space X and if $T : K \to K$ is nonexpansive, then T has a fixed point. Moreover the fixed point set of T is a closed convex subset of K.

While studying a paper by Brodskii and Milman [30], Kirk [120] was able to discover the above stated result. Let us mention that the normal structure property is an old concept, not directly related to nonexpansive mappings (for instance it is known that Hilbert space and uniformly convex Banach spaces have the normal structure property). In fact, the proofs given independently by Browder [32] and Göhde [85] of their results do not use this property at all, but refer directly to the concept of the uniform convexity.

2.5.1 Classical Existence Results

Let K be a bounded closed convex subset of a Banach space $(X, \|.\|)$ and $T : K \to K$ be a nonexpansive mapping. Let $\varepsilon \in (0, 1)$ and $x_0 \in K$. Define T_ε by

$$T_\varepsilon(x) = \varepsilon x_0 + (1 - \varepsilon)T(x), \quad \text{for any } x \in K.$$

It is easy to check that $T_\varepsilon(K) \subset K$ (since K is convex) and it is a contraction. The Banach Contraction Principal implies the existence of a unique point $x_\varepsilon \in K$ such that

$$x_\varepsilon = \varepsilon x_0 + (1 - \varepsilon)T(x_\varepsilon).$$

which implies

$$\|x_\varepsilon - T(x_\varepsilon)\| = \varepsilon \|x_0 - T(x_\varepsilon)\| \leq \varepsilon \, \text{diam}(K) = \varepsilon \, \sup\{\|x - y\| : x, y \in K\}.$$

Therefore we have

$$\inf_{x \in K} \|x - T(x)\| = 0.$$

This property is known as the *approximate fixed point property*. Therefore, any nonexpansive mapping defined on a bounded closed convex subset of Banach space has this property. By taking a sequence $\{\varepsilon_n\}$ which goes to 0, we generate a sequence of points $\{x_n\}$ from K such that

$$\lim_{n \to \infty} \|x_n - T(x_n)\| = 0.$$

Such a sequence is called an *approximate fixed point sequence* (a.f.p.s.). In order to grasp the ideas behind the proofs of the existence of fixed point for nonexpansive mappings in the Hilbert and uniformly convex spaces, we need the concept of *asymptotic center* of a sequence discovered by Edelstein [71]. This concept is very useful whenever one deals with sequential approximations in Banach spaces. Let $\{x_n\}$ be a bounded sequence in a Banach space X, and let C be a closed convex subset of X. Consider the functional $\tau : C \to [0, \infty)$ defined by

$$\tau(x) = \limsup_{n \to \infty} \|x_n - x\|.$$

Usually we use the notation $\tau(x) = r(x, \{x_n\})$. The infimum of $\tau(x)$ over C is called the *asymptotic radius* of $\{x_n\}$ and denoted by $r(C, \{x_n\})$, i.e.,

$$r(C, \{x_n\}) = \inf_{x \in C} r(x, \{x_n\}) = \inf_{x \in C} \left(\limsup_{n \to \infty} \|x_n - x\| \right).$$

The set

$$A(C, \{x_n\}) = \Big\{ x \in C : r(x, \{x_n\}) = r(C, \{x_n\}) \Big\}$$

is the set of all asymptotic centers of $\{x_n\}$. If $\{x_n\}$ converges to $x \in C$, then $A(C, \{x_n\}) = \{x\}$. In general, the set $A(C, \{x_n\})$ is not reduced to one point.

Theorem 2.9. *Every bounded sequence in a uniformly convex Banach space X has a unique asymptotic center with respect to any closed convex subset of X.*

Proof. Let X be a uniformly convex Banach space. Let $\{x_n\}$ be a bounded sequence in X, and let C be a closed convex subset of X. Set $R = r(C, \{x_n\})$. Without loss of generality, we may assume $R > 0$. For any $\varepsilon > 0$, set

$$C_\varepsilon = \{x \in C : r(x, \{x_n\}) \leq R + \varepsilon\}.$$

Since the function $x \to r(x, \{x_n\})$ is continuous and convex, we conclude that C_ε is a nonempty closed convex subset of C. For any $x, y \in C_\varepsilon$, we have $\|x - y\| \leq 2(R + \varepsilon)$. Hence C_ε is bounded. Since X is uniformly convex, it is reflexive. Hence C_ε is weakly compact. Set $C_n = C_{1/n}$, for $n \geq 1$. The sequence $\{C_n\}$ is decreasing, which implies by weak compactness that $\bigcap_{n \geq 1} C_n$ is a nonempty closed bounded convex subset of C. It is easy to check that

$$A(C, \{x_n\}) = \bigcap_{n \geq 1} C_n.$$

Let us finish the proof of Theorem 2.9 by showing that $A(C, \{x_n\})$ has at most one point. Let $x, y \in A(C, \{x_n\})$. Then we have

$$\limsup_{n \to +\infty} \|x - x_n\| = \limsup_{n \to +\infty} \|y - x_n\| = R.$$

Let $\eta > 0$. Then there exists $n_0 \geq 1$ such that

$$\sup_{n \geq n_0} \|x - x_n\| \leq R + \eta, \quad \text{and} \quad \sup_{n \geq n_0} \|y - y_n\| \leq R + \eta.$$

Assume that $x \neq y$. Set $\varepsilon = \|x - y\| > 0$. Since X is uniformly convex, then we have

$$\left\| \frac{x+y}{2} - x_n \right\| \leq (R + \eta) \left(1 - \delta_X \left(\frac{\varepsilon}{R + \eta} \right) \right)$$

for any $n \geq n_0$, where δ_X is the modulus of uniform convexity of X. Since $A(C, \{x_n\})$ is convex, we get

$$R = r \left(\frac{x+y}{2}, \{x_n\} \right) = \limsup_{n \to +\infty} \left\| \frac{x+y}{2} - x_n \right\| \leq (R + \eta) \left(1 - \delta_X \left(\frac{\varepsilon}{R + \eta} \right) \right),$$

which implies

$$\delta_X \left(\frac{\varepsilon}{R + \eta} \right) \leq \frac{\eta}{R + \eta},$$

for any $\eta > 0$. Fix $\eta_0 > 0$. Since δ_X is nondecreasing, we get

$$0 < \delta_X \left(\frac{\varepsilon}{R + \eta_0} \right) \leq \delta_X \left(\frac{\varepsilon}{R + \eta} \right) \leq \frac{\eta}{R + \eta},$$

for any $\eta < \eta_0$. This is clearly a contradiction which forces $x = y$. $\quad\square$

In the Hilbert case, we can prove even more.

Theorem 2.10. *In a Hilbert space H, the weak limit of a weakly convergent sequence coincides with its asymptotic center with respect to H.*

Proof. Let H be a Hilbert space. Denote by $<,>$ the scalar product of H. Let $\{x_n\}$ be a sequence which converges weakly to x in H. For any $y \in H$, we have

$$\|y - x_n\|^2 = \|y - x\|^2 + 2 < y - x, x - x_n > + \|x - x_n\|^2,$$

for any $n \geq 1$. Since $\{x_n\}$ converges weakly to x, we get $\lim_{n \to +\infty} <y - x, x - x_n >= 0$. Hence

$$\limsup_{n \to +\infty} \|y - x_n\|^2 = \|y - x\|^2 + \limsup_{n \to +\infty} \|x - x_n\|^2. \qquad \text{(HE)}$$

Therefore we have

$$\limsup_{n \to +\infty} \|x - x_n\| = \inf \left\{ \limsup_{n \to +\infty} \|y - x_n\| : y \in H \right\},$$

i.e., $x \in A(H, \{x_n\})$. Since H is uniformly convex, then we must have $A(H, \{x_n\}) = \{x\}$.
$\square$

Remark 2.2. In any Hilbert space H, the asymptotic center of a bounded sequence $\{x_n\}$ with respect to H belongs to the closed convex hull of $\{x_n\}$. Indeed, let $\{x_n\}$ be a bounded sequence in a Hilbert space H. Let x be the asymptotic center of $\{x_n\}$ with respect to H. Let $C = \overline{conv}(\{x_n\})$ be the closed convex hull of $\{x_n\}$. Let $P : H \to C$ be the nearest point projection. Using the nonexpansive behavior of P in any Hilbert space, we get

$$\|x_n - P(x)\| \leq \|x_n - x\|,$$

for any $n \geq 1$. Hence

$$\limsup_{n \to +\infty} \|x_n - P(x)\| \leq \limsup_{n \to +\infty} \|x_n - x\| = r(H, \{x_n\}).$$

Since x is the asymptotic center of $\{x_n\}$, we must have $x = P(x)$. Note that this property does not necessarily hold in (even uniformly convex) Banach spaces.

Theorem 2.11. *Let C be a bounded closed convex subset of a uniformly convex Banach space X. Let $T : C \to C$ be a nonexpansive mapping.*

(a) Let $\{x_n\}$ be an a.f.p.s. of T in C. Let $z \in C$ be its asymptotic center with respect to C, then it is quite easy to check that $T(z)$ is also an asymptotic center of $\{x_n\}$ with respect to C. By the uniqueness of the asymptotic center, we get $T(z) = z$.
(b) Let $x \in X$ and consider the orbit of x under T, i.e., $\{T^n(x)\}$. Let $z \in C$ be its asymptotic center with respect to C, then $T(z)$ is also an asymptotic center of $\{T^n(x)\}$ with respect to C. By the uniqueness of the asymptotic center, we get $T(z) = z$.

2.5.2 The Normal Structure Property

The reason behind separating the Normal Structure Property from the other results is the important role it played during the first 20 years (since 1965) of the theory. In order to appreciate this property, let us give more information on nonexpansive mappings in Banach spaces. Indeed, let C be a weakly compact convex subset of a Banach space X. Let $T : C \to C$ be nonexpansive. The first pioneers of the metric fixed point theory were mostly concerned about the existence of fixed points. So assume that T fails to have a fixed point in C. Since C is weakly compact, there exists (by the Zorn Lemma) a minimal nonempty closed convex subset K of C invariant under T, i.e., $T(K) \subset K$. The central research of the metric fixed point theory in Banach spaces always orbited around the discovery of new properties of such minimal convex sets. The first properties are due to Kirk.

Theorem 2.12. *[120] Under the above assumptions and notations, we have*

(a) $\overline{conv}(T(K)) = K$.
(b) $\sup_{z \in K} \|z - x\| = diam(K)$, for any $x \in K$.

Proof. Let C be a weakly compact convex subset of a Banach space X. Let $T : C \to C$ be nonexpansive. Let K be a minimal T-invariant closed convex subset of C. Set $K_0 = \overline{conv}(T(K))$. Since K is T-invariant and convex, then we have $K_0 \subset K \subset C$. Hence $T(K_0) \subset T(K) \subset K_0$, i.e., K_0 is T-invariant. By minimality, we get $K = K_0 = \overline{conv}(T(K))$. Next we prove (b). Let $x \in K$. Set $r = \sup_{z \in K} \|z - x\|$. Set

$$K_0 = \{y \in K : \sup_{z \in K} \|z - y\| \leq r\}.$$

Then we have

$$K_0 = K \cap \left(\bigcap_{z \in K} B(z, r) \right),$$

where $B(z, r)$ is the closed ball in X centered at z with radius r. Since $x \in K_0$, we deduce that K_0 is a nonempty closed convex subset of K. Let us show that K_0 is T-invariant. Let $y \in K_0$. Then $K \subset B(y, r)$. Since T is nonexpansive, we get $T(K) \subset B(T(y), r)$. Since balls are convex, we get $\overline{conv}(T(K)) \subset B(T(y), r)$. The conclusion of (a) implies that $K \subset B(T(y), r)$. Therefore, we must have $T(y) \in K_0$, i.e., K_0 is T-invariant. By minimality of K, we conclude that $K = K_0$. Hence

$$\sup_{z \in K} \|y - z\| \leq \sup_{z \in K} \|z - x\|,$$

for any $y \in K$. Since x was arbitrarily taken in K, we get

$$\sup_{z \in K} \|y - z\| = \sup_{z \in K} \|z - x\|,$$

for any $y \in K$. Hence

$$\operatorname{diam}(K) = \sup_{y,z \in K} \|y - z\| = \sup_{y \in K} \left(\sup_{z \in K} \|y - z\| \right) = \sup_{z \in K} \|z - x\|,$$

for any $x \in K$. This completes the proof of Theorem 2.12.
$\square$

Definition 2.5. A closed convex subset C of a Banach space X is said to have the *normal structure property* if any bounded convex subset K of C which contains more than one point, contains a *nondiametral point* , i.e., there exists a point $x_0 \in K$ such that

$$\sup_{x \in K} \|x_0 - x\| < \operatorname{diam}(K) .$$

We will also say that X has the *normal structure property* if any bounded closed convex subset has the normal structure property.

The normal structure property forbids the conclusion (b) in Theorem 2.12 to hold (see [21]).

Throughout this section we will use the following notations:

- $\text{diam}(C) = \sup\limits_{x,y \in C} \|x - y\|$
- $r_z(C) = \sup\limits_{x \in C} \|z - x\|$
- $R(C) = \inf\limits_{x \in C} r_x(C)$
- $\mathscr{C}(C) = \{z \in C : r_z(C) = r(C)\}$

The number $R(C)$ and the set $\mathscr{C}(C)$ are called the *Chebyshev radius* and *Chebyshev center* of C, respectively.

In light of the discussion above and Theorem 2.12, we get the following fundamental theorem [120].

Theorem 2.13 (Kirk's Fixed Point Theorem). *Let X be a Banach space and suppose that C is a nonempty weakly compact convex subset of X which has the normal structure property. Then any nonexpansive mapping $T : C \to C$ has a fixed point.*

Remark 2.3. When X is strictly convex, and under the assumptions of Theorem 2.13, the fixed point set of T, $F(T)$, is convex. Indeed, let $x, y \in F(T)$. Then we have

$$\left\| x - T\left(\frac{x+y}{2}\right) \right\| = \left\| T(x) - T\left(\frac{x+y}{2}\right) \right\| \leq \left\| x - \frac{x+y}{2} \right\| = \frac{\|x-y\|}{2}.$$

Similarly we have

$$\left\| y - T\left(\frac{x+y}{2}\right) \right\| \leq \frac{\|x-y\|}{2}.$$

Hence

$$\left\| x - T\left(\frac{x+y}{2}\right) \right\| = \left\| y - T\left(\frac{x+y}{2}\right) \right\| = \frac{\|x-y\|}{2}.$$

Since X is strictly convex, we must have

$$T\left(\frac{x+y}{2}\right) = \frac{x+y}{2},$$

i.e., $\dfrac{x+y}{2} \in F(T)$. In fact, the convexity of the fixed point set of nonexpansive mappings characterizes Banach spaces which are strictly convex (see [104]).

Since the publication of Kirk's fixed point theorem in 1965, people intensified their investigation of the normal structure property. For example, it was proved that uniformly convex Banach spaces have this property. Let X be a Banach space. Define the *characteristic (or coefficient) of uniform convexity* of the Banach space X by

$$\varepsilon_0(X) = \sup\{\varepsilon \in [0,2] : \delta_X(\varepsilon) = 0\},$$

where

$$\delta_X(\varepsilon) = \inf\left\{1 - \left\|\frac{x+y}{2}\right\| : \|x\| \le 1, \|y\| \le 1, \|x-y\| \ge \varepsilon,\, x,y \in X\right\},$$

is the modulus of uniform convexity of X. Recall that X is uniformly convex if and only if $\varepsilon_0(X) = 0$, i.e., $\delta_X(\varepsilon) > 0$, for any $\varepsilon \in (0,2]$. And X is said to be *uniformly non-square* if and only if $\varepsilon_0(X) < 2$. James [99] studied these spaces extensively (see also [76]). Assume that $\varepsilon_0(X) < 1$, and let $\varepsilon \in (\varepsilon_0(X), 1)$. Then for any nonempty bounded closed convex subset C of X not reduced to one point, there exists a point $x \in C$ such that

$$r(x,C) \le (1 - \delta(\varepsilon))\, \text{diam}(C).$$

So we have

$$N(X) = \sup\left\{\frac{r(C)}{\text{diam}(C)}\right\} \le 1 - \delta(\varepsilon_0(X))$$

where the supremum is taken over all nonempty bounded closed convex subset of X with more than one point. The number $N(X)$, introduced by Bynum [39], is known as the *coefficient of uniform normal structure property*. So we have the normal structure property and even more, uniform normal structure property, i.e., $N(X) < 1$. Therefore, uniformly convex Banach spaces have a stronger geometric property.

Since its early connection with the fixed point property, the normal structure property had occupied a central role in the study of metric fixed point theory. At one point, it was asked whether the two properties are equivalent. In order to show that it is not the case, one has to dig deeper into the properties of the minimal convex sets associated to nonexpansive mappings. Independently, Goebel [78] and Karlovitz [102] proved the following:

Theorem 2.14. *[78, 102] Let K be a subset of a Banach space X which is minimal with respect to being nonempty, weakly compact, convex, and T-invariant for some nonexpansive mapping T, and suppose $\{x_n\} \subseteq K$ is an a.f.p.s., that is, $\lim_{n\to\infty} \|x_n - T(x_n)\| = 0$. Then for each $x \in K$, we have*

$$\lim_{n\to\infty} \|x - x_n\| = diam(K).$$

Proof. Let K be a subset of a Banach space X which is minimal with respect to being nonempty, weakly compact, convex, and T-invariant for some nonexpansive mapping T. Let $\{x_n\}$ be an a.f.p.s of T in K. Let $\mathscr{U}$ be a nontrivial ultrafilter over $\mathbb{N}$. Since K is bounded, then $\lim_{\mathscr{U}} \|x - x_n\| = l(x)$ exists for any $x \in K$. It is obvious that the function $l : K \to [0, +\infty)$ is continuous convex function. Moreover we have

$$\lim_{\mathscr{U}} \|x - x_n\| = \lim_{\mathscr{U}} \|x - T(x_n)\|,$$

since $\{x_n\}$ is an a.f.p.s. of T. Since T is nonexpansive, we get $l(T(x)) \le l(x)$, for any $x \in K$. Fix $x \in K$. Set $K_x = \{y \in K : l(y) \le l(x)\}$. Obviously we have $x \in K_x$. Using the properties of l and T, we conclude that K_x is a closed nonempty convex subset of K which is T-invariant. Using the minimality of K, we conclude that $K_x = K$, for any $x \in K$. Clearly this will force the function l to be constant, i.e., there exists $r \geq 0$ such that

$$\lim_{\mathcal{U}} \|x - x_n\| = r,$$

for any $x \in K$. Clearly we have $r \leq \mathrm{diam}(K)$. Assume that $r \neq \mathrm{diam}(K)$. Let $\varepsilon \in (0, \mathrm{diam}(K) - r)$. Since $\mathcal{U}$ is not a trivial ultrafilter, then for any $y_1, y_2, \cdots, y_p \in K$, there exists $n \geq 1$ such that

$$\|y_i - x_n\| \leq r + \varepsilon, \quad i = 1, 2, \cdots, p,$$

for any $p \geq 1$. Hence the family $\{B(y, r + \varepsilon)\}$ of closed balls in K has the finite intersection property. Since K is weakly compact, then $K_\infty = K \bigcap \left(\bigcap_{x \in K} B(x, r + \varepsilon) \right)$ is not empty. For any $x \in K_\infty$, we have $\sup_{z \in K} \|x - z\| \leq r + \varepsilon < \mathrm{diam}(K)$. This is the sought contradiction with the property (b) of Theorem 2.12. Hence $r = \mathrm{diam}(K)$. Since $\mathcal{U}$ was an arbitrary nontrivial ultrafilter over $\mathbb{N}$, we conclude that $\lim_{n \to \infty} \|x - x_n\|$ exists and

$$\lim_{n \to \infty} \|x - x_n\| = \mathrm{diam}(K).$$

$\square$

Recall again that it was this property that allowed Karlovitz to prove that the normal structure property and the fixed point property are not equivalent.

2.5.3 The Demiclosedness Principle

A fundamental result in the theory of nonexpansive mappings is Browder-Göhde's demiclosedness principle [32, 85].

Definition 2.6. Let X be a Banach space and $K \subseteq X$. A mapping $f : K \to X$ is *demiclosed* (at y) if the conditions $\{x_j\}$ converges weakly to x and $\{f(x_j)\}$ converges strongly to y then $x \in K$ and $f(x) = y$.

The demiclosedness principle asserts that if K is a closed convex subset of a uniformly convex Banach space X and if $T : K \to X$ is nonexpansive, then $I - T$ is demiclosed. This fact is proved in [32] using sharp properties of the modulus of convexity and a clever subsequential thinning process. In this section, we give the proof of this result in the case of Hilbert spaces.

Theorem 2.15. *Let K be a bounded closed convex subset of a Hilbert space H. Suppose $T : K \to K$ is nonexpansive. Then $I - T$ is demiclosed at 0.*

Proof. Let $\{x_n\}$ be a sequence in K weakly convergent to x which is an a.f.p.s. for T, i.e., $\lim_{n\to+\infty} \|x_n - T(x_n)\| = 0$. Since K is weakly closed, $x \in K$. Let us prove that $x \in F(T)$, i.e., $T(x) = x$. Using the equation (HE) in the proof of Theorem 2.10, we get

$$\limsup_{n\to+\infty} \|y - x_n\|^2 = \|y - x\|^2 + \limsup_{n\to+\infty} \|x - x_n\|^2, \tag{HE}$$

for any $y \in H$. In particular, we have

$$\limsup_{n\to+\infty} \|T(x) - x_n\|^2 = \|T(x) - x\|^2 + \limsup_{n\to+\infty} \|x - x_n\|^2.$$

Since $\{x_n\}$ is an a.f.p.s. for T, then we have

$$\limsup_{n\to+\infty} \|T(x) - x_n\|^2 = \limsup_{n\to+\infty} \|T(x) - T(x_n)\|^2 \le \limsup_{n\to+\infty} \|x - x_n\|^2,$$

because T is nonexpansive. Hence

$$\|T(x) - x\|^2 + \limsup_{n\to+\infty} \|x - x_n\|^2 \le \limsup_{n\to+\infty} \|x - x_n\|^2,$$

which implies $\|T(x) - x\|^2 \le 0$, i.e., $T(x) = x$.
$\square$

Remark 2.4. The equation (HE) is fundamental and is at the heart of the proof of Theorem 2.15. So it is of no surprise that whenever a similar formula holds, we have a similar conclusion. In fact, this theorem may be adapted to other topologies than the weak topology. Indeed, in [27] it is proved that if $\{f_n\}$ is a sequence of L^p-uniformly bounded functions on a measure space, and if $f_n \to f$ almost everywhere, then

$$\liminf_{n\to\infty} \|f_n - g\|^p = \liminf_{n\to\infty} \|f_n - f\|^p + \|f - g\|^p, \tag{AEQ}$$

for all $p \in (0, \infty)$ and $g \in L^p$. Let K be a nonempty subset of L^p which is closed almost everywhere. Let $T : K \to K$ be nonexpansive. Let $\{f_n\}$ be a sequence of L^p-uniformly bounded functions on a measure space such that $f_n \to f$ almost everywhere. Assume that $\{f_n\}$ is an a.f.p.s. for T. Then we have $f \in K$. Moreover from the equation (AEQ), we get

$$\liminf_{n\to\infty} \|f_n - T(f)\|^p = \liminf_{n\to\infty} \|f_n - f\|^p + \|f - T(f)\|^p.$$

Since $\{f_n\}$ is is an a.f.p.s. for T and T is nonexpansive, we get

$$\liminf_{n\to\infty} \|f_n - T(f)\|^p = \liminf_{n\to\infty} \|T(f_n) - T(f)\|^p \le \liminf_{n\to\infty} \|f_n - f\|^p.$$

As in the proof of Theorem 2.15, we will get $T(f) = f$, i.e., T is demiclosed at 0.

The fundamental application of Browder-Göhde's demiclosedness principle in fixed point theory may now be stated.

Theorem 2.16. *[32, 85] If K is a bounded closed convex subset of a uniformly convex Banach space X and if $T : K \to K$ is nonexpansive, then T has a fixed point. Moreover, the fixed point set of T is a closed convex subset of K.*

Proof. Since X is strictly convex, then $F(T)$ is convex. Moreover, since T is continuous, then $F(T)$ is closed. We are only left to prove that $F(T)$ is not empty. Since K is bounded and convex, then T has an a.f.p.s. $\{x_n\}$ in K. Since X is reflexive, then K is weakly compact. Hence, a subsequence $\{x_{\phi(n)}\}$ of $\{x_n\}$ exists which is weakly convergent to some $x \in K$. The demiclosedness principle will then imply $T(x) = x$. □

The equations (HE), (AEQ) and similar equations [110] play a major role in the geometry of Banach spaces in connection with the fixed point theory. This was initially noted by Radon [180] and Riesz [184, 185] who introduced a property known as Radon-Riesz property, also known as Kadec-Klee property, which will be a subject of our discussion in the next section.

2.5.4 Opial and Kadec-Klee Properties

The Opial property finds its roots in [173], where Opial strengthened a previous result of Browder and Petryshyn [33] by proving the weak convergence of the iterate of a nonexpansive mapping defined within Hilbert spaces. The Kadec-Klee property, also known as Radon-Riesz [180, 184, 185] property or property (H) [94], deals with the problem of when the weak convergence implies the strong convergence in Banach spaces. In particular, this property is satisfied on the unit sphere of Hilbert spaces. Both properties continue to play a major role in the study of the metric fixed point property. The original definitions of both properties are closely linked to the weak topology.

Let X be a Banach space. Let τ be a topological vector space topology on X that is weaker than the norm topology. Throughout this section, we will assume that the topology τ is lower semi continuous with respect to the norm, that is, if $\{x_n\}$ τ-converges to $x \in X$, then

$$||x|| \leq \liminf_{n \to \infty} ||x_n||.$$

It is easy to see that this will happen if and only if the closed balls are sequentially τ-closed.

Definition 2.7. We will say that X satisfies τ-Opial condition if for every bounded $\{x_n\} \in X$ that τ-converges to $x \in X$, then

$$\liminf_{n \to \infty} ||x_n - x|| < \liminf_{n \to \infty} ||x_n - y||,$$

for every $y \neq x$. We will say that X satisfies the uniform τ-Opial condition if for every $R > 0$, for every $\varepsilon > 0$ there exists $\eta > 0$ such that for every $\{x_n\} \in X$ which

τ-converges to $x \in X$ and for every $y \in X$, we have

$$\liminf_{n \to \infty} ||x_n - x|| + \eta \leq \liminf_{n \to \infty} ||x_n - y||,$$

provided $\liminf_{n \to \infty} ||x_n - x|| \leq R$ and $||x - y|| \geq \varepsilon$.

Opial's property plays an important role in the study of τ-convergence of iterates nonexpansive mappings and of the asymptotic behavior of nonlinear semigroups [148, 173, 177]. Clearly uniform τ-Opial condition implies τ-Opial condition. In his original paper, Opial [173] showed the demiclosedness principle for nonexpansive mappings in Banach spaces which satisfy the Opial condition.

Theorem 2.17. *Let X be a Banach space which satisfies the Opial condition for the weak topology. Let C be a nonempty closed convex subset of X, and $T : C \to X$ be a nonexpansive mapping. Then T is demiclosed, i.e., for any $\{x_n\} \subset C$ such that $\{x_n\}$ is weakly convergent to $x_0 \in C$, and $\{x_n - T(x_n)\}$ is strongly convergent to $y_0 \in X$, we have $x_0 - T(x_0) = y_0$.*

Proof. Let $\{x_n\} \subset C$ be such that $\{x_n\}$ is weakly convergent to $x_0 \in C$, and $\{x_n - T(x_n)\}$ is strongly convergent to $y_0 \in X$. Since T is nonexpansive, we get

$$\liminf_{n \to \infty} ||x_n - x_0|| \geq \liminf_{n \to \infty} ||T(x_n) - T(x_0)|| = \liminf_{n \to \infty} ||x_n - y_0 - T(x_0)||.$$

Since X satisfies the Opial condition, we get $x_0 = y_0 + T(x_0)$, or $x_0 - T(x_0) = y_0$.
□

Note that this proof can be easily adapted to any topology τ as described above. As a direct application of this result, we get the Opial property version of Theorem 2.16, which we mentioned in the previous sections.

Theorem 2.18. *[32, 85] Let X be a Banach space which satisfies the Opial condition for the weak topology. Let C be a nonempty weakly compact convex subset of X, and $T : C \to C$ be a nonexpansive mapping. Then T has a fixed point.*

Another important application of Theorem 2.17 is the answer to a long-standing question whether there exists a larger class of mappings for which the orbits converge, for some topology, to a fixed point. Recall that a map T is called asymptotically regular whenever we have $\lim_{n \to \infty} ||T^{n+1}(x) - T^n(x)|| = 0$. Opial [173] proved the following result.

Theorem 2.19. *Let X be a uniformly convex Banach space which satisfies the Opial condition for the weak topology. Let C be a nonempty closed convex subset of X, and $T : C \to C$ be a nonexpansive asymptotically regular mapping with a nonempty fixed point set. Then for any $x \in C$, the sequence $\{T^n(x)\}$ is weakly convergent to a fixed point of T.*

Proof. Let $x \in C$. Set

$$\tau(z) = \limsup_{n \to +\infty} \|T^n(x) - z\|, \ z \in C.$$

Since T has a nonempty fixed points set, then $\{T^n(x)\}$ is bounded. Indeed, let us fix $y \in F(T)$, then by the nonexpansiveness of T, $\|T^n(x) - y\| \leq \|x - y\|$, which means that $\{T^n(x)\} \subset B(y, \|x - y\|)$. Since X is uniformly convex, then it is reflexive which implies that C is weakly compact. Hence $\{T^n(x)\}$ has a subsequence which converges weakly. Let $z_0 \in C$ be a weak cluster point of $\{T^n(x)\}$, i.e., there exists $\{T^{\psi(n)}(x)\}$ a subsequence which converges weakly to z_0. We claim that $z_0 \in F(T)$. Indeed, since T is asymptotically regular, we have $\lim_{n \to +\infty} \|T^{\psi(n)}(x) - T^{\psi(n)+1}(x)\| = 0$. Using Theorem 2.17, we get $T(z_0) = z_0$. We will show that z_0 is an asymptotic center of $\{T^n(x)\}$. Indeed since $z_0 \in F(T)$, we have

$$\tau(z_0) = \limsup_{n \to +\infty} \|T^n(x) - z_0\| = \lim_{n \to +\infty} \|T^n(x) - z_0\| = \liminf_{n \to +\infty} \|T^{\psi(n)}(x) - z_0\|.$$

Using the Opial property, we get

$$\tau(z_0) \leq \liminf_{n \to +\infty} \|T^{\psi(n)}(x) - z\|$$
$$\leq \limsup_{n \to +\infty} \|T^{\psi(n)}(x) - z\|$$
$$\leq \limsup_{n \to +\infty} \|T^n(x) - z\| = \tau(z),$$

for any $z \in C$. Hence $\tau(z_0) = \min_{z \in C} \tau(z)$, i.e., z_0 is an asymptotic center of $\{T^n(x)\}$. Using Theorem 2.9, z_0 is the unique asymptotic center of $\{T^n(x)\}$. Therefore, $\{T^n(x)\}$ has one weak-cluster point, i.e., $\{T^n(x)\}$ is weakly convergent. Note that its weak limit is its asymptotic center which is a fixed point of T. $\square$

The first nontrivial example of asymptotically regular mappings was given by Ishikawa [98]. Indeed, Ishikawa noted that if D is bounded and convex, and if $T : D \to D$ is nonexpansive, then the mapping $f = (I + T)/2$ has the same fixed point set as T and is asymptotically regular. Before we give proof of such a statement, we need the following technical lemma.

Lemma 2.2. *[111] Suppose $\{x_n\}$ and $\{y_n\}$ are sequences in a Banach space X which satisfy for all $n \in \mathbb{N}$:*

(i) $x_{n+1} = \dfrac{1}{2}(x_n + y_n)$,

(ii) $\|y_{n+1} - y_n\| \leq \|x_{n+1} - x_n\|$.

Then for all $i, n \in \mathbb{N}$,

$$\left(1 + \frac{n}{2}\right) \|y_i - x_i\| \leq 2^n \left[\|y_i - x_i\| - \|y_{i+n} - x_{i+n}\|\right] + \|y_{i+n} - x_i\|.$$

Proof. The proof is by induction on n. Observe that the lemma is trivially true for all i if $n = 0$, and make the inductive assumption that the theorem holds for a given $n \in \mathbb{N}$ and all i. This gives (upon replacing i with $i+1$ in the inequality)

$$\left(1 + \frac{n}{2}\right) \|y_{i+1} - x_{i+1}\| \leq 2^n \left[\|y_{i+1} - x_{i+1}\| - \|y_{i+n+1} - x_{i+n+1}\|\right] + \|y_{i+n+1} - x_{i+1}\|.$$

Next observe that by (i), the triangle inequality, and (ii) respectively:

$$\|y_{i+n+1} - x_{i+1}\| \leq (1/2)\left[\|y_{i+n+1} - x_i\| + \|y_{i+n+1} - y_i\|\right]$$
$$\leq (1/2)\left[\|y_{i+n+1} - x_i\| + \sum_{k=0}^{n} \|y_{i+k+1} - y_{i+k}\|\right]$$
$$\leq (1/2)\left[\|y_{i+n+1} - x_i\| + \sum_{k=0}^{n} \|x_{i+k+1} - x_{i+k}\|\right].$$

Combined with the previous inequality, this gives

$$\left(1 + \frac{n}{2}\right) \|y_{i+1} - x_{i+1}\| \leq 2^n \left[\|y_{i+1} - x_{i+1}\| - \|y_{i+n+1} - x_{i+n+1}\|\right]$$
$$+ (1/2)\left[\|y_{i+n+1} - x_i\| + \sum_{k=0}^{n} \|x_{i+k+1} - x_{i+k}\|\right].$$

Thus

$$2\left(1 + \frac{n}{2}\right) \|y_{i+1} - x_{i+1}\| \leq 2^{n+1} \left[\|y_{i+1} - x_{i+1}\| - \|y_{i+n+1} - x_{i+n+1}\|\right]$$
$$+ \left[\|y_{i+n+1} - x_i\| + \sum_{k=0}^{n} \|x_{i+k+1} - x_{i+k}\|\right].$$

Next we observe that (i) and (ii) imply

$$\|y_{n+1} - x_{n+1}\| \leq \|y_{n+1} - y_n\| + \|y_n - x_{n+1}\|$$
$$\leq \|x_{n+1} - x_n\| + \|y_n - x_{n+1}\|$$
$$= \|y_n - x_n\|.$$

This means that the sequence $\{\|y_n - x_n\|\}$ is monotone nonincreasing. Thus

$$\sum_{k=0}^{n} \|x_{i+k+1} - x_{i+k}\| = \frac{1}{2} \sum_{k=0}^{n} \|y_{i+k} - x_{i+k}\| \leq \frac{(n+1)}{2} \|y_i - x_i\|,$$

and we have

$$2\left(1+\frac{n}{2}\right)\|y_{i+1}-x_{i+1}\| \leq 2^{n+1}\Big[\|y_{i+1}-x_{i+1}\| - \|y_{i+n+1}-x_{i+n+1}\|\Big]$$
$$+\Big[\|y_{i+n+1}-x_i\| + ((n+1)/2)\|y_i-x_i\|\Big]$$

$$= 2^{n+1}\Big[\|y_i-x_i\| - \|y_{i+n+1}-x_{i+n+1}\|\Big]$$
$$+2^{n+1}\|y_{i+1}-x_{i+1}\|$$
$$+[((n+1)/2)-2^{n+1}]\|y_i-x_i\| + \|y_{i+n+1}-x_i\|$$

Since $2\left(1+\frac{n}{2}\right)-2^{n+1} \leq 0$,

$$\Big[2\left(1+\frac{n}{2}\right)-2^{n+1}\Big]\|y_{i+1}-x_{i+1}\| + \Big[2^{n+1}-((n+1)/2)\Big]\|y_i-x_i\|$$
$$\geq \Big[2\left(1+\frac{n}{2}\right)-2^{n+1}\Big]\|y_i-x_i\| + \Big[2^{n+1}-((n+1)/2)\Big]\|y_i-x_i\|$$
$$= \Big[1+(n+1)/2\Big]\|y_i-x_i\|,$$

and we have

$$\Big[1+(n+1)/2\Big]\|y_i-x_i\| \leq 2^{n+1}\Big[\|y_{i+1}-x_{i+1}\| - \|y_{i+n+1}-x_{i+n+1}\|\Big]$$
$$+\|y_{i+n+1}-x_i\|.$$

This completes the proof.
□

Using Lemma 2.2, we get Ishikawa's result on asymptotically regular mappings.

Theorem 2.20. *Let K be a bounded convex subset of a Banach space and suppose $T : K \to K$ is nonexpansive. Then the mapping $f : K \to K$ defined by setting $f(x) = (x+T(x))/2$ is asymptotically regular.*

Proof. Select $x_0 \in K$, let $y_0 = f(x_0)$, and having defined x_n take $y_n = f(x_n)$ and set $x_{n+1} = (x_n+y_n)/2, n = 1,2,\cdots$. It is easy to see that the resulting sequences $\{x_n\},\{y_n\}$ satisfy the assumptions of Lemma 2.2. Thus for each $i,n \in \mathbb{N}$,

$$\left(1+\frac{n}{2}\right)\|y_i-x_i\| \leq 2^n\Big[\|y_i-x_i\| - \|y_{i+n}-x_{i+n}\|\Big]$$
$$+\|y_{i+n}-x_i\|.$$

Since $\{\|x_n-y_n\|\} = \{\|x_n-f(x_n)\|\}$ is monotone nonincreasing, there exists a number $r \geq 0$ such that $\lim\limits_{n\to\infty}\|x_{n+1}-f(x_{n+1})\| = r$. Now let $i \to \infty$ in the above inequality. Then

$$\left(1+\frac{n}{2}\right)r \leq diam(K).$$

Clearly this implies $r = 0$.
□

Remark 2.5. A slight modification of the preceding argument yields the fact that each of the mappings $f_\alpha = (1 - \alpha)I + \alpha T$ are asymptotically regular for each $\alpha \in (0,1)$. Moreover, it can be shown that the convergence in Theorem 2.20 is uniform in the sense that the rate of convergence does not depend on the initial point x_0, nor indeed on the mapping T. Details may be found in [81].

Let us introduce another important property which played a major role in the study of metric fixed point theory [68, 69, 94].

Definition 2.8. We will say that X satisfies τ-Kadec-Klee property if for some $\varepsilon > 0$ there exists $\eta > 0$ such that for every $\{x_n\}$ in the unit ball of X which is τ-convergent to x, we have

$$||x|| \leq 1 - \eta,$$

provided that

$$sep\{x_n\} = \inf\{||x_n - x_m|| : n \neq m\} > \varepsilon.$$

We will say that X satisfies τ- uniform Kadec-Klee property if the above holds for every ε.

The most popular topology that one thinks of in Banach spaces, which satisfies most of the above desired properties, is the weak topology. In fact, the ideas described in this section should be extended to a larger class of topologies.

Example 2.1. We have seen in Remark 2.4 that if $\{f_n\}$ is a sequence of L^p-uniformly bounded functions on a measure space, and if $f_n \to f$ almost everywhere, then

$$\liminf_{n \to \infty} ||f_n||^p = \liminf_{n \to \infty} ||f_n - f||^p + ||f||^p,$$

for all $p \in (0, \infty)$. It is easy to check that L^p satisfy the uniform Opial condition and the uniform Kadec-Klee property [110] for the topology of convergence almost everywhere. It is not hard to see that this conclusion still holds for convergence in measure. But $L^p([0, 1], dx)$, where dx is the Lebesgue measure, for $p \in [1, \infty)$, $p \neq 2$, fails to satisfy the Opial condition for the weak topology [173].

The connection between the metric fixed point property, Opial, and Kadec-Klee properties is given in the following theorem.

Theorem 2.21. *Let X be a Banach space and τ be a topology as described above.*

(i) *Assume that X satisfies the τ-Opial condition. Let D be a τ-sequentially compact bounded convex subset of X. Then D has the fixed point property for nonexpansive mappings, i.e., any $T : D \to D$ nonexpansive has a fixed point.*

(ii) *Assume that X satisfies the τ-uniform Kadec-Klee property. Then X has τ-normal structure property.*

2.6 Ishikawa and Mann Iterations in Banach spaces

Another aspect in the study of metric fixed point theory is the theory of approximation of fixed point, when it exists. This theory is very important for the applications. Very often, the solution to a problem may be translated to be a fixed point. Its existence may be insured by, for example, physical inspections. Therefore, for such problems, the question of existence of the fixed point is not an issue. Since the explicit solution may not be available, one has to consider a method of approximating such a solution. As Banach Contraction Principle suggests, one may think of the iterates of the given map. Unfortunately, such iterates do not converge in general. Because of this, mathematicians started to think of other type of iterations. Here we focus on two of them that attracted most of the attention.

Let X be a Banach space. Let C be a nonempty closed convex subset of X. Let $T : C \to C$ be a map. One of the first iteration processes was introduced by Halpern [91]

$$x_{n+1} = \alpha_{n+1} x_0 + (1 - \alpha_{n+1}) T(x_n), \ n \geq 0, \tag{HI}$$

where the initial guess $x_0 \in C$ is taken arbitrarily and the sequence $\{\alpha_n\}$ is in the interval $[0, 1]$. Then another iteration was introduced by Mann [161] which is defined by

$$x_{n+1} = \alpha_n x_n + (1 - \alpha_n) T(x_n), \ n \geq 0, \tag{MI}$$

where the initial guess $x_0 \in C$ is taken arbitrarily and the sequence $\{\alpha_n\}$ is in the interval $[0, 1]$. Yet another iteration process is referred to as Ishikawa's [97] process which is defined recursively by

$$x_{n+1} = \alpha_n x_n + (1 - \alpha_n) T \left(\beta_n x_n + (1 - \beta_n) T(x_n) \right), \ n \geq 0, \tag{II}$$

where the initial guess $x_0 \in C$ is taken arbitrarily and the sequences $\{\alpha_n\}$ and $\{\beta_n\}$ are in the interval $[0, 1]$.

In general, not much is known regarding the convergence of the iterations (MI) and (II) unless the underlying space X has some convenient geometric properties.

Theorem 2.22. *[91, 200] Let $\{b_n\}_{n \geq 0}$ be a sequence in $[0, +\infty)$ such that*

(i) $\quad B_n = \displaystyle\sum_{i=0}^{i=n} b_i \to +\infty \text{ as } n \to +\infty;$

(ii) $\quad \dfrac{1}{B_n} \displaystyle\sum_{i=1}^{i=n} |b_i - b_{i-1}| \to 0 \text{ as } n \to +\infty.$

Let K be a closed convex subset of a Hilbert space H and $T : K \to K$ be nonexpansive such that $F(T)$ is not empty. Then for any $x \in K$, the sequence $\{x_n\}$ generated by *the iteration (HI), with $\alpha_n = \dfrac{b_n}{B_n}$ and $x_0 = x$, is convergent in the norm to the unique nearest point $P(x) \in F(T)$.*

Proof. Without loss of generality, we may assume that $0 \in F(T)$. We know that in a strictly convex Banach space, the set of fixed points of a nonexpansive mapping is convex. Hence $F(T)$ is a closed convex subset of H. In this case, the nearest point projection map $P : H \to F(T)$ exists. In particular, we have

$$\|x - P(x)\| \le \|x - y\|, \ \ y \in F(T).$$

Recall that $P(x)$ is characterized by the inequality

$$\langle y - P(x), x - P(x) \rangle \le 0, \ \ y \in F(T). \tag{NPP}$$

where $\langle \cdot, \cdot \rangle$ is the scalar product of H. Fix $x \in H$. Since $0 \in F(T)$, then we have $\|T(z)\| \le \|z\|$, for any $z \in H$. Using this inequality, we see that $\|x_n\| \le \|x\|$, for any $n \ge 0$, where $\{x_n\}$ is generated by the iteration (HI), with $x_0 = x$. By induction, we prove that

$$\|x_n - x_{n-1}\| \le 2 \frac{\|x\|}{B_n} \left(\sum_{i=0}^{i=n} |b_i - b_{i-1}| + \sum_{i=1}^{i=n} \frac{b_i b_{i-1}}{B_{i-1}} \right), \ \ n \ge 0 \tag{HW}$$

where $b_{-1} = 0$ and $x_{-1} = 0$. Indeed, it is clear that the inequality (HW) holds for $n = 0$. Assume it holds for $n \ge 0$. Then we have

$$\begin{aligned}
\|x_{n+1} - x_n\| &\le |\alpha_{n+1} - \alpha_n| \, \|x\| + |\alpha_{n+1} - \alpha_n| \, \|T(x_{n-1})\| \\
&\quad + (1 - \alpha_{n+1}) \, \|T(x_n) - T(x_{n-1})\| \\
&\le |\alpha_{n+1} - \alpha_n| \, \|x\| + |\alpha_{n+1} - \alpha_n| \, \|x_{n-1}\| \\
&\quad + (1 - \alpha_{n+1}) \, \|x_n - x_{n-1}\| \\
&\le 2|\alpha_{n+1} - \alpha_n| \, \|x\| + (1 - \alpha_{n+1}) \, \|x_n - x_{n-1}\| \\
&\le 2 \left(\frac{|b_{n+1} - b_n|}{B_{n+1}} + \frac{b_n b_{n+1}}{B_n B_{n+1}} \right) \|x\| \\
&\quad + \frac{B_n}{B_{n+1}} \frac{2\|x\|}{B_n} \left(\sum_{i=0}^{i=n} |b_i - b_{i-1}| + \sum_{i=1}^{i=n} \frac{b_i b_{i-1}}{B_{i-1}} \right) \\
&= \frac{2\|x\|}{B_{n+1}} \left(\sum_{i=0}^{i=n+1} |b_i - b_{i-1}| + \sum_{i=1}^{i=n+1} \frac{b_i b_{i-1}}{B_{i-1}} \right).
\end{aligned}$$

Hence the inequality (HW) holds for $n + 1$. Therefore by induction it holds for any $n \ge 0$. From our assumptions on $\{b_n\}$, we deduce that $\alpha_n \to 0$ when $n \to +\infty$, and

$$\lim_{n \to \infty} \|x_n - x_{n-1}\| = 0.$$

Since

$$\|x_n - T(x_n)\| \le \|x_n - (1-\alpha_n)\,T(x_{n-1})\| +$$
$$(1-\alpha_n)\|T(x_{n-1}) - T(x_n)\| + \alpha_n\|T(x_n)\|$$
$$\le \alpha_n\,\|x\| + \|x_n - x_{n-1}\| + \alpha_n\,\|x\|,$$

we obtain

$$\lim_{n\to\infty}\|x_n - T(x_n)\| = 0,$$

i.e., $\{x_n\}$ is an a.f.p.s. for T. Next we prove that

$$\limsup_{n\to\infty}\ \langle T(x_n) - P(x), x - P(x)\rangle\ \le 0.$$

Indeed assume not. Then there exist $D > 0$ and a subsequence $\{x_{\psi(n)}\}$ of $\{x_n\}$ such that

$$\langle T(x_{\psi(n)}) - P(x), x - P(x)\rangle\ \ge D,\ n \ge 1.$$

We may assume that $\{x_{\psi(n)}\}$ weakly converges to $y \in K$. Since $\{x_n\}$ is an a.f.p.s., then T is demiclosed, i.e., $y \in F(T)$ and $\{T(x_{\psi(n)})\}$ also weakly converges to y. So we must have

$$\langle y - P(x), x - P(x)\rangle\ \ge D > 0.$$

This is a contradiction with the inequality (NPP). So we have

$$\limsup_{n\to\infty}\ \langle T(x_n) - P(x), x - P(x)\rangle\ \le 0.$$

Moreover using the Abel-Dini theorem [129], we conclude that $\sum_{n=0}^{\infty}\alpha_n = +\infty$ which implies

$$\lim_{n\to+\infty}\ \prod_{i=m}^{i=n}(1-\alpha_i) = 0, \tag{E1}$$

for any $m \in \mathbb{N}$. Fix $\varepsilon > 0$. Then there exists $n_0 \ge 0$ such that

$$\langle T(x_n) - P(x), x - P(x)\rangle\ \le \varepsilon,$$

and $\alpha_n\,\|x - P(x)\|^2 \le \varepsilon$, for any $n \ge n_0$. Then for any $n \ge n_0$, we have

$$\|x_{n+1} - P(x)\|^2 = \alpha_{n+1}^2\,\|x - P(x)\|^2$$
$$2\alpha_{n+1}(1-\alpha_{n+1})\ <T(x_n) - P(x), x - P(x)>$$
$$+ (1-\alpha_{n+1})^2\,\|T(x_n) - P(x)\|^2$$
$$\le \alpha_{n+1}^2\,\|x - P(x)\|^2 + 2\alpha_{n+1}\,\varepsilon + (1-\alpha_{n+1})\,\|T(x_n) - P(x)\|^2$$
$$\le \alpha_{n+1}\,3\varepsilon + (1-\alpha_{n+1})\,\|T(x_n) - P(x)\|^2$$
$$\le 3\alpha_{n+1}\,\varepsilon + (1-\alpha_{n+1})\,\|x_n - P(x)\|^2.$$

Hence

$$\|x_m - P(x)\|^2 \le 3\varepsilon + \|x_{n_0} - P(x)\|^2 \prod_{i=n_0+1}^{m} (1 - \alpha_i), \quad m > n_0.$$

Using the limit (E1), we get

$$\limsup_{n \to \infty} \|x_n - P(x)\|^2 \le 3\varepsilon.$$

Since ε was taken arbitrarily positive, we conclude that $\lim_{n \to \infty} \|x_n - P(x)\| = 0$.
$\square$

Remark 2.6. The proof of Theorem 2.22 shows how difficult such conclusions can be. Moreover, one has to keep in mind that moving from Hilbert spaces to Banach spaces adds more to the difficulty. Halpern's iteration (HI) finds its roots in the approximation used to find approximate fixed point sequences for a nonexpansive mapping.

It is worth mentioning that the Ishikawa process is more general than the Mann process. But research has been concentrated on the latter due probably to the reasons that the formulation of the Mann process is simpler, and that a convergence theorem for the Mann process may possibly lead to a convergence theorem for the Ishikawa process, provided the sequence $\{\beta_n\}$ satisfies certain appropriate conditions. However, the introduction of the Ishikawa process has value on its own merit. As a matter of fact, the Mann process may fail to converge while the Ishikawa process can still converge for some mappings in a Hilbert space, [45]. Reich [182] shows that if the underlying space X is uniformly convex and has a Fréchet differentiable norm, T is nonexpansive, and if the sequence $\{\alpha_n\}$ is such that $\sum_{n=0}^{\infty} \alpha_n(1 - \alpha_n) = +\infty$, then the sequence generated by the Mann processes converges weakly to a fixed point of T.

For a good reference to the theory of successive iterations, one may consult the book by Berinde [24]. Recently, Kozlowski proved the weak convergence of generalized Mann and Ishikawa processes for asymptotic pointwise nonexpansive mappings acting in Banach spaces with Opial property [133] and in uniformly smooth Banach spaces [137].

2.7 Metric Convexity and Convexity Structures

If (M,d) is a metric space and if $p,q \in M$ then a point $r \in M$ is said to be *metrically between* p and q if $p \ne r \ne q$ and $d(p,q) = d(p,r) + d(r,q)$. A metric space is said to be *metrically convex* if, given any two points $p,q \in M$, there exists at least one point $r \in M$ such that r is metrically between p and q.

The concept of metric convexity is fundamental in the study of abstract metric spaces. However, in its most general form, it fails to satisfy one of the fundamental properties of convexity in the algebraic sense (in a linear space); namely, if A and B are two (metrically) convex subsets of a metric space M, then *it need not be the case that $A \cap B$ is convex*. To see this, it is sufficient to consider the ordinary unit circle S in $\mathbb{R}^2$, where the distance between two points is taken to be the length of the shortest arc joining them. The upper half circle and lower half circle of S are each metrically convex but the intersection of these two sets is a set consisting of just two antipodal points.

A subset S of a metric space M is called a *metric segment* joining $p, q \in S$ if there exists a closed interval $[a, b]$ in the real line $\mathbb{R}$ and an isometry φ which maps $[a, b]$ onto S with $\varphi(a) = p$ and $\varphi(b) = q$. If a metric space M has the property that each two of its points are the endpoints of a metric segment, then clearly M is metrically convex, and it is natural to ask when the converse is true. The answer is surprisingly often. This is a fundamental result in the theory due to Karl Menger [162], one of the pioneers in the study of abstract metric spaces.

Theorem 2.23. *If M is a complete and metrically convex metric space, then each two points of M are joined by at least one metric segment of M.*

We shall be concerned almost entirely with complete metric spaces; hence with the stronger notion of convexity assured by Menger's Theorem whenever the concept arises. For reasons that will be clear later, the usefulness of metric convexity in a fixed point context is hindered by the fact that the family of all convex subsets of a metrically convex metric space need not be stable under intersections. A quick way to remedy this defect is to turn to a more abstract notion of convexity.

Definition 2.9. A family $\mathscr{C}$ of subsets of a set X is called an (abstract) *convexity structure* if

(1) Both $\emptyset$ and X are in $\mathscr{C}$.
(2) $\mathscr{C}$ is stable under intersections; that is, if $\{D_\alpha\}_{\alpha \in I}$ is any nonempty subfamily of $\mathscr{C}$ then $\bigcap_{\alpha \in I} D_\alpha \in \mathscr{C}$.
(3) $\mathscr{C}$ is stable for nested unions; that is, if $\{D_\alpha\}_{\alpha \in I}$ is any nonempty subfamily of $\mathscr{C}$ which is totally ordered by set inclusion, then $\bigcup_{\alpha \in I} D_\alpha \in \mathscr{C}$.

In the study of metric fixed point theory, closed balls turn out to play a fundamental role. A subset A of a bounded metric space M will be said to be *admissible* of A, and can be written as the intersection of a family of closed balls centered at points of M. The family $\mathscr{A}(M)$ of all admissible subsets of M enters into the study of metric fixed point theory in a very natural way, and this is therefore the obvious candidate for the needed underlying convexity structure. However, while this family satisfies assumptions (1) and (2), in general property (3) does not hold. (This fact cannot be rectified even if the collection is enlarged to include all ball intersections rather than just intersections of *closed* balls). Nonetheless, the family $\mathscr{A}(M)$ plays a major role in the metric fixed point theory. Thus we shall exchange property (3) with the assumption that the convexity structure under consideration contains all the closed balls of M.

The structure $\mathscr{A}(M)$ has one very distinctive advantage when compared with metric convexity. Any subset A of a metric space is contained in the set

$$cov(A) := \bigcap \{B : B \text{ is a closed ball in } M \text{ and } B \supseteq A\}.$$

Moreover $cov(A) \in \mathscr{A}(M)$. This provides a convenient analog to the concept of a "closed convex hull" in functional analysis, with the family $\mathscr{A}(M)$ replacing the family of all closed (algebraically) convex sets. This analogy does not quite work with metric convexity. It is natural to define the *convex hull* of a subset of A of a convex metric space M to be a set which is closed, convex, contains A, and for which none of its proper subsets has those properties. With this definition, the existence of a unique convex hull for each subset A of a complete convex metric space *which has unique metric segments* is immediate. Merely take the "convex hull" of A to be the intersection of all closed convex sets which contain A. However, problems arise in the general case since, as we have observed, the intersection of two convex sets need not be a convex set. On the other hand, if M is any compact metric space and if $A \subseteq M$, then A does have a convex hull, although the convex hull of A need not be unique. This is a classical result proved by Karl Menger in 1931. It rests on the following lemma.

Lemma 2.3. *Let $C_1 \supseteq C_2 \supseteq \cdots$ be a descending sequence of nonempty closed convex subsets of a compact metric space (M,d). Then $\bigcap_{n=1}^{\infty} C_n$ is nonempty and convex.*

Proof. The fact that the intersection is nonempty is immediate from compactness. Suppose $x, y \in \bigcap_{n=1}^{\infty} C_n$ with $x \neq y$. Then in each of the sets C_n there exists a point z_n such that

$$d(x,z_n) = d(y,z_n) = \frac{1}{2}d(x,y),$$

(where we used the fact that x and y are actually joined by a metric segment which lies in C_n.) By compactness of M the sequence $\{z_n\}$ has a subsequence $\{z_{n_k}\}$ which converges to a point $z \in M$ and since each of the sets C_n is closed, $z \in \bigcap_{n=1}^{\infty} C_n$. Since the metric d is continuous,

$$d(x,z) = d(y,z) = \frac{1}{2}d(x,y).$$

This proves that $\bigcap_{n=1}^{\infty} C_n$ is convex.

$\square$

Theorem 2.24. *If A is a nonempty subset of a compact and convex metric space M, then A has a convex hull.*

Proof. Let Σ denote the collection of all closed convex subsets of M which contain A. Observe that $\Sigma \neq \emptyset$ since $M \in \Sigma$. Since M is compact, for each $n \in \mathbb{N}$ there is a finite collection $\mathscr{U}_n := \{U_{n(1)}, U_{n(2)}, \cdots U_{n(k_n)}\}$ of open balls of radius $1/n$ which covers M. For each $K \in \Sigma$ there is a well defined integer $\mu_n(K)$ $(\mu_n(K) \leq k(n))$ such that K intersects $\mu_n(K)$ members of $\mathscr{U}_n$. Choose $C_1 \in \Sigma$ so that

$$\mu_1(C_1) = \inf\{\mu_1(C) : C \in \Sigma\}.$$

Having defined C_n choose $C_{n+1} \subseteq C_n$ so that

$$\mu_{n+1}(C_{n+1}) = \inf\{\mu_{n+1}(C) : C \in \Sigma \text{ and } C \subseteq C_n\}.$$

Then clearly $C_1 \supseteq C_2 \supseteq \cdots$ and $A \subseteq C := \bigcap_{n=1}^{\infty} C_n$. Also, by Lemma 2.3, C is convex. Suppose some $H \in \Sigma$ is a proper subset of C. Then there exists a point $c \in C \backslash H$ and in turn there exists an integer n for which the open ball $U(c; 1/n) \cap C \subset C \backslash H$. In particular, some member U of $\mathscr{U}_{2n}$ lies in $U(c; 1/n)$; hence

$$\mu_{2n}(H) < \mu_{2n}(C) \leq \mu_{2n}(C_{2n}),$$

contradicting the definition of $\mu_{2n}(C_{2n})$.
□

It is possible to give a much quicker proof of the above theorem using Zorn's Lemma. (The proof given here is Menger's original. It predates the discovery of Zorn's Lemma.) In fact, the following is true.

Theorem 2.25. *Let M be a convex metric space, and suppose the intersection every descending chain of closed convex subsets of M is itself convex. Then every nonempty subset A of M has a convex hull.*

Proof. As above, let Σ denote the family of all closed convex subsets of M which contain A, and order Σ by set inclusion . $\Sigma \neq \emptyset$ since $M \in \Sigma$, and by assumption every descending chain in $(\Sigma, \supseteq)$ is bounded below by its intersection. By Zorn's lemma Σ has a minimal element, completing the proof.
□

2.8 Uniformly Convex Metric Spaces

Let (M, d) be a metric space. Suppose that there exists a family $\mathscr{F}$ of metric segments such that any two points x, y in M are endpoints of a unique metric segment $[x, y] \in \mathscr{F}$ ($[x, y]$ is an isometric image of the real line interval $[0, d(x, y)]$). In this case, there exists a mapping $W : M \times M \times [0, 1] \to M$ such that for each $(x, y, \lambda) \in M \times M \times [0, 1]$ and $z \in M$, we have

$$d(z, W(x, y, \lambda)) \leq \lambda \, d(z, x) + (1 - \lambda) d(z, y).$$

Throughout let us write $W(x,y,\lambda) = \lambda x \oplus (1-\lambda)y$, whenever the choice of the mapping W is irrelevant. Moreover, if we have

$$d\left(\frac{1}{2}p \oplus \frac{1}{2}x, \frac{1}{2}p \oplus \frac{1}{2}y\right) \leq \frac{1}{2}d(x,y),$$

for all p,x,y in M, then M is said to be a *hyperbolic metric space* (see [82]). We will say that a subset C of a hyperbolic metric space M is convex if $[x,y] \subset C$ whenever x,y are in C. It is not hard to check that balls in hyperbolic metric spaces are convex.

Definition 2.10. Let (M,d) be a hyperbolic metric space. We say that M is *uniformly convex* (in short, UC) if for any $a \in M$, for every $r > 0$, and for each $\varepsilon > 0$

$$\delta(r,\varepsilon) = \inf\left\{1 - \frac{1}{r}\, d\left(\frac{1}{2}x \oplus \frac{1}{2}y, a\right) : d(x,a) \leq r, d(y,a) \leq r, d(x,y) \geq r\varepsilon\right\} > 0.$$

The definition of uniform convexity finds its origin in Banach spaces [50]. To the best of our knowledge, the first attempt to generalize this concept to metric spaces was done in [84]. The reader may also consult [82]. We have the following properties [113]:

(a) Let us observe that $\delta(r,0) = 0$, and $\delta(r,\varepsilon)$ is an increasing function of ε for every fixed r. Also, for $r_1 \leq r_2$ there holds

$$1 - \frac{r_2}{r_1}\left(1 - \delta\left(r_2, \varepsilon\frac{r_1}{r_2}\right)\right) \leq \delta(r_1,\varepsilon).$$

(b) If (M,d) is uniformly convex, then (M,d) is strictly convex, i.e., whenever

$$d\left(\frac{1}{2}x \oplus \frac{1}{2}y, a\right) = d(x,a) = d(y,a)$$

for any $x,y,a \in M$, then we must have $x = y$.

Obviously, normed linear spaces are hyperbolic spaces. As nonlinear examples, one can consider the Hadamard manifolds [38], the Hilbert open unit ball equipped with the hyperbolic metric [82], and the CAT(0) spaces [122, 120, 126, 156] (see example 2.3).

Example 2.2. [82] Let B be the open unit ball of the infinite Hilbert space H. On B, we consider the Poincaré hyperbolic metric ρ:

$$\rho(x,y) = \inf_{\gamma} \int_0^1 \alpha(\gamma(t), \gamma'(t))dt$$

where the infimum is taken over all piecewise differentiable curves such that $\gamma(0) = x$ and $\gamma(1) = y$, and where

$$\alpha(x,v) = \sup_{f \in \mathscr{F}} \|Df(x)(v)\|$$

with $\mathscr{F} = \{f : B \rightarrow B : \text{holomorphic}\}$. Note that (B,ρ) is a complete metric space (unbounded). Using the Mobius transformations, one can prove that

$$\rho(x,y) = \text{Argth}\left(1 - \sigma(x,y)\right)^{1/2} \quad \text{with} \quad \sigma(x,y) = \frac{(1-|x|^2)(1-|y|^2)}{|1- <x,y> |^2}.$$

Using several complicated computations, one may prove that for any x,y in B, and $\lambda \in [0,1]$, there exists a unique $z \in B$ such that

$$\rho(z,w) \leq \lambda \rho(x,w) + (1-\lambda)\rho(y,w)$$

for any $w \in B$. In other words, we have $z = W(x,y,\lambda) = \lambda x \oplus (1-\lambda)y$. In fact, the metric space (B,ρ) enjoys some geometric properties similar to uniformly convex Banach spaces. Moreover, we have

$$\begin{cases} \rho(a,x) \leq r \\ \rho(a,y) \leq r \\ \rho(x,y) \geq r\varepsilon \end{cases} \Longrightarrow \rho\left(a, \frac{1}{2}x \oplus \frac{1}{2}y\right) \leq r(1 - \delta(r,\varepsilon)),$$

for any a,x,y in B and any positive r and ε, where

$$\delta(r,\varepsilon) = 1 - \frac{1}{r}\text{Argth}\left(\frac{\sinh(r(1+\varepsilon/2))\sinh(r(1-\varepsilon/2))}{\cosh(r)}\right)^{1/2}.$$

It is easy to check that for $r > 0$ and $\varepsilon > 0$, we have $\delta(r,\varepsilon) > 0$, i.e., the open unit ball (B,ρ) is uniformly convex. Note that holomorphic mappings are nonexpansive mappings for the Poincaré distance ρ.

Example 2.3. [28] Let (X,d) be a metric space. A *geodesic* from x to y in X is a mapping c from a closed interval $[0,l] \subset \mathbb{R}$ to X such that $c(0) = x$, $c(l) = y$, and $d(c(t),c(t')) = |t-t'|$ for all $t,t' \in [0,l]$. In particular, c is an isometry and $d(x,y) = l$. The image α of c is called a *geodesic (or metric) segment* joining x and y. The space (X,d) is said to be a *geodesic space* if every two points of X are joined by a geodesic and X is said to be *uniquely geodesic* if there is exactly one geodesic joining x and y for each $x,y \in X$, which will be denoted by $[x,y]$, and called the segment joining x to y.

A *geodesic triangle* $\Delta(x_1,x_2,x_3)$ in a geodesic metric space (X,d) consists of three points x_1,x_2,x_3 in X (the *vertices* of Δ) and a geodesic segment between each pair of vertices (the *edges* of Δ). A *comparison triangle* for geodesic triangle $\Delta(x_1,x_2,x_3)$ in (X,d) is a triangle $\overline{\Delta}(x_1,x_2,x_3) := \Delta(\bar{x}_1,\bar{x}_2,\bar{x}_3)$ in $\mathbb{R}^2$ such that $d_{\mathbb{R}^2}(\bar{x}_i,\bar{x}_j) = d(x_i,x_j)$ for $i,j \in \{1,2,3\}$. Such a triangle always exists (see [28]).

A geodesic metric space is said to be a *CAT* (0) space if all geodesic triangles of appropriate size satisfy the following *CAT* (0) *comparison axiom:*

Let Δ be a geodesic triangle in X and let $\overline{\Delta} \subset \mathbb{R}^2$ be a comparison triangle for Δ. Then Δ is said to satisfy the *CAT* (0) *inequality* if for all $x, y \in \Delta$ and all comparison points $\bar{x}, \bar{y} \in \overline{\Delta}$,

$$d(x, y) \leq d(\bar{x}, \bar{y}).$$

Complete *CAT* (0) spaces are often called *Hadamard spaces (see [120])*. If x, y_1, y_2 are points of a *CAT* (0) space and y_0 is the midpoint of the segment $[y_1, y_2]$, i.e., $y_0 = \dfrac{y_1 \oplus y_2}{2}$, then the *CAT* (0) inequality implies:

$$d^2\left(x, \frac{y_1 \oplus y_2}{2}\right) \leq \frac{1}{2} d^2(x, y_1) + \frac{1}{2} d^2(x, y_2) - \frac{1}{4} d^2(y_1, y_2).$$

This inequality is the (CN) inequality of Bruhat and Tits [37]. As for the Hilbert space, the (CN) inequality implies that CAT(0) spaces are uniformly convex with

$$\delta(r, \varepsilon) = 1 - \sqrt{1 - \frac{\varepsilon^2}{4}}.$$

Next we give some results of uniformly convex metric spaces which are useful in the study of metric fixed point theory in these metric spaces. We start with the following technical lemma which will help establish a property in uniformly convex metric spaces equivalent to reflexivity in Banach spaces.

Lemma 2.4. *[113] Assume that (M, d) is uniformly convex. Let $\{C_n\} \subset M$ be a sequence of nonempty, nonincreasing, convex, bounded, and closed sets. Let $x \in M$ be such that*

$$0 < d = \lim_{n \to \infty} d(x, C_n) < \infty.$$

Let $x_n \in C_n$ be such that $d(x, x_n) \to d$. Then $\{x_n\}$ is a Cauchy sequence.

Recall that a hyperbolic metric space (M, d) is said to have the property (R) [105] if any nonincreasing sequence of nonempty, convex, bounded, and closed sets has a nonempty intersection. Our next result deals with the existence and the uniqueness of the best approximants of convex, closed, and bounded sets in a uniformly convex metric space. This result is of interest by itself as uniform convexity implies the property (R), which reduces to reflexivity in the linear case.

Theorem 2.26. *Assume that (M, d) is complete and uniformly convex. Let $C \subset M$ be nonempty, convex, and closed. Let $x \in M$ be such that $d(x, C) < \infty$. Then there exists a unique best approximant of x in C, i.e., there exists a unique $x_0 \in C$ such that*

$$d(x, x_0) = d(x, C).$$

The following result gives the analog result to the well-known theorem that states that any uniformly convex Banach space is reflexive. For a reference, the reader may read Theorem 2.1 in [82].

Theorem 2.27. *If (M,d) is complete and uniformly convex, then (M,d) has the property (R).*

Remark 2.7. Note that any hyperbolic metric space M which satisfies the property (R) is complete. Indeed, let $\{x_n\}$ be a Cauchy sequence in M. Denote

$$\varepsilon_n = \sup\{d(x_m, x_s) : m, s \geq n\}, \ n = 1, \cdots$$

Our assumption implies that $\lim_{n \to \infty} \varepsilon_n = 0$. In hyperbolic metric spaces, closed balls are convex. Therefore the property (R) implies that $\bigcap_{n \geq 1} B(x_n, \varepsilon_n) \neq \emptyset$. It is easy to check that this intersection is reduced to one point which is the limit of $\{x_n\}$.

The following technical lemma is useful to establish a metric version of the main results in [201] proved in the setting of Banach spaces.

Lemma 2.5. *[113] Let (M,d) be a uniformly convex metric space. Assume that there exists $R \in [0, +\infty)$ such that*

$$\limsup_{n \to \infty} d(x_n, a) \leq R, \ \limsup_{n \to \infty} d(y_n, a) \leq R, \ and \ \lim_{n \to \infty} d\left(a, \frac{1}{2}x_n \oplus \frac{1}{2}y_n\right) = R.$$

Then $\lim_{n \to \infty} d(x_n, y_n) = 0$.

A metric version of the parallelogram identity goes as follows (see [18, 82, 201]).

Theorem 2.28 (Parallelogram Inequality). *[113] Let (M,d) be uniformly convex. Fix $a \in M$. For each $0 < r$ and for each $\varepsilon > 0$ denote*

$$\Psi(r, \varepsilon) = \inf\left\{\frac{1}{2}d^2(a,x) + \frac{1}{2}d^2(a,y) - d^2\left(a, \frac{1}{2}x \oplus \frac{1}{2}y\right)\right\},$$

where the infimum is taken over all $x, y \in M$ such that $d(a,x) \leq r$, $d(a,y) \leq r$, and $d(x,y) \geq r\varepsilon$. Then $\Psi(r, \varepsilon) > 0$ for any $0 < r$ and for each $\varepsilon > 0$. Moreover, for a fixed $r > 0$, we have

(a) $\Psi(r, 0) = 0$
(b) $\Psi(r, \varepsilon)$ is a nondecreasing function of ε;
(c) if $\lim_{n \to \infty} \Psi(r, t_n) = 0$, then $\lim_{n \to \infty} t_n = 0$.

The concept of p-uniform convexity was used extensively by Xu [201] (see also [16] p. 310); its nonlinear version for $p = 2$ is given below [113].

Definition 2.11. We will say that (M,d) is *2-uniformly convex* if

$$c_M = \inf\left\{\frac{\Psi(r, \varepsilon)}{r^2 \varepsilon^2} : r > 0, \ \varepsilon > 0\right\} > 0.$$

Note that (M,d) is 2-uniformly convex if and only if

$$\inf\left\{\frac{\delta(r,\varepsilon)}{\varepsilon^2} : r > 0,\ \varepsilon > 0\right\} > 0.$$

Example 2.4. Let $M,d)$ be $CAT(0)$ geodesic metric space. The (CN) inequality implies that

$$\Psi(r,\varepsilon) = \frac{r^2\varepsilon^2}{4}.$$

This clearly implies that any CAT(0) space is 2-uniformly convex with $c_M = \dfrac{1}{4}$ [113].

As an application of the previous technical results, we discuss the existence of fixed points of uniformly Lipschitzian mappings [80].

Definition 2.12. A mapping $T : C \rightarrow C$ (a subset of M) is said to be *uniformly Lipschitzian* if there exists a non-negative number k such that

$$d(T^n(x), T^n(y)) \leq k\, d(x,y),$$

for all x and y in C, and $n \geq 1$. The smallest such constant k will be denoted by $\lambda(T)$.

It is well known that if a mapping is uniformly Lipschitzian, then one may find an equivalent distance for which the mapping is nonexpansive (see [82, pages 34–38]). Indeed, let $T : C \rightarrow C$ be uniformly Lipschitzian. Denote

$$\rho(x,y) = \sup\{d(T^n x, T^n y) : n = 0,1,2,\ldots\}$$

for all $x,y \in C$. Then ρ is an equivalent distance to the distance d and relative to which T is nonexpansive. In this context, it is natural to ask the question; if a set C has the fixed point property (fpp) for nonexpansive mappings with respect to the metric d, then does C also have (fpp) for mappings which are nonexpansive relative to an equivalent metric? This is known as the stability of (fpp). The first result in this direction is due to Goebel and Kirk [80]. Motivated by such questions, many investigated the fixed point property of uniformly Lipschitzian mappings in uniformly convex Banach spaces and hyperbolic metric spaces.

Recall that the normal structure coefficient $N(M)$ of the hyperbolic metric space M is defined (see [39]):

$$N(M) = \inf\left\{\frac{\mathrm{diam}(C)}{R(C)} : C \text{ bounded convex subset of } M \text{ with diam}(C) > 0\right\},$$

where $\mathrm{diam}(C)$ is the diameter of C, and $R(C)$ is the Chebyshev radius of C. Recall that we have

$$\mathrm{diam}(C) = \sup\{d(x,y) : x,y \in C\}, \text{ and } R(C) = \inf\{\sup_{y\in C} d(x,y); x \in C\}.$$

The equation (HE) satisfied in Hilbert spaces has a nonlinear version in uniformly convex hyperbolic metric spaces [113]:

Theorem 2.29. *Let (M,d) be hyperbolic metric space and let C be a nonempty, closed, and convex subset of M. Assume that M is 2-uniformly convex. Let $\{x_n\}$ be a bounded sequence in C. Then there exists a unique point $z \in C$ such that*

$$\limsup_{n\to\infty} d^2(x_n,z) + 2c_M\, d^2(z,x) \le \limsup_{n\to\infty} d^2(x_n,x), \qquad \text{(HEM)}$$

for any $x \in C$.

The next result is the nonlinear version of Theorem 3 of [201].

Theorem 2.30. *[113] Let (M,d) be hyperbolic metric space which is 2-uniformly convex. Let C be a nonempty, closed, convex, and bounded subset of M. Let $T : C \to C$ be uniformly Lipschitzian with*

$$\lambda(T) < \left(\frac{1 + \sqrt{1 + 8c_M N(M)^2}}{2} \right)^{1/2}.$$

Then T has a fixed point in C.

2.9 More on Convexity Structures

Kirk's fixed point theorem involves some kind of compactness and the normal structure property. Both properties are easy to define in metric spaces. The first attempt to extend Kirk's theorem to metric spaces was done by Takahashi [197]. He considered compact metric spaces in his generalization, which was very restrictive since Kirk's theorem assumes weak compactness, not strong compactness. Penot [176], on the other hand, defined compactness for convexity structures, which leads to weak compactness in the linear case. Indeed, a convexity structure Σ is said to be compact if and only if every family of subsets of Σ which has the finite intersection property has a nonempty intersection, i.e., if $(A_i)_{i\in I}$, with $A_i \in \Sigma$, then $\bigcap_{i\in I} A_i \ne \emptyset$ provided $\bigcap_{i\in I_f} A_i \ne \emptyset$ for any finite subset I_f of I. As for the normal structure property, Penot introduced such a property in convexity structure as follows:

Definition 2.13. The convexity structure Σ is said to be normal if and only if for any nonempty and bounded $A \in \Sigma$ not reduced to one point, there exists $a \in A$ such that

$$\sup\{d(a,x): x \in A\} < \sup\{d(x,y): x,y \in A\} = diam(A).$$

Penot's formulation of Kirk's fixed point theorem becomes

Theorem 2.31. *[176] Let (M,d) be a nonempty bounded metric space which possesses a convexity structure which is compact and normal. Then every nonexpansive mapping $T : M \to M$ has a fixed point.*

In the original Kirk's theorem, the Banach space is supposed to have a normal structure which means that the family of all convex sets is normal. However, this family is a large one, and contains the admissible sets. For example, the Banach space l^∞ is a good example which illustrates the power behind Penot's formulation. Indeed, l^∞ fails to have the normal structure property but $\mathscr{A}(l^\infty)$ is compact and normal, which implies the following theorem discovered separately by Sine [190] and Soardi [191] in 1979.

Theorem 2.32. *Let A be a nonempty admissible subset of l^∞. Then every nonexpansive mapping $T : A \to A$ has a fixed point.*

Remark 2.8. In the original proof of Kirk's theorem, the weak compactness is used to prove the existence of a minimal invariant set via Zorn's lemma. Gillespie and Williams [77] showed that a constructive proof may be found which uses only countable compactness. In other words, the convexity structure Σ is assumed to satisfy a countable intersection property, i.e., for any $(A_n)_{1 \le n}$, with $A_n \in \Sigma$, then $\bigcap_{n=1}^{\infty} A_n \neq \emptyset$ provided $\bigcap_{n=1}^{m} A_n \neq \emptyset$ for any $m \ge 1$. This weakening is very important, since in many practical cases we do not have a compactness generated by a topology, but a compactness defined sequentially. The latter usually generates some kind of countable compactness. Note also, that if the convexity structure Σ is uniformly normal, then it is countably compact [12, 105]. This is an amazing metric translation of a well known similar result in Banach spaces due to Maluta [160]. It is natural to ask whether a convexity structure which is countably compact is actually compact. The answer is yes if we are dealing with uniform normal structure [149]. In fact, if $\mathscr{A}(M)$ is countably compact and normal, then $\mathscr{A}(M)$ is compact.

The special role played by l^∞ above is not unique. In fact, it is a special case of a more general theory known as *hyperconvex metric spaces* [10].

2.10 Common Fixed Points

Since the early stages of development of the metric fixed point theory, many asked whether any results that dealt with the existence of the fixed point for one mapping extends to a family of mappings. In this case, we talk about the common fixed point. In particular, we deal with the question of whether the conclusion of Kirk's fixed point theorem holds for a family of nonexpansive mappings. Since the fixed point set remained invariant when the mappings involved were commutative, it was natural to assume that the given family is commutative. Since the beginning it was clear that in order to obtain a common fixed point result, one has to study the structure of the fixed point set. This opened a new area of study in metric fixed point theory.

Definition 2.14. [108] Let (M,d) be a metric space. A subset A of M is said to be a *one-local retract* of M if for every family $\{B_i\}_{i\in I}$ of closed balls centered in A with nonempty intersection, it is the case that $A \cap \left(\bigcap_{i\in I} B_i \right) \neq \emptyset.$

It is immediate that each nonexpansive retract of M is a one-local retract (but not conversely). Recall that $A \subset M$ is a nonexpansive retract of M if there exists a nonexpansive map $r : M \to A$ such that $r(a) = a$, for every $a \in A$.

The following theorem will shed some light on this notion.

Theorem 2.33. *[108] Let (M,d) be a metric space and N be a nonempty subset of M. The following are equivalent.*

(i) N is a 1-local retract of M.
(ii) N is a nonexpansive retract of $N \cup \{x\}$, for every $x \in M$.
(iii) Let H be a metric space and $T : N \to H$ be a Lipschitzian map, i.e., there exists $\alpha \geq 0$ such that $d_H(T(a),T(b)) \leq \alpha\, d_M(a,b)$, for every $a,b \in N$, where d_H and d_M stand respectively for the distances on H and M. Then, for every $x \in M$, there exists an extension $T^* : N \cup \{x\} \to H$, such that $d_H(T^*(a),T^*(b)) \leq \alpha\, d_M(a,b)$, for every $a,b \in N \cup \{x\}$, and $T^*(a) = T(a)$ for every $a \in N$.

Let (M,d) be a metric space. If we assume that $\mathscr{A}(M)$ is compact and normal, then it is not the case that these assumptions remain valid for $\mathscr{A}(N)$, where $N \subset M$. Recall that these assumptions are at the core of the Kirk's fixed point theorem in metric spaces.

Theorem 2.34. *[108] Let (M,d) be a metric space and N be a nonempty subset one-local retract of M. Assume that $\mathscr{A}(M)$ is compact and normal. Then $\mathscr{A}(N)$ is compact and normal.*

The idea of one-local retract was inspired by the beautiful work of Baillon [13]. The following result finds its roots in Baillon's work.

Theorem 2.35. *[108] Let (M,d) be a metric space for which $\mathscr{A}(M)$ is compact and normal. Let $(M_\beta)_{\beta\in\Gamma}$ be a decreasing family of nonempty one-local retracts of M, where Γ is totally ordered. Then $\bigcap_{\beta\in\Gamma} M_\beta$ is not empty and is a one-local retract of M.*

Let us point out that under Kirk's fixed point theorem, the fixed point set of a nonexpansive mapping is a one-local retract. Indeed let $M,d)$ be a metric space for which $\mathscr{A}(M)$ is compact and normal. Let $T : M \to M$ be a nonexpansive map. We know that $F(T)$ is not empty. Let $\{B_i\}_{i\in I}$ be a family of closed balls centered in $F(T)$ with a nonempty intersection. Since T is nonexpansive, then $S = \bigcap_{i\in I} B_i$ is T-invariant, i.e.,

$T(S) \subset S$. The set S belongs to $\mathscr{A}(M)$, which implies that $\mathscr{A}(S)$ is compact and normal. Therefore Kirk's fixed point theorem implies that T has a fixed point in S, i.e., $S \cap F(T) \neq \emptyset$. This completes the proof of our claim.

The next results discuss the existence of fixed points of commutative family of nonexpansive mappings and the structure of their common fixed point set.

Theorem 2.36. *Let (M,d) be a bounded metric space such that $\mathscr{A}(M)$ is compact and normal. Then any commuting family of nonexpansive mappings $\{T_i\}_{i \in I}$, $T_i :$ $M \to M$, has a common fixed point. Moreover if we denote by $F\left(\{T_i\}_{i \in I}\right)$ the set of the common fixed points, then $F\left(\{T_i\}_{i \in I}\right)$ is a one-local retract of M.*

Proof. Let M and $\{T_i\}_{i \in I}$ be as assumed in the theorem. Let I_f be a nonempty finite subset of I. For any $i, j \in I_f$, we have $F(T_i)$ is nonempty and is a one-local retract of M. Hence, $\mathscr{A}(F(T_i))$ is normal and compact. Since T_j commutes with T_i, we have $T_j(F(T_i)) \subset F(T_i)$. Hence, T_j has a fixed point in $F(T_i)$ and its fixed points set in $F(T_i)$ is a one-local retract, i.e., $F(T_i) \cap F(T_j)$ is a one-local retract. Hence, $\bigcap_{i \in I_f} F(T_i)$ is not empty and is a one-local retract. Therefore, the conclusion of Theorem 2.36 follows from Theorem 2.35.

$\square$

Chapter 3
Modular Function Spaces

In this chapter we recall the foundations of the theory of modular function spaces defined by convex regular function modulars. The results discussed in this chapter will be used throughout the rest of the book. This chapter also presents the reader with an exhaustive list of examples that will frequently reoccur in later parts of the book. In Sect. 3.4 "Generalizations and Special Cases", we summarize a more general approach, based on [132], which allows nonconvex function modulars and extends the setting to spaces of function with values in Banach spaces. While these results are not necessary for understanding the contents of the remaining chapters, the general framework introduced in this section may help to appreciate the full potential of the theory of modular function spaces.

3.1 Foundations

Let Ω be a nonempty set and Σ be a nontrivial σ-algebra of subsets of Ω. Let $\mathscr{P}$ be a nontrivial δ-ring of subsets of Ω which means that $\mathscr{P}$ is closed with respect to forming of countable intersections, and finite unions and differences. Assume further that $E \cap A \in \mathscr{P}$ for any $E \in \mathscr{P}$ and $A \in \Sigma$. Let us assume that there exists an increasing sequence of sets $K_n \in \mathscr{P}$ such that $\Omega = \bigcup K_n$. By $\mathscr{E}$ we denote the linear space of all simple functions with supports from $\mathscr{P}$. By $\mathscr{M}_\infty$ we denote the space of all extended measurable functions, that is, all functions $f : \Omega \to [-\infty, \infty]$ such that there exists a sequence $\{g_n\} \subset \mathscr{E}$, $|g_n| \leq |f|$ and $g_n(\omega) \to f(\omega)$ for all $\omega \in \Omega$. By 1_A we denote the characteristic function of the set A.

Definition 3.1. Let $\rho : \mathscr{M}_\infty \to [0, \infty]$ be a nontrivial, convex, and even function. We say that ρ is a *regular convex function pseudomodular* if:

(a) $\rho(0) = 0$;
(b) ρ is *monotone* , that is, $|f(\omega)| \leq |g(\omega)|$ for all $\omega \in \Omega$ implies $\rho(f) \leq \rho(g)$, where $f, g \in \mathscr{M}_\infty$;

© Springer International Publishing Switzerland 2015
M. A. Khamsi, W. M. Kozlowski, *Fixed Point Theory in Modular Function Spaces*,
DOI 10.1007/978-3-319-14051-3_3

(c) ρ is *orthogonally subadditive* , that is, $\rho(f1_{A\cup B}) \leq \rho(f1_A) + \rho(f1_B)$ for any
  $A, B \in \Sigma$ such that $A \cap B \neq \emptyset$, where $f \in \mathcal{M}_\infty$;
(d) ρ has the *Fatou property*: $|f_n(\omega)| \uparrow |f(\omega)|$ for all $\omega \in \Omega$ implies $\rho(f_n) \uparrow \rho(f)$,
  where $f \in \mathcal{M}_\infty$;
(e) ρ is *order continuous* in $\mathcal{E}$, that is, $g_n \in \mathcal{E}$ and $|g_n(\omega)| \downarrow 0$ implies $\rho(g_n) \downarrow 0$.

Similarly as in the case of measure spaces, we say that a set $A \in \Sigma$ is ρ-*null* if
$\rho(g1_A) = 0$ for every $g \in \mathcal{E}$. We say that a property holds ρ-*almost everywhere* if
the exceptional set is ρ-null. As usual we identify any pair of measurable sets whose
symmetric difference is ρ-null as well as any pair of measurable functions differing
only on a ρ-null set. With this in mind we define

$$\mathcal{M} = \{f \in \mathcal{M}_\infty : |f(\omega)| < \infty \, \rho - a.e.\},$$

where each $f \in \mathcal{M}$ is actually an equivalence class of functions equal ρ-a.e. rather
than an individual function.

Definition 3.2. Let ρ be a regular convex function pseudomodular.

(a) We say that ρ is a *regular convex function semimodular* if $\rho(\alpha f) = 0$ for every
  $\alpha > 0$ implies that $f = 0 \, \rho - a.e.$;
(b) We say that ρ is a *regular convex function modular* if $\rho(f) = 0$ implies that
  $f = 0 \, \rho - a.e.$

The class of all nonzero regular convex function modulars defined on Ω is denoted
by $\mathfrak{R}$.

Remark 3.1. Let us denote $\rho(f, E) = \rho(f1_E)$ for $f \in \mathcal{M}$, $E \in \Sigma$. Also, by convention for $\alpha > 0$ we will write $\rho(\alpha, E)$ instead of $\rho(\alpha 1_E)$. We will use these
notations when convenient. It is easy to prove that $\rho(f, E)$ is a convex function
modular in the sense of Definition 2.1.1 in [132] (more precisely, it is a convex function modular with the Fatou property). Therefore, we can use results of
the standard theory of modular function spaces as per the framework defined by
Kozlowski in [130, 131, 132, 136] and sketched in Sect. 3.6.1 of this book; see also
[166, 167, 169, 170] for the basics of the general modular theory.

Remark 3.2. Note that if ρ is a regular convex function modular, then to verify that
a set E is ρ-null it suffices to prove that there exists $\alpha > 0$ such that $\rho(\alpha, E) = 0$.

Definition 3.3. [130, 131, 132] Let ρ be a convex function modular.

(a) A *modular function space* is the vector space $L_\rho(\Omega, \Sigma)$, or briefly L_ρ, defined
  by
$$L_\rho = \{f \in \mathcal{M} : \rho(\lambda f) \to 0 \text{ as } \lambda \to 0\}.$$

(b) The following formula defines a norm in L_ρ (frequently called the *Luxemburg*
  norm) :
$$\|f\|_\rho = \inf\{\alpha > 0 : \rho(f/\alpha) \leq 1\}.$$

Remark 3.3. It is not difficult to prove that $\|\cdot\|_\rho$ defines actually a norm such that $\|f\|_\rho \le \|g\|_\rho$ whenever $|f)| \le |g|$ ρ-a.e. It is also straightforward to demonstrate that $\|f_n\|_\rho \to 0$ if and only if $\rho(\alpha f_n) \to 0$ for every $\alpha > 0$. While it is interesting to investigate the Luxemburg norm related properties of modular function spaces, we focus our attention on those fixed point results that can be formulated purely in the term of function modulars. The relationship between modulars and respective Luxemburg norms is discussed in Propositions 3.7, 3.12, 3.13 below, see also Example 5.4.

In the fixed point theory in modular function spaces, we use often a notion of ρ-convergence. It is described, among other important terms, in the following definition.

Definition 3.4. Let $\rho \in \mathfrak{R}$.

(a) We say that $\{f_n\}$ is ρ-*convergent* to f and write $f_n \to f$ (ρ) if $\rho(f_n - f) \to 0$.
(b) A sequence $\{f_n\}$ where $f_n \in L_\rho$ is called a ρ-*Cauchy* sequence if $\rho(f_n - f_m) \to 0$ as $n, m \to \infty$.
(c) A set $B \subset L_\rho$ is called ρ-*closed* if for any sequence of $f_n \in B$, the convergence $f_n \to f$ (ρ) implies that f belongs to B.
(d) A set $B \subset L_\rho$ is called ρ-*bounded* if its ρ-diameter is finite; the ρ-diameter of B is defined as $\delta_\rho(B) = \sup\{\rho(f - g) : f \in B, g \in B\}$.
(e) A set $B \subset L_\rho$ is called *strongly ρ-bounded* if there exists $\beta > 1$ such that $M_\beta(B) = \sup\{\rho(\beta(f - g)) : f \in B, g \in B\} < \infty$.
(f) A set $B \subset L_\rho$ is called ρ-*compact* if for any $\{f_n\}$ in B, there exists a subsequence $\{f_{n_k}\}$ and an $f \in B$ such that $\rho(f_{n_k} - f) \to 0$.
(g) A set $B \subset L_\rho$ is called ρ-*a.e. closed* if for any $\{f_n\}$ in B, which ρ-a.e. converges to some f, we have $f \in B$.
(h) A set $B \subset L_\rho$ is called ρ-*a.e. compact* if for any $\{f_n\}$ in B, there exists a subsequence $\{f_{n_k}\}$, which ρ-a.e. converges to some $f \in B$.
(i) Let $f \in L_\rho$ and $B \subset L_\rho$. The ρ-*distance* between f and B is defined as

$$d_\rho(f, B) = \inf\{\rho(f - g) : g \in B\}.$$

Let us start with the following easy but important fact.

Proposition 3.1. *Assume that a sequence of $f_n \in L_\rho$ ρ-converges to an $f \in \mathcal{M}$, that is, $\rho(f_n - f) \to 0$. Then $f \in L_\rho$.*

Proof. Fix $\varepsilon > 0$ and take any $0 < \lambda < \dfrac{1}{2}$. Observe that

$$\rho(\lambda f) \le \frac{1}{2}\rho(2\lambda(f - f_n)) + \frac{1}{2}\rho(\lambda f_n) \le \frac{1}{2}\rho(f - f_n) + \frac{1}{2}\rho(\lambda f_n) \le \frac{\varepsilon}{2} + \frac{1}{2}\rho(\lambda f_n),$$

for $n \ge n_0$. Since $f_{n_0} \in L_\rho$, it follows that there exists $\lambda_0 > 0$ such that $\rho(\lambda f_{n_0}) < \varepsilon$ for $\lambda < \lambda_0$. Hence, $\rho(\lambda f) \le \varepsilon$ for $\lambda < \lambda_0$, which means that $f \in L_\rho$.
$\square$

It is very important to know that the ρ-limit is actually uniquely defined for a given sequence of $f_n \in L(\rho)$.

Proposition 3.2. *Assume that we have two sequences of functions $f_n \in L_\rho$ and $g_n \in L_\rho$ such that $\rho(f_n - f) \to 0$ and $\rho(g_n - f) \to 0$. Then $f = g$ ρ-a.e.*

Proof. Since

$$\rho\left(\frac{f-g}{2}\right) \le \frac{1}{2}\rho(f - f_n) + \frac{1}{2}\rho(g - g_n) \to 0,$$

we have $\rho\left(\dfrac{f-g}{2}\right) = 0$, which implies that $f = g$ ρ-a.e. $\square$

Let us note, however, that ρ-convergence does not necessarily imply ρ-Cauchy condition. Also, $f_n \to f$ (ρ) does not imply in general $\lambda f_n \to \lambda f$ (ρ), for $\lambda > 1$.

The following useful result is a direct consequence of the order continuity property of function modulars, see part (e) of Definition 3.1.

Proposition 3.3. *If $\{f_n\}$ converges uniformly to f on a set $E \in \mathscr{P}$, then for every $\alpha > 0$,*

$$\lim_{n\to\infty} \rho\left(\alpha(f_n - f), E\right) = 0.$$

The next result follows from the Fatou property of ρ.

Proposition 3.4. *Let $\rho \in \mathfrak{R}$ and $f, f_n \in \mathscr{M}_\infty$. If $f_n \to f$ $\rho - a.e.$, then*

$$\rho(f) \le \liminf_{n\to\infty} \rho(f_n).$$

Proof. Set $g_n = \inf_{k \ge n} |f_k|$. Hence,

$$\liminf_{n\to\infty} \rho(g_n) \le \liminf_{n\to\infty} \rho(f_n).$$

Note that, in virtue of the Fatou property, the left-hand side of the above inequality is equal to $\rho(f)$, since $|g_n| \uparrow |f|$ almost everywhere with respect to ρ. $\square$

The next theorem describes an important relationship between the ρ-convergence and the convergence ρ-a.e.

Theorem 3.1. *Let $\rho \in \mathfrak{R}$. Assume that $\{f_n\}$ satisfies ρ-Cauchy condition. Then, there exists a subsequence $\{f_{n_k}\}$ of $\{f_n\}$ and a function $f \in \mathscr{M}$ such that $f_{n_k} \to f$ ρ-a.e.*

Proof. Denoting $E_{n,m}(\varepsilon) = \{\omega \in \Omega : |f_n(\omega) - f_m(\omega)| \ge \varepsilon\}$, we get

$$\rho(\varepsilon, E_{n,m}(\varepsilon)) \le \rho(f_n - f_m, E_{n,m}(\varepsilon)) \le \rho(f_n - f_m) \to 0.$$

Hence, for any $k \in \mathbb{N}$, there exists $n_0(k) \in \mathbb{N}$ such that if $n \ge n_0(k)$ and $m \ge n_0(k)$, then

$$\rho\left(2^{-k}, E_{n,m}(2^{-k})\right) < 2^{-k}.$$

Taking $n_1 = n_0(1)$, $n_k = \max\{n_{k-1}, n_0(k)\}$, we get the sequence $g_k = f_{n_k}$. Define

$$E_k = \{\omega \in \Omega : |g_k(\omega) - g_{k+1}(\omega)| \geq 2^{-k}\},$$

and observe that $\rho(2^{-k}, E_k) < 2^{-k}$. Denote $G_j = \bigcup_{m=j}^{\infty} E_m$ and observe that for every $j \geq i \geq k$ and $\omega \in \Omega \setminus G_k$, there holds

$$|g_i(\omega) - g_j(\omega)| \leq \sum_{m=i}^{\infty} |g_m(\omega) - g_{m+1}(\omega)| \leq 2^{1-i}.$$

Define a measurable function $h = \sum_{j=1}^{\infty} 2^{-j} 1_{G_j \setminus G_{j+1}}$ and calculate

$$\rho(h, G_k) = \rho\left(h, \bigcup_{n=k}^{\infty} (G_n \setminus G_{n+1})\right) \leq$$

$$\sum_{n=k}^{\infty} \rho(h_n, G_n \setminus G_{n+1}) \leq \sum_{n=k}^{\infty} \rho(2^{-n}, E_n) \leq 2^{1-k}.$$

Hence, $\rho(h, G) = 0$, where $G = \bigcap_{j=1}^{\infty} G_j$. Observe that if ω does not belong to G, then there exists k such that for any $j \geq i \geq k$, there holds $|g_i(\omega) - g_j(\omega)| < 2^{1-i}$. Therefore, there exists a real number $f(\omega)$ such that $f_{n_i}(\omega) = g_i(\omega) \to f(\omega)$. Clearly, the function f is measurable. It remains to be shown that G is ρ-null. Note that $G = \bigcup_{n=1}^{\infty} D_n$, where $D_n = G \cap \{\omega \in \Omega : h(\omega) \geq n^{-1}\}$. Therefore,

$$0 = \rho(h, G) \geq \rho(h, D_n) \geq \rho(n^{-1}, D_n).$$

Hence, D_n is ρ-null for every $n \in \mathbb{N}$, which implies that G is ρ-null. □

We will use the above result to prove the fundamental completeness theorem.

Theorem 3.2. *Let $\rho \in \mathfrak{R}$. L_ρ is complete with respect to the ρ-convergence.*

Proof. Let a sequence $\{f_n\} \subset L_\rho$ be ρ-Cauchy. From Theorem 3.1, there exist a subsequence $\{f_{n_k}\}$ and a function $f \in \mathcal{M}$ such that $f_{n_k} \to f$ ρ-a.e. Fix $\varepsilon > 0$ and note that there exists $n_0 \in \mathbb{N}$ such that $\rho(f_m - f_n) < \varepsilon$ provided $m, n \geq n_0$. Using Proposition 3.4, we get

$$\rho(f - f_n) \leq \liminf_{k \to \infty} \rho(f_{n_k} - f_n) \leq \varepsilon$$

for $n \geq n_0$. It remains to be proved that $f \in L_\rho$. Let $\lambda_n \to 0$. Fix $\varepsilon > 0$ and take $k \in \mathbb{N}$ such that $\rho(f - f_k) < \varepsilon$. For n sufficiently large, we obtain

$$\rho(\lambda_n(f - f_k)) \leq \rho(f - f_k) < \varepsilon,$$

which implies that $\rho(\lambda_n(f - f_k)) \to 0$ as $n \to \infty$. Thus, $f_k - f$ belongs to L_ρ. Since $f_k \in L_\rho$ and L_ρ is a linear space, then f belongs to L_ρ as claimed. $\quad\square$

Let us follow [132] and prove a modular version of the *Egoroff theorem*, which plays a critical role in the theory of modular function spaces.

Theorem 3.3. *Let $\rho \in \mathfrak{R}$ and let $f_n \to f$ ρ − a.e. where f, $f_n \in \mathcal{M}$. There exists a nondecreasing sequence of sets $H_k \in \mathscr{P}$ such that $H_k \uparrow \Omega$ and $\{f_n\}$ converges uniformly to f on every H_k.*

Proof. Let us fix $\alpha > 0$ and recall that there exists an nondecreasing sequence of sets $K_n \in \mathscr{P}$ such that $\Omega = \bigcup K_n$. Let us fix temporarily $n \in \mathbb{N}$. We shall prove that there exist sets $F_n \in \mathscr{P}$ such that $F_n \subset K_n$ and

$$\rho(\alpha, K_n \setminus F_n) < \frac{1}{2^n} \tag{3.1}$$

and $\{f_m\}$ converges uniformly to f on each F_n. By omitting, if necessary, a ρ-null set from Ω, we may assume that the sequence $\{f_n\}$ pointwise converges to f everywhere. Set

$$E_k^m = \bigcap_{i=k}^{\infty} \left\{ \omega \in K_n : |f_i(\omega) - f(\omega)| < \frac{1}{m} \right\}$$

and observe that $E_1^m \subset E_2^m \subset \ldots$ and, since $\{f_n\}$ pointwise converges to f everywhere, that

$$K_n \subset \bigcup_{k=1}^{\infty} E_k^m$$

for every $m \in \mathbb{N}$. Then there exists a positive integer $n_0 = n_0(m)$ such that

$$\rho(\alpha, K_n \setminus E_{n_0(m)}^m) < \frac{1}{2^{m+n}}.$$

Denoting

$$F_n = \bigcap_{m=1}^{\infty} E_{n_0(m)}^m,$$

we have

$$\rho(\alpha, K_n \setminus F_n) = \rho\left(\alpha, K_n \setminus \bigcap_{m=1}^{\infty} E_{n_0(m)}^m\right)$$

$$= \rho\left(\alpha, \bigcup_{m=1}^{\infty} \left(K_n \setminus E_{n_0(m)}^m\right)\right) \le \sum_{m=1}^{\infty} \rho\left(\alpha, K_n \setminus E_{n_0(m)}^m\right) < \frac{1}{n},$$

proving (3.1). It follows from the definition of F_n that to every $m \in \mathbb{N}$ there corresponds $k_m \in \mathbb{N}$ such that for $i \ge k_m$,

$$|f_i(\omega) - f(\omega)| < \frac{1}{m},$$

for all $\omega \in F_n$, which means that $\{f_i\}$ converges uniformly to f on F_n. Let us define now

$$H_k = \bigcup_{n=1}^{k} F_n, \ H = \bigcup_{k=1}^{\infty} H_k.$$

Clearly, $\{H_k\}$ is a nondecreasing sequence of sets from $\mathscr{P}$ such that

$$\rho(\alpha, K_k \setminus H) \le \rho(\alpha, K_k \setminus H_k) \le \rho(\alpha, K_k \setminus F_k) < \frac{1}{2^k}$$

and $\{f_m\}$ converges uniformly to f on every H_k. It remains to be proved that $\Omega \setminus H$ is ρ-null. Indeed,

$$\rho(\alpha, \Omega \setminus H) \le \rho(\alpha, K_k \setminus H) + \sum_{n=k+1}^{\infty} \rho(\alpha, K_n \setminus H) < \frac{1}{2^k} + \sum_{n=k+1}^{\infty} \frac{1}{2^n}$$

for every $k \in \mathbb{N}$. Thus, $\rho(\alpha, \Omega \setminus H) = 0$. The proof is complete. $\square$

The following useful result is an easy corollary from Theorems 3.1 and 3.3.

Proposition 3.5. *Let $\rho \in \mathfrak{R}$. Assume that $\rho(f_n - f) \to 0$, then there exists a subsequence $\{f_{n_k}\}$ of $\{f_n\}$ such that $f_{n_k} \to f$ ρ-a.e.*

Proof. Arguing like in the proof of Theorem 3.1, there exists a subsequence $\{f_{n_k}\}$ of $\{f_n\}$ and a measurable function g such that $f_{n_k} \to g$ ρ-a.e. We need to show that $f = g$ ρ-a.e. This is not obvious as the modular ρ does not need to satisfy the triangle property. However, by the Egoroff theorem 3.3, there exists an increasing sequence of sets $\{H_m\}$ from $\mathscr{P}$ such that $\bigcup_{m=1}^{\infty} H_m = \Omega$ and f_n converges to g uniformly on each H_m. Hence,

$$\rho\left(\frac{f-g}{2}, H_m\right) \le \rho(f - f_{n_k}, H_m) + \rho(f_{n_k} - g, H_m) \to 0, \quad \text{as } k \to \infty.$$

Hence, $\rho(f - g, H_m) = 0$ for every m, which implies that $f = g$ ρ-a.e. $\square$

As an immediate consequence of Propositions 3.4 and 3.5, we get the following important result.

Proposition 3.6. *Let* $\rho \in \mathfrak{R}$. *Then* ρ-balls $B_\rho(f,r) = \{g \in L_\rho : \rho(f-g) \leq r\}$ are ρ-closed and ρ-a.e. closed.

Let us come back to the question of the relationship between the modular and norm convergence in modular function spaces.

Proposition 3.7. *Let* $\rho \in \mathfrak{R}$. *The following assertions are true:*

(a) If $\|f\|_\rho < 1$ *then* $\rho(f) \leq \|f\|_\rho$
(b) $\|f\|_\rho \leq 1$ *if and only if* $\rho(f) \leq 1$.

Proof. Part (a): Let α be any number such that $\|f\|_\rho < \alpha < 1$. Then it follows from the definition of the Luxemburg norm that

$$\rho\left(\frac{f}{\alpha}\right) \leq 1,$$

hence

$$\rho(f) = \rho\left(\alpha\frac{f}{\alpha}\right) \leq \alpha\rho\left(\frac{f}{\alpha}\right) \leq \alpha.$$

Since α was chosen arbitrarily, it follows that $\rho(f) \leq \|f\|_\rho$, as claimed.
Part (b): Assume first that $\|f\|_\rho \leq 1$ and fix arbitrarily $0 < \lambda < 1$. Hence $\|\lambda f\|_\rho < 1$. By Part (a) we get

$$\rho(\lambda f) \leq \|\lambda f\|_\rho = \lambda\|f\|_\rho.$$

Taking $\lambda \to 1$ and using the Fatou property, we obtain the desired inequality. Let us prove now the other implication. Assume now that $\rho(f) \leq 1$. If $\rho(f) = 1$ then by definition, $\|f\|_\rho \leq 1$, so we may assume that $\rho(f) < 1$. Assume to the contrary that $\|f\|_\rho > 1$. There exists then a sequence of positive numbers $\{\lambda_n\}$ such that $\lambda_n \uparrow 1$ and $\|\lambda_n f\|_\rho > 1$ which, by the definition of the Luxemburg norm, implies that $\rho(\lambda_n f) > 1$. By passing with $n \to \infty$ and using the Fatou property we get $\rho(f) \geq 1$, which contradicts our assumption. $\square$

From part (a) of Proposition 3.7, we conclude immediately the following very important fact.

Proposition 3.8. *Let* $\rho \in \mathfrak{R}$, *let* $f_n, f \in L_\rho$. *Then* $\|f_n - f\|_\rho \to 0$ *implies also the* ρ-convergence, that is $\rho(f_n - f) \to 0$.

The next two facts are actually true in any modular space. We reproduce them here for the sake of completeness.

Proposition 3.9. *Let* $\rho \in \mathfrak{R}$ *and let* $f \in L_\rho$. *Then* $\|f\|_\rho > 1$ *implies* $\rho(f) \geq \|f\|_\rho$.

Proof. Let us fix temporarily $1 < \alpha < \|f\|_\rho$. It follows form the definition of the Luxemburg norm that $\rho(f/\alpha) > 1$ and therefore by the convexity of ρ that

$$1 < \rho\left(\frac{f}{\alpha}\right) \leq \frac{1}{\alpha}\rho(f).$$

Hence $\rho(f) \geq \alpha > 1$, which implies that $\rho(f) \geq \|f\|_\rho$ since $\alpha < \|f\|_\rho$ was chosen arbitrarily. $\square$

Proposition 3.10. *Let $\rho \in \mathfrak{R}$, let f_n, $f \in L_\rho$. Then $\|f_n - f\|_\rho \to 0$ if and only if $\rho(\lambda(f_n - f)) \to 0$ for all $\lambda > 0$.*

Proof. If $\|f_n - f\|_\rho \to 0$ then $\|\lambda(f_n - f)\|_\rho \to 0$ for any $\lambda > 0$ and hence $\rho(\lambda(f_n - f)) \to 0$ by Proposition 3.8. Conversely, assume that $\rho(\lambda(f_n - f)) \to 0$ for all $\lambda > 0$ and assume to the contrary that $\|f_n - f\|_\rho$ does not tend to zero. Passing to a subsequence, if necessary, we can assume that there exists an $\varepsilon > 0$ such that $\|f_n - f\|_\rho > \varepsilon$. By multiplying both sides of this inequality by a suitable $\lambda > 0$, we have

$$\|\lambda(f_n - f)\|_\rho > 1.$$

By Proposition 3.9,

$$\rho(\lambda(f_n - f)) > \|\lambda(f_n - f)\|_\rho > 1,$$

which contradicts the fact that $\rho(\lambda(f_n - f)) \to 0$. $\square$

In the sequel, we will use the following result that describes the relation between the norm and the modular convergence.

Proposition 3.11. *Let $\rho \in \mathfrak{R}$. If $C \subset L_\rho$ is ρ-bounded then C is bounded with respect to the Luxemburg norm.*

Proof. Assume to the contrary that C is not $\|\cdot\|_\rho$-bounded. There exist therefore $h_n, h \in C$ such that $\|h_n - h\|_\rho \to \infty$. Without loss of generality we can assume that $\|h_n - h\|_\rho > 1$ for all $n \in \mathbb{N}$. By Proposition 3.9,

$$\rho(h_n - h) \geq \|h_n - h\|_\rho \to \infty, \tag{3.2}$$

which contradicts the fact that $\sup_{f,g \in C} \rho(f - g) < \infty$ by ρ-boundedness of C.

$\square$

The following definition plays an important role in the theory of modular function spaces.

Definition 3.5. *Let $\rho \in \mathfrak{R}$. We say that ρ has the Δ_2 property if $\rho(2f_n) \to 0$ whenever $\rho(f_n) \to 0$.*

The next result is an immediate consequence of the above definition and of Proposition 3.7.

Proposition 3.12. *Let $\rho \in \mathfrak{R}$. The convergence with respect to the function modular ρ is equivalent to the convergence with respect to the Luxemburg norm $\|\cdot\|_\rho$ if and only if ρ has the Δ_2-property.*

The following comparison of various modular and norm notions is an immediate consequence of the above results.

Proposition 3.13. *Let $\rho \in \mathfrak{R}$. The following relationships hold for sets $C \subset L_\rho$:*

(a) If C is ρ-closed then C is $\|\cdot\|_\rho$-closed.
(b) If C is ρ-bounded then C is $\|\cdot\|_\rho$-bounded.
(c) If C is ρ-compact then C is ρ-a.e. compact.
(d) If C is $\|\cdot\|_\rho$-compact then C is ρ-compact.
(e) Convergence with respect to $\|\cdot\|_\rho$ always implies ρ-convergence.
(f) $\|f_n - f\|_\rho \to 0$ if and only if $\rho(\lambda(f_n - f)) \to 0$ for all $\lambda > 0$.
(g) If ρ satisfies Δ_2 then $\|\cdot\|_\rho$-convergence and ρ-convergence are equivalent in L_ρ.
(h) If ρ satisfies Δ_2 then $\|\cdot\|_\rho$-compactness and ρ-compactness are equivalent in L_ρ.

A stronger condition than the Δ_2 property is sometimes used in the literature, as defined below.

Definition 3.6. Let $\rho \in \mathfrak{R}$. We say that ρ satisfies the Δ_2-*type condition* if there exists a constant $0 < K < \infty$ such that for every $f \in L_\rho$ we have $\rho(2f) \leq K\rho(f)$.

Note that if ρ satisfies the Δ_2-type condition then for every $f \in L_\rho$, we have $\rho(f) < \infty$. It is also clear that if ρ satisfies the Δ_2-type condition then ρ has the Δ_2-property, and that the converse is not true. In this book we refer mainly to the Δ_2 property as defined in Definition 3.5. In some cases, especially in Sect. 5.3.5, we will discuss function modulars satisfying the Δ_2-type condition in the sense of Definition 3.6.

It is important to be able to measure the growth of the function modular in the sense of the following definition.

Definition 3.7. Let $\rho \in \mathfrak{R}$. Define the *growth function* ω_ρ as follows:

$$\omega_\rho(t) = \sup \left\{ \frac{\rho(tf)}{\rho(f)} : 0 < \rho(f) < \infty \right\},$$

for all $t \geq 0$.

Note that in general the growth function $\omega_\rho(t)$ may be equal to infinity for some $t > 1$, hence it is not very useful in such cases. However, the growth function ω_ρ has some interesting properties provided ρ satisfies the Δ_2-type condition.

Lemma 3.1. [63] *Let $\rho \in \mathscr{R}$. Assume that ρ is convex and satisfies the Δ_2-type condition. The growth function ω_ρ has the following properties:*

(i) $\omega_\rho(t) < \infty$, for any $t \in [0, \infty)$.
(ii) $\omega_\rho : [0, \infty) \to [0, \infty)$ is a convex, strictly increasing function. Hence, it is continuous.
(iii) $\omega_\rho(\alpha\beta) \leq \omega_\rho(\alpha)\omega_\rho(\beta)$, for any $\alpha, \beta \in [0, \infty)$.
(iv) $\omega_\rho^{-1}(\alpha)\omega_\rho^{-1}(\beta) \leq \omega_\rho^{-1}(\alpha\beta)$, for any $\alpha, \beta \in [0, \infty)$, where ω_ρ^{-1} is the inverse function of ω_ρ.
(v) $\|f\| \leq \dfrac{1}{\omega_\rho^{-1}\left(\frac{1}{\rho(f)}\right)}$, for any $f \in L_\rho$.

Proof. The proofs of (i)–(iv) follow straightforward from the definition. To prove (v), let us assume, $0 < \alpha < ||f||_\rho$, hence $1 < \rho\left(\frac{f}{\alpha}\right)$, which implies

$$\frac{1}{\rho(f)} < \omega_\rho\left(\frac{1}{\alpha}\right)$$

and so

$$\omega_\rho^{-1}\left(\frac{1}{\rho(f)}\right) < \frac{1}{\alpha}.$$

Letting $\alpha \to ||f||_\rho^-$ we obtain

$$||f||_\rho \leq \frac{1}{\omega_\rho^{-1}\left(\frac{1}{\rho(f)}\right)}.$$

$\square$

Obviously when ρ satisfies the Δ_2-type condition, then the Luxemburg norm convergence and ρ-convergence are equivalent. Using the growth function, we get a better understanding of this result.

Lemma 3.2. *Let $\rho \in \mathscr{R}$. Assume that ρ satisfies the Δ_2-type condition. For every $\varepsilon > 0$, there exists $\delta > 0$ such that $\rho(f) < \delta$ implies $||f|| < \varepsilon$.*

Proof. Let $\varepsilon > 0$. Set $\delta = \dfrac{1}{\omega_\rho(1/\varepsilon)}$. Since ω^{-1} is an increasing function, property (v) of Lemma 3.1 implies that $||f|| < \varepsilon$ whenever $\rho(f) < \delta$. $\square$

In the sequel we will be using the following notion of the ρ-lower semicontinuity.

Definition 3.8. A function $\lambda : C \to [0, \infty]$, where $C \subset L_\rho$ is nonempty and ρ-closed, is called ρ-*lower semicontinuous* if for any $\alpha > 0$, the set $C_\alpha = \{f \in C : \lambda(f) \leq \alpha\}$ is ρ-closed.

Remark 3.4. It can be easily proved that ρ-lower semicontinuity is equivalent to the condition

$$\lambda(f) \leq \liminf_{n \to \infty} \lambda(f_n) \text{ provided } f, f_n \in C, \text{ and } \rho(f - f_n) \to 0.$$

The modular version of the property (R), defined previously for metric spaces, will be essential for the proof of several fixed point theorems in modular function spaces. Following [117], let us formally define this property.

Definition 3.9. We say that L_ρ has *property* (R) if and only if every nonincreasing sequence $\{C_n\}$ of nonempty, ρ-bounded, ρ-closed, convex subsets of L_ρ has nonempty intersection.

In some cases we will be referring to a somewhat stronger notion of the property (R'), defined below.

Definition 3.10. We say that L_ρ has *property* (R') if and only if for every ρ-bounded sequence of $f_n \in L_\rho$ there exists a subsequence $\{f_{n_k}\}$ such that

$$\bigcap_{n \in \mathbb{N}} cl_\rho(conv\{f_{n_k} : k \geq n\})$$

is nonempty and reduced to one point. By $cl_\rho(conv(A))$, we understand the ρ-closure of the smallest convex subset of L_ρ containing A.

Remark 3.5. It is easy to verify that in any modular function space property (R') implies property (R).

The ρ-lower semicontinuous real-valued convex functions play an important role in modular function spaces because they attain their infimum provided ρ has property (R), as shown in next result.

Lemma 3.3. *Let us assume that $\rho \in \mathfrak{R}$ has property (R). Let $K \subset L_\rho$ be nonempty, convex, ρ-closed, and ρ-bounded. If $\varphi : K \to [0, \infty)$ is a ρ-lower semicontinuous convex function, then there exists $x_0 \in K$ such that*

$$\varphi(x_0) = \inf\{\varphi(x) : x \in K\}.$$

Proof. Let $m = \inf\{\varphi(x) : x \in K\}$. The assumptions on φ imply $m < \infty$. For any $n \geq 1$, set

$$K_n = \left\{ x \in K : \varphi(x) \leq m + \frac{1}{n} \right\}.$$

Clearly K_n is not empty and is a convex set because φ is a convex function. Also, K_n is ρ-closed since φ is ρ-lower semicontinuous. Since ρ satisfies property (R), then

$$K_\infty = \bigcap_{n \geq 1} K_n \neq \emptyset.$$

For $x_0 \in K_\infty$, there holds

$$\varphi(x_0) = \inf\{\varphi(x) : x \in K\},$$

as claimed. □

3.2 Space E_ρ, Convergence Theorems, and Vitali Property

Definition 3.11. Define $L_\rho^0 = \{f \in L_\rho : \rho(f, \cdot) \text{ is order continuous}\}$ and the linear space $E_\rho = \{f \in L_\rho : \lambda f \in L_\rho^0 \text{ for every } \lambda > 0\}$.

The position of E_ρ with respect to L_ρ is characterized in the following theorem.

Theorem 3.4. [130, 131, 132] *Let $\rho \in \mathfrak{R}$. Then E_ρ is a $\|\cdot\|_\rho$-closed subspace of L_ρ.*

Proof. First, we will prove that E_ρ is a linear subspace of L_ρ. Clearly E_ρ is a linear space; it suffices, therefore, to prove that $E_\rho \subset L_\rho$. Let $f \in E_\rho$ and $0 \le \lambda_n \to 0$. Since $f \in \mathcal{M}$ it follows that there exists a sequence $\{s_m\}$ of $\mathcal{P}$-simple functions such that $|s_m| \uparrow |f|$. By the Egoroff theorem, we get a sequence $\{H_i\}$ such that $H_i \in \mathcal{P}, H_i \uparrow \Omega$, and $\{s_m\}$ converges uniformly to f on every H_k. Let us choose a number $\lambda > 0$ such that $\lambda_n \le \lambda$ for all natural numbers n, and fix an arbitrary $\varepsilon > 0$. Since $f \in E_\rho$, there exists an index $k_0 \in \mathbb{N}$ such that

$$\rho(\lambda f, \Omega \setminus H_{k_0}) < \frac{\varepsilon}{3}.$$

By the uniform convergence of $\{s_m\}$ on H_{k_0} we can choose $m_0 \in \mathbb{N}$ such that

$$\rho(2\lambda(s_{m_0} - f), H_{k_0}) < \frac{\varepsilon}{3}.$$

Remember that s_{m_0} belongs to $\mathcal{E} \subset L_\rho$ and hence there exists $n_0 \in \mathbb{N}$ such that

$$\rho(2\lambda_n s_{m_0}, H_{k_0}) < \frac{\varepsilon}{3}$$

for $n \ge n_0$. Finally, for $n \ge n_0$, we have

$$\rho(\lambda_n f) \le \rho(\lambda_n f, \Omega \setminus H_{k_0}) + \rho(\lambda_n f, H_{k_0})$$
$$\le \rho(\lambda f, \Omega \setminus H_{k_0}) + \rho(2\lambda(s_{m_0} - f), H_{k_0}) + \rho(2\lambda_n s_{m_0}, H_{k_0}) < \varepsilon.$$

This means that the function f is a member of the space L_ρ. We will prove now that E_ρ is closed. Let $f_n \in E_\rho$, $\|f_n - f\|_\rho \to 0$, $f \in L_\rho$. Take a sequence of sets $E_n \in \Sigma$ with $E_n \downarrow \emptyset$. Fix positive numbers ε, $\alpha > 0$ and choose $k_0 \in \mathbb{N}$ such that

$$\rho(2\alpha(f - f_{k_0})) < \frac{\varepsilon}{2}.$$

Since $f_{k_0} \in E_\rho$, it follows that for n sufficiently large,

$$\rho(2\alpha f_{k_0}, E_n) < \frac{\varepsilon}{2}.$$

Thus, for n sufficiently large, we have

$$\rho(\alpha f, E_n) \le \rho(2\alpha(f - f_{k_0}), E_n) + \rho(2\alpha f_{k_0}, E_n)$$
$$\le \rho(2\alpha(f - f_{k_0})) + \rho(2\alpha f_{k_0}, E_n) < \varepsilon,$$

which implies that $f \in E_\rho$. Therefore, E_ρ is a closed subspace of L_ρ, as claimed. $\square$

According to Proposition 3.12, the ρ-convergence and the $\|\cdot\|_\rho$-convergence are in general equivalent if only if ρ satisfies Δ_2. However, it is legitimate to ask on what subsets of L_ρ such equivalence may hold even ρ does not have Δ_2. We introduce a concept of sets with the Vitali property that will play this role, see [141]. The reference to Giuseppe Vitali is justified by the following version of the *Vitali convergence theorem*, which was proved in the context of modular function spaces in [132].

Theorem 3.5. [132] *Let $\rho \in \mathfrak{R}$. Let $f_n \in E_\rho$, $f \in L_\rho$, and $f_n \to f$ ρ-a.e. Then the following conditions are equivalent:*

(i) $f \in E_\rho$ and $\|f_n - f\|_\rho \to 0$.
(ii) For every $\alpha > 0$, the subadditive measures $\rho(\alpha f_n, \cdot)$ are order equicontinuous, that is, if $E_k \in \Sigma$ are such that $E_k \downarrow \emptyset$ then $\lim\limits_{k \to \infty} \sup\limits_{n \in \mathbb{N}} \rho(\alpha f_n, E_k) = 0$.

Proof. $(i) \Rightarrow (ii)$. Let us choose a sequence of sets $E_k \in \Sigma$ such that $E_k \downarrow \emptyset$, and fix arbitrarily numbers $\varepsilon > 0$ and $\alpha > 0$. Since $\|f_n - f\|_\rho \to 0$, it follows that there exists $n_0 \in \mathbb{N}$ such that for all $n \geq n_0$,

$$\rho(2\alpha(f_n - f)) < \frac{\varepsilon}{2}.$$

From the definition of the space E_ρ, it follows that there exists a natural number $k_0 \in \mathbb{N}$ with

$$\rho(\alpha f_n, E_k) < \varepsilon \tag{3.3}$$

for $k \geq k_0$ and $n = 1, 2, \ldots, n_0 - 1$. Similarly, since $f \in E_\rho$, there exists $k_1 > k_0$ such that for any $k \geq k_1$,

$$\rho(2\alpha f, E_k) < \frac{\varepsilon}{2}.$$

Thus,

$$\rho(\alpha f_n, E_k) \leq \rho(2\alpha(f_n - f)) + \rho(2\alpha f, E_k) < \varepsilon \tag{3.4}$$

for all natural $k_1 > k_0$ and $n \geq n_0$. The inequalities (3.3) and (3.4) give the desired result.

$(ii) \Rightarrow (i)$. Let us fix two numbers $\varepsilon > 0$ and $\alpha > 0$. Since $f_n \to f$ ρ-a.e., it follows from the Egoroff theorem (Theorem 3.3) that there exists a sequence of sets $H_k \in \mathscr{P}$ such that $H_k \uparrow \Omega$ and $\{f_n\}$ converge uniformly to f on every H_k. By the assumed order equicontinuity of function modulars, we can pick up an index $k_0 \in \mathbb{N}$ such that for all $n \in \mathbb{N}$,

$$\rho(2\alpha f_n, \Omega \setminus H_{k_0}) < \frac{\varepsilon}{4}.$$

Since $\{f_n\}$ converges uniformly to f on $H_{k_0} \in \mathscr{P}$, one can find $n_0 \in \mathbb{N}$ such that

$$\rho(2\alpha(f_n - f)) < \frac{\varepsilon}{4}$$

for $n \geq n_0$. Let $n, m \geq n_0$. We have

$$\rho(\alpha(f_n - f_m)) \leq \rho(\alpha(f_n - f_m), \Omega \setminus H_{k_0}) + \rho(\alpha(f_n - f_m), H_{k_0})$$
$$\leq \rho(2\alpha f_n, \Omega \setminus H_{k_0}) + \rho(2\alpha f_m, \Omega \setminus H_{k_0})$$
$$+ \rho(2\alpha(f_n - f)) + \rho(2\alpha(f_m - f)) < \varepsilon.$$

Hence the sequence $\{\alpha f_n\}$ satisfies the ρ-Cauchy condition, which implies that $\{f_n\}$ satisfies the Cauchy condition in the sense of the Luxemburg norm $\|\cdot\|_\rho$. Since $f_n \in E_\rho$, which is complete in the norm sense as the closed subspace of a complete metric space L_ρ, it follows that there exists a function $g \in E_\rho$ such that $\|f_n - g\|_\rho \to 0$. Therefore, in view of Proposition 3.5 there exists a subsequence of $\{f_n\}$ that converges ρ-a.e. to g. On the other hand, $f_n \to f$ ρ-a.e. Thus, $f = g$ ρ-a.e. and finally $\|f_n - f\|_\rho \to 0$, as claimed. $\square$

As an immediate consequence of the Vitali convergence theorem, we get a result that is of great importance for applications, namely the following version of the *Lebesgue dominated convergence theorem*.

Theorem 3.6. [132] *Let $\rho \in \mathfrak{R}$. Let $f_n, f \in \mathcal{M}$, $f_n \to f$ ρ-a.e. Assume that there exists a function $g \in E_\rho$ such that $|f_n(\omega)| \leq |g(\omega)|$ for every $n \in \mathbb{N}$. Then $\|f_n - f\|_\rho \to 0$.*

Theorem 3.7. [130, 131, 132] *Let $\rho \in \mathfrak{R}$. Then E_ρ is the $\|\cdot\|_\rho$ closure of the space of all $\mathcal{P}$ simple functions $\mathcal{E}$.*

Proof. We will prove first that $cl_{\|\cdot\|_\rho}(\mathcal{E}) \subset E_\rho$. Take a simple function $g \in \mathcal{E}$, a positive number λ, and a sequence $E_n \downarrow \emptyset$, $E_n \in \Sigma$. Denote

$$\alpha = \sup_{\omega \in \Omega} |\lambda g(\omega)|,$$

and observe that

$$\rho(\lambda g, E_n) = \rho(\lambda g, supp(g) \cap E_n) \leq \rho(\alpha, supp(g) \cap E_n) \to 0,$$

because $supp(g) \cap E_n \in \mathbb{P}$. Hence, $\mathcal{E} \subset E_\rho$ and consequently $cl_{\|\cdot\|_\rho}(\mathcal{E}) \subset E_\rho$, since E_ρ is $\|\cdot\|_\rho$-closed. To prove the inverse inclusion, let us choose arbitrarily $f \in E_\rho$ and $\varepsilon > 0$. Since $f \in E_\rho$ and Ω is a countable union of sets from $\mathcal{P}$, there exists a set $E \in \mathcal{P}$ such that, denoting by $E' = \Omega \setminus E$, we have $\|f 1_{E'}\|_\rho < \frac{\varepsilon}{2}$. Let us select a sequence $\{g_n\}$ of $\mathcal{P}$-simple functions such that for all $n \in \mathbb{N}$ and all $\omega \in E$,

$$supp(g_n) \subset E,$$
$$|g_n(\omega)| \leq |f(x)|,$$
$$g_n(\omega) \to f(x).$$

By the Lebesgue dominated convergence theorem, $\|g_n - f1_E\|_\rho \to 0$. Hence,

$$\|g_{n_0} - f1_E\|_\rho < \frac{\varepsilon}{2}$$

for a certain $n_0 \in \mathbb{N}$, and

$$\|g_{n_0} - f\|_\rho \le \|g_{n_0} - f1_E\|_\rho + \|g_{n_0}1_{E'}\|_\rho + \|f1_{E'}\|_\rho \le \varepsilon,$$

that is $E_\rho \subset cl_{\|\cdot\|_\rho}(\mathscr{E})$. This completes the proof. $\square$

Theorem 3.8. [130, 131, 132] *Let $\rho \in \mathfrak{R}$. Then the following conditions are equivalent:*

(a) ρ has the Δ_2-property;
(b) L_ρ^0 is a linear subspace of L_ρ;
(c) $E_\rho = L_\rho^0 = L_\rho$;
(b) $\|\cdot\|_\rho$-convergence and ρ-convergence are equivalent.

Proof. $(a) \Rightarrow (b)$. Let $f \in L_\rho^0$ and $E_k \downarrow \emptyset$, $E_k \in \Sigma$. Since $f \in L_\rho^0$, it follows that $\rho(f, E_k) \to 0$. By Δ_2 we get that $\rho(2f, E_k) \to 0$. Hence, $2f \in L_\rho^0$ and, consequently, L_ρ^0 is a linear space.

$(b) \Rightarrow (c)$. We have $E_\rho \subset L_\rho^0 \subset L_\rho$, and L_ρ is the smallest linear space containing L_ρ^0. E_ρ is the largest space contained in L_ρ. Since L_ρ^0 is linear, then $E_\rho = L_\rho^0 = L_\rho$.

$(c) \Rightarrow (a)$. Suppose to the contrary that there exists a sequence of functions $\{f_n\}$ from L_ρ such that $\rho(f_n, \cdot)$ are order continuous while $\rho(2f_n, \cdot)$ are not. It follows then that $\rho(2f_n, \cdot)$ are not uniformly exhaustive, that is, there exists a sequence of disjoint sets $\{D_k\}$ from Σ and a constant $\eta > 0$ such that for every $k \in \mathbb{N}$,

$$\rho(2f_k, D_k) > \eta.$$

Observe that $\rho(f_k, D_k) \to 0$ because

$$\rho(f_k, D_k) \le \sup_{n \in \mathbb{N}} \rho(f_n, D_k) \to 0,$$

which is a consequence of the fact that $\sup_{n \in \mathbb{N}} \rho(f_n, \cdot)$ is order continuous and hence exhaustive. Let $\{f_{k_i}\}$ be a subsequence of $\{f_n\}$ such that

$$\sum_{i=1}^{\infty} \rho(f_{k_i}, D_{k_i}) \le 1.$$

Denoting

$$s_m = \sum_{i=1}^{m} f_{k_i}1_{D_{k_i}}, \ f = \sum_{i=1}^{\infty} f_{k_i}1_{D_{k_i}},$$

we get $s_m \in L_\rho = E_\rho$ and

$$\rho(f - s_m) \le \sum_{i=m+1}^{\infty} \rho(f_{k_i}, D_{k_i}) \to 0.$$

The function f is a member of $L_\rho = E_\rho$. Indeed, let $\varepsilon > 0$ be chosen arbitrarily. Moreover, let $0 < \varepsilon_k < \frac{1}{2}$, $\varepsilon_k \to 0$, $\lambda_k = 2\varepsilon_k < 1$. Then we have

$$\rho(\varepsilon_k f) \le \rho(\lambda_k(f - s_m)) + \rho(\lambda_k s_m) \le \rho(f - s_m) + \rho(\lambda_k s_m). \tag{3.5}$$

We may take $m_0 \in \mathbb{N}$ such that $\rho(f - s_{m_0}) < \frac{\varepsilon}{2}$ and $k_0 \in \mathbb{N}$ such that $\rho(\lambda_k s_{m_0}) < \frac{\varepsilon}{2}$ for natural $k \ge k_0$. Hence, $\rho(\varepsilon_k f) < \varepsilon$ for $k \ge k_0$, and then $f \in L_\rho = E_\rho$. Since $L_\rho = E_\rho$ is linear, it follows that $2f$ is a member of E_ρ as well. However, it was observed above that $\rho(2f, D_k) > \eta > 0$. Hence $\rho(2f, \cdot)$ is a σ-subadditive measure, which is not exhaustive, i.e., $\rho(2f, \cdot)$ must not be order continuous, which means that the function $2f$ does not belong to E_ρ. This contradiction completes this part of the proof.

$(a) \Rightarrow (d)$. It suffices to prove that $\rho(f_n) \to 0$ implies $\rho(2f_n) \to 0$ for $f_n \in L_\rho$. Assume, therefore, that $f_n \in L_\rho$ and $\rho(f_n) \to 0$. Let us take an arbitrary subsequence $\{h_n\}$ of $\{f_n\}$. We can choose a subsequence $\{g_n\}$ of $\{h_n\}$ such that $g_n \to 0$ ρ-a.e. As we proved, it follows from (a) that $E_\rho = L_\rho^0 = L_\rho$. It follows from the Vitali convergence theorem that $\rho(g_n, \cdot)$ are order equicontinuous. Hence, $\rho(2g_n, \cdot)$ are also equicontinuous by Δ_2. Using the Vitali convergence theorem again, we get $\rho(2g_n) \to 0$. By the arbitrariness of $\{h_n\}$ we have that $\rho(2f_n) \to 0$.

$(d) \Rightarrow (a)$. Let $f_n \in L_\rho$ and let $\rho(f_n, \cdot)$ be order equicontinuous. Assume to the contrary that there exists a sequence of sets $E_k \in \Sigma$ such that $E_k \downarrow \emptyset$, an $\varepsilon > 0$, and a subsequence $\{g_n\}$ of $\{f_n\}$ such that

$$\rho(2g_k, E_k) = \rho(2g_k 1_{E_k}) > \varepsilon. \tag{3.6}$$

On the other hand,

$$\rho(g_k, E_k) \le \sup_{n \in \mathbb{N}} \rho(f_n, E_k) \to 0.$$

By (d) then,

$$\rho(2g_k, E_k) = \rho(2g_k 1_{E_k}) \to 0,$$

which contradicts (3.6) and completes the proof. $\quad\square$

Definition 3.12. A set $C \subset L_\rho$ is said to posses the *Vitali property* if $C \subset E_\rho$, and for any $g \in L_\rho$ and $g_n \in C$ with $\rho(g_n - g) \to 0$, there exists a subsequence $\{g_{n_k}\}$ of $\{g_n\}$ such that for every $\alpha > 0$ the subadditive measures $\rho(\alpha g_{n_k}, \cdot)$ are order equicontinuous.

Our next result characterizes sets with the Vitali property as those subsets of E_ρ on which the ρ-convergence and the $\| \cdot \|_\rho$-convergence are indeed equivalent.

Theorem 3.9. *Let $\rho \in \mathfrak{R}$. A set $C \subset L_\rho$ has the Vitali property if and only the following two conditions are satisfied:*

(i) $C \subset E_\rho$.
(ii) If $g \in L_\rho$ and $g_n \in C$ with $\rho(g_n - g) \to 0$, then $\|g_n - g\|_\rho \to 0$.

Proof. Let us assume first that $C \subset L_\rho$ has the Vitali property. Hence, by definition, the condition (i) is satisfied. To prove (ii), take a sequence $\{g_n\}$, and a function $g \in L_\rho$ such that $\rho(g_n - g) \to 0$. By Proposition 3.5, there exists a subsequence $\{g_{n_k}\}$ such that

$$g_{n_k} \to g \; \rho - a.e. \tag{3.7}$$

Because C has the Vitali property, there exists a subsequence $\{g_{n_{k_l}}\}$ such that for every $\alpha > 0$ the submeasures $\rho(\alpha g_{n_{k_l}}, \cdot)$ are order equicontinuous. Hence, using (3.7) and the Vitali theorem (Theorem 3.5), we conclude that $\|g_{n_{k_l}} - g\|_\rho \to 0$. Since this reasoning can be repeated for any subsequence of $\{g_n\}$, every subsequence of $\{g_n\}$ contains a subsequence converging in $\|\cdot\|_\rho$ to g, proving (ii).

Conversely, let us assume that (i) and (ii) are satisfied. Assume to the contrary that C does not have the Vitali property. Then there exists a sequence $\{g_k\}$ of elements of C, $g \in L_\rho$, and $\alpha > 0$ such that the submeasures $\rho(\alpha g_n, \cdot)$ are not order equicontinuous while $\rho(g_k - g) \to 0$. The latter fact implies by (ii) that $\|g_n - g\|_\rho \to 0$, which also implies, via closedness of E_ρ, that $g \in E_\rho$. By Proposition 3.5 we can assume, without loosing generality, that $g_{n_k} \to g$ ρ-a.e. From Theorem 3.5, we conclude that $\|g_n - g\|_\rho$ cannot tend to zero. Contradiction completes the proof. □

Remark 3.6. Combining Proposition 3.13 with Theorem 3.9, we can easily see that a set with the Vitali property is ρ-closed if and only if it is $\|\cdot\|_\rho$-closed.

Remark 3.7. Let $C \subset L_\rho$ be a set with the Vitali property and let a, $b \in \mathbb{R}$. Let $u : [a,b] \to C$ be a ρ-continuous function, that is, $\rho(u(t_n) - u(t)) \to 0$ provided $t_n \to t$. It follows immediately from Theorem 3.9 that u is $\|\cdot\|_\rho$-continuous.

As an immediate corollary to Remark 3.7, we obtain the following important result.

Remark 3.8. Let Z be a separable linear subspace of $(E_\rho, \|\cdot\|_\rho)$ and let $C \subset Z$ have the Vitali property. Assume that the function $u : [a,b] \to C$ is ρ-continuous. Then u is the Bochner integrable function with respect to the the Lebesgue measure m on $[a,b]$, i.e., $u \in L^1(\Omega, Z, m)$.

Let us discuss the Fatou property in the context of the Vitali property.

Proposition 3.14. *Let $\rho \in \mathfrak{R}$. Assume that $C \subset E_\rho$ has the Vitali property and that*

$$\rho(f_n - f) \to 0, \; \rho(g_n - g) \to 0 \tag{3.8}$$

as $n \to \infty$, with $f_n, f, g_n, g \in C$. Then,

$$\rho(f - g) \leq \liminf_{n \to \infty} \rho(f_n - g_n). \tag{3.9}$$

Proof. Since C has the Vitali property, it follows from (3.8) that

$$\rho((f_n - g_n) - (f - g)) \to 0. \tag{3.10}$$

Using (3.10) and the Fatou property of ρ, it is easy to get required inequality (3.9). $\square$

Let us finish this section with few examples of sets with the Vitali property.

Example 3.1. If ρ has Δ_2 property then every set $C \subset L_\rho$ has the Vitali property.

Example 3.2. Let $C \subset E_\rho$. If there exists $g \in E_\rho$ such that $|f(\omega)| \le |g(\omega)|$ ρ-a.e. for every $f \in C$, then C has the Vitali property.

Example 3.3. Let $C \subset E_\rho$ be $\|\cdot\|_\rho$-conditionally compact. Then C has the Vitali property, see Theorem 3.11 in Sect. 3.4.

3.3 An Equivalent Topology

The concept of ρ-a.e. closed, and ρ-a.e. compact sets has been defined using sequences. One of the problems that many authors have found hard to circumvent is whether these notions are related to a topology. In this section we will discuss this problem. In particular, we will construct a topology τ for which the ρ-a.e. compactness is equivalent to the usual compactness for τ. This is crucial when we try to use Zorn's lemma. Throughout this section, we assume that ρ satisfies the Δ_2-type condition and it is σ-finite, i.e., there exists an increasing sequence of sets $K_n \in \mathscr{P}$ such that $0 < \rho(1_{K_n}) < \infty$ and $\Omega = \bigcup_{n \ge 1} K_n$.

Proposition 3.15. *Assume that ρ satisfies the Δ_2-type condition. The functional $d : L_\rho \times L_\rho \to [0, +\infty)$ defined*

$$d(f,g) = \sum_{k=1}^{\infty} \frac{1}{2^k} \frac{1}{\rho(1_{K_k})} \rho\left(\frac{|f-g|}{1+|f-g|} 1_{K_k}\right)$$

satisfies the following:

1. $d(f,g) = 0$ if and only if $f = g$ ρ-a.e.;
2. $d(f,g) = d(g,f)$;
3. $d(f,g) \le \dfrac{\omega_\rho(2)}{2}(d(f,h) + d(h,g))$;

for any f, g, and h in L_ρ.

Proof. (1) and (2) are obvious. To prove (3) we only need to recall the inequality

$$\frac{|a+b|}{1+|a+b|} \le \frac{|a|}{1+|a|} + \frac{|b|}{1+|b|}$$

for all positive numbers a, b and use the definition of the growth function ω_ρ. $\square$

Remark 3.9. The functional d is not a distance because of (3). But there are many mathematical objects that fail the triangle inequality but are very useful tools. That is the case with d.

In the next proposition, we discuss the relationship between ρ-a.e. convergence and the convergence for the functional d.

Proposition 3.16. *Let ρ be a convex, σ-finite modular satisfying the Δ_2-type condition and $\{f_n\}_n$ be a sequence in L_ρ. If $\{f_n\}_n$ is ρ-a.e. convergent to f, then*

$$\lim_{n\to\infty} d(f_n, f) = 0.$$

Moreover, if $\lim_{n\to\infty} d(f_n, f) = 0$, then there exists a subsequence $\{f_{n_k}\}_k$ that converges ρ-a.e. to f.

Proof. Assume that $\{f_n\}_n$ ρ-a.e. converges to f. We will show that $\lim_{n\to\infty} d(f_n, f) = 0$.

Let $\varepsilon > 0$, and choose $N \in \mathbb{N}$ such that $\displaystyle\sum_{k=N+1}^{\infty} \frac{1}{2^k} < \varepsilon$. We have

$$\lim_{n\to\infty} d(f_n, f) \leq \lim_{n\to\infty} \sum_{k=1}^{N} \frac{1}{2^k} \frac{1}{\rho(1_{K_k})} \rho\left(\frac{|f_n - f|}{1 + |f_n - f|} 1_{K_k}\right) + \varepsilon$$

$$= \sum_{k=1}^{N} \frac{1}{2^k} \frac{1}{\rho(1_{K_k})} \lim_{n\to\infty} \rho\left(\frac{|f_n - f|}{1 + |f_n - f|} 1_{K_k}\right) + \varepsilon.$$

Since $\dfrac{|f_n - f|}{1 + |f_n - f|} 1_{K_k} \leq 1_{K_k}$, and $\dfrac{|f_n - f|}{1 + |f_n - f|} 1_{K_k} \overset{\rho\text{-a.e.}}{\longrightarrow} 0$ as $n \to \infty$, for any $k \geq 1$, Lebesgue convergence theorem will then imply

$$\lim_{n\to\infty} \rho\left(\frac{|f_n - f|}{1 + |f_n - f|} 1_{K_k}\right) = 0, \quad k \geq 1.$$

Thus $\lim_{n\to\infty} d(f_n, f) \leq \varepsilon$, for any $\varepsilon > 0$, i.e., $\lim_{n\to\infty} d(f_n, f) = 0$. Conversely, assume now that $\lim_{n\to\infty} d(f_n, f) = 0$. For any $k \geq 1$, we have

$$\lim_{n\to\infty} \rho\left(\frac{|f_n - f|}{1 + |f_n - f|} 1_{K_k}\right) = 0.$$

Thus there exists a subsequence $\{f_n^1\}_n$ of $\{f_n\}_n$ such that $\dfrac{|f_n^1 - f|}{1 + |f_n^1 - f|} 1_{K_1} \overset{\rho\text{-a.e.}}{\longrightarrow} 0$ and so $f_n^1 \overset{\rho\text{-a.e.}}{\longrightarrow} f$ in K_1 i.e., $\lim_{n\to\infty} f_n^1(x) = f(x)$ whenever $x \in K_1 \setminus A_1$ where $A_1 \subset K_1$ and $\rho(1_{A_1}) = 0$. By induction and using a diagonal argument we obtain a subsequence of $\{f_{\varphi(n)}\}_n$, which converges ρ-a.e. to f. $\qquad\square$

Next we use the functional d to define the topology τ.

Definition 3.13. Let C be a subset of L_ρ.

(a) C is said to be d-closed if and only if for any sequence $\{f_n\}_n$ in C which d-converges to f, we have $f \in C$.
(b) C is d-open if and only if $L_\rho \backslash C$ is d-closed.
(c) C is said to be d-sequentially compact if for each sequence $\{f_n\}_n$ there exists a subsequence $\{f_{\varphi(n)}\}_n$, which d-converges to a point in C.

It is easily seen that the family of all d-open subsets of L_ρ forms a topology on L_ρ. Furthermore from Proposition 3.16, d-sequentially compact sets and ρ-a.e. compact sets are identical. Even though d satisfies (3) instead of the triangular inequality, the usual arguments, which prove that sequential compactness and compactness are identical in metric spaces, hold in this setting. We also have d-sequential compactness and d-compactness identical.

3.4 Compactness and Separability

In the context of the previous sections, let us discuss the separability of $(E_\rho, \|\cdot\|_\rho)$. First, we need the following definition.

Definition 3.14. The function modular $\rho \in \mathfrak{R}$ is called *separable* if $\|f 1_{(\cdot)}\|_\rho)$ is a *separable set function* for each $f \in \mathscr{E}$, which means that there exists a countable $\mathscr{A} \subset \mathscr{P}$ such that to every $A \in \mathscr{P}$ there corresponds a sequence $\{A_k\}$ of elements of $\mathscr{A}$ with

$$\rho(\alpha f, A \Delta A_k) \to 0 \tag{3.11}$$

for every $\alpha > 0$, where Δ denotes the symmetric set difference.

We are now ready to prove the following theorem that characterizes separable $(E_\rho, \|\cdot\|_\rho)$ spaces.

Theorem 3.10. [132] *Let* $\rho \in \mathfrak{R}$. *The space* $(E_\rho, \|\cdot\|_\rho)$ *is a separable Banach space if and only if* ρ *is separable.*

Proof. Assume first that $(E_\rho, \|\cdot\|_\rho)$ is separable. Hence the space of all $\mathscr{P}$-simple functions $\mathscr{E} \subset E_\rho$ is separable. Suppose to the contrary that ρ is not a separable function modular. Then there exists a function $f \in \mathscr{E}$, a number $\varepsilon > 0$, and an uncountable family of sets $\mathscr{B} \subset \mathscr{P}$ such that $\|f 1_{B \Delta B'}\|_\rho > \varepsilon$ for arbitrary $B, B' \in \mathscr{B}$. Hence,

$$\varepsilon < \|f 1_{B \Delta B'}\|_\rho = \|f 1_B - f 1_{B'}\|_\rho.$$

Put $W = \{f 1_B : B \in \mathscr{B}\}$, observe that W is an uncountable subset of $\mathscr{E}$ and that $\|g - h\|_{\rho|} > \varepsilon$ for $g, h \in W$, $g \neq h$. This fact contradicts the separability of $\mathscr{E}$. Let us prove the converse. For $n \in \mathbb{N}$ and $r \in \mathbb{Q}$, let us denote by $\{\mathscr{B}_{n,r}\}$ a countable subfamily of sets dense in $\{\Omega_n \cap A : A \in \mathscr{P}\}$ with respect to the pseudometric

$$(A, B) \mapsto \|r 1_{\Omega_n \cap (A \Delta B)}\|_\rho,$$

where $A, B \in \mathscr{B}$. Denote

$$\mathscr{B} = \bigcup_{n \in \mathbb{N},\ r \in \mathbb{Q}} B_{n,r}.$$

It is enough to prove that $\{r1_B : r \in \mathbb{Q},\ B \in \mathscr{B}\}$ is dense in $\{s1_A : s \in \mathbb{R},\ A \in \mathscr{P}\}$. Fix $s \in \mathbb{R}$ and $\varepsilon > 0$. By the density of $\mathbb{Q}$ in $\mathbb{R}$ and by Proposition 3.3, we can choose an $r \in \mathbb{Q}$ such that

$$\|s1_A - r1_A\|_\rho < \frac{\varepsilon}{3}.$$

Since A, $\Omega_n \in \mathscr{P}$ and $\Omega_n \uparrow \Omega$, it follows that $A \setminus \Omega_n \downarrow \emptyset$ and that $\|r1_{A\setminus\Omega_n}\|_\rho < \frac{\varepsilon}{3}$ for n sufficiently large. From the assumption of the separability of ρ we conclude that there exists a set $B \in \mathscr{B}_{n,r}$ such that

$$\|r1_{A\cap\Omega_n} - r1_B\|_\rho = \|r1_{\Omega_n\cap(A\Delta B)}\|_\rho < \frac{\varepsilon}{3}.$$

Finally,

$$\|s1_A - r1_B\|_\rho \leq \|s1_A - r1_A\|_\rho + \|r1_{A\cap\Omega_n} - r1_B\|_\rho + \|r1_{A\setminus\Omega_n}\|_\rho < \varepsilon.$$

This completes the proof. □

Finally, we can combine Theorem 3.10 with Remark 3.8 into the following very useful statement.

Proposition 3.17. *Let* $\rho \in \mathfrak{R}$ *be a separable function modular. If* $u : [a,b] \to C$ *is* ρ*-continuous, where* $C \subset E_\rho$ *has the Vitali property, then* $u \in L^1(\Omega, Z, m)$.

Next, let us characterize the compact subsets of the space E_ρ.

Theorem 3.11. [132] *Let* $\rho \in \mathfrak{R}$. *A set* $D \subset E_\rho$ *is conditionally compact in the sense of the Luxemburg norm* $\|\cdot\|_\rho$ *if and only if the following conditions are satisfied:*

 (i) For every $\alpha > 0$ *the set function* $\sup\{\rho(\alpha f, \cdot) : f \in D\}$ *is order continuous;*
 (ii) For every sequence $\{f_n\}$ *of elements from* D *there exist a subsequence* $\{f_{n_k}\}$ *and a function* $f \in E_\rho$ *such that* $\rho(f_{n_k} - f) \to 0$.

Proof. Sufficiency: Let $f_n \in D$ for $n \in \mathbb{N}$. By (ii) we can choose a subsequence $\{f_{n_k}\}$ and a function $f \in E_\rho$ such that $f_{n_k} \to f$ ρ-a.e. Combining this with (i) and using the Vitali theorem, we get that $\|f_{n_k} - f\|_\rho \to 0$.

Necessity: To prove (i), let us fix an $\varepsilon > 0$ and sets $E_k \downarrow \emptyset$. By the conditional compactness of D we may find a finite set $\{f_1, \ldots, f_n\}$ of elements of E_ρ such that to every $f \in D$ there corresponds an index $i \in \{1, \ldots, n\}$ for which $\|f - f_i\|_\rho < \frac{\varepsilon}{2}$. Since all functions f_i belong to E_ρ it follows that there exists $k_0 \in \mathbb{N}$ such that $\|f_i 1_{E_k}\|_\rho < \frac{\varepsilon}{2}$ for $k \geq k_0$, $1 \leq i \leq n$. This completes the proof of (i). To prove (ii), let us choose a sequence $\{f_n\}$ of elements of D. Since D is conditionally compact, it follows that there exists a subsequence $\{f_{n_k}\}$ of $\{f_n\}$ and a function $f \in E_\rho$ such that $\|f_{n_k} - f\|_\rho \to 0$, which implies that $\rho(f_{n_k} - f) \to 0$. □

3.5 Examples

Let us illustrate the notion of modular function spaces with few examples.

Example 3.4. L^p-space, $p \geq 1$, is a modular function space generated by

$$\rho(f) = \int_{\mathbb{R}} |f(t)|^p dm(t).$$

Example 3.5. l^p-space, $p \geq 1$, is a modular function space generated by

$$\rho(f) = \sum_{i=1}^{\infty} |f_i|^p.$$

Example 3.6. Orlicz space is a modular function space generated by

$$\rho(f) = \int_{\mathbb{R}} \varphi(|f(t)|) dm(t),$$

provided φ is an Orlicz function, that is, φ is a nonnegative, convex function such that $\varphi(0) = 0$.

Example 3.7. Musielak–Orlicz space is a modular function space generated by

$$\rho(f) = \int_{\mathbb{R}} \phi(t, |f(t)|) dm(t),$$

provided φ is nonnegative, convex function in the second variable such that $\varphi(x, 0) = 0$, and φ is a measurable, locally integrable function in the first variable such that $\varphi(\cdot, u) > 0$ $m - a.e.$ for every $u > 0$.

Example 3.8. Lorentz p-space is a modular function space generated by

$$\rho(f) = \sup_{\tau \in \mathscr{T}} \int_{\mathbb{R}} |f(t)|^p d\mu_\tau(t),$$

where μ is a σ-finite measure, $\mathscr{T}$ is a group of all measure preserving transformations $\tau : \mathbb{R} \to \mathbb{R}$ and $\mu_\tau(E) = \mu(\tau^{-1}(E))$.

Example 3.9. Orlicz–Lorentz space is a modular function space generated by

$$\rho(f) = \sup_{\tau \in \mathscr{T}} \int_{\mathbb{R}} \varphi(|f(t)|) d\mu_\tau(t),$$

where φ is an Orlicz function, μ is a σ-finite measure, $\mathscr{T}$ is a group of all measure preserving transformations $\tau : \mathbb{R} \to \mathbb{R}$ and $\mu_\tau(E) = \mu(\tau^{-1}(E))$.

Example 3.10. Let μ be a finite measure on Ω. Denote by $(\mathscr{M}, \mu)$ the space of all μ-measurable real-valued functions. Let $E \subset (\mathscr{M}, \mu)$ be an order continuous Banach function lattice with the Fatou property, that is, a Banach space with a norm $\| \cdot \|_E$ such that

1. If $f \in (\mathcal{M}, \mu)$, $g \in E$ and $|f| \leq |g|$ μ-a.e., then $f \in E$ and $\|f\|_E \leq \|g\|_E$,
2. There exists a function in E that is strictly positive μ-a.e.,
3. If $f, f_n \in E$, $|f_n| \leq |f|$ and $f_n \to 0$ μ-a.e., then $\|f_n\|_E \to 0$,
4. If $0 \leq f_n \in E$ and $f \in (\mathcal{M}, \mu)$ and $f_n \uparrow f$, then $\|f_n\|_E \uparrow \|f\|_E$.

Let φ be an Orlicz function. Define

$$\rho(f) = \begin{cases} \|\varphi \circ f\|_E, & \text{if } \varphi \circ f \in E, \\ \infty, & \text{otherwise.} \end{cases}$$

It is a straightforward calculation to show that ρ is a regular convex function modular. The corresponding modular function space is called the Calderon–Lozanovskii space defined by the space E and the Orlicz function φ and denoted by E_φ. Note that if $E = L^1$ then E_φ becomes the Orlicz space L^φ. Similarly, if $E = L^p$ then E_φ is the Banach space of all measurable functions such that

$$\rho(f) = \left(\int_\Omega |\varphi(f(\omega))|^p d\mu \right)^{\frac{1}{p}} < \infty.$$

For more general discussion of Calderon–Lozanovskii spaces, see for example [43] or [95].

Example 3.11. Let $\Omega = [0,1]$, $\{\Omega_n\}$ be a countable disjoint partition of Ω such that $m(\Omega_p) = 2^{-p}$, where m is the Lebesgue measure on $[0,1]$. Let $\mathscr{P}$ be a δ-ring generated by the sets of the form $A \cap \Omega_p$ for all $p \in \mathbb{N}$ and all measurable sets $A \subset [0,1]$. For a measurable function $f : \Omega \to \mathbb{R}$, we put

$$\rho(f) = \sum_{p=1}^{\infty} \left(\int_{\Omega_p} |f(t)|^p dm(t) \right)^{\frac{1}{p}} + \sup_{p \in \mathbb{N}} \int_{\Omega_p} |f(t)|^p dm(t).$$

Example 3.12. An extremal flexibility gained by using the apparatus of the modular function spaces can be illustrated as follows: the operator itself is used for the construction of a modular and hence a space in which this operator has required properties. Let us consider for instance the following Uryshon integral operator:

$$T(f)(x) = \int_0^1 k(x, y, |f(y)|) dy + f_0(x),$$

where f_0 is a fixed function and $f : [0,1] \to \mathbb{R}$ is Lebesgue measurable. For the kernel k we assume that

(a) $k : [0,1] \times [0,1] \times \mathbb{R}_+ \to \mathbb{R}_+$ is Lebesgue measurable,
(b) $k(x, y, 0) = 0$,
(c) $k(x, y, .)$ is continuous, convex and increasing to $+\infty$,
(d) $\int_0^1 k(x, y, t) dx > 0$ for $t > 0$ and $y \in (0, 1)$.

Assume in addition that for almost all $t \in [0,1]$ and measurable functions f,g there holds

$$\int_0^1 \left\{ \int_0^1 k(t,u,|k(u,v,|f(v)|) - k(u,v,|g(v)|)|)dv \right\}du \leq \int_0^1 k(t,u,|f(u) - g(u)|)du.$$

Setting $\rho(f) = \int_0^1 \left\{ \int_0^1 k(x,y,|f(y)|)dy \right\}dx$ and using Jensen's inequality, it is easy to show that ρ is a nonnegative, even, convex nonlinear functional on the space of measurable functions, $L_\rho = \{f : [0,1] \to \mathbb{R} : \exists \lambda > 0, \ \rho(\lambda f) < \infty\}$, and that $\rho(T(f) - T(g)) \leq \rho(f - g)$, that is, T is nonexpansive with respect to ρ. We will come back to this example in Chap. 4 (Example 5.1) and Chap. 5 (Example 7.6).

3.6 Generalizations and Special Cases

3.6.1 General Definition of Function Modular

While in this book we will always restrict our considerations to the modular function spaces of scalar-valued measurable functions defined by convex regular function modulars, it is worthwhile to mention a more general framework and, in this context, to point to some interesting applications and possible research directions. The general framework presented in this section follows the book [132] and, as it was mentioned earlier in this chapter, the notions used in this book are special cases of the general theory, hence all the results of the general theory of modular function spaces are applicable to convex regular function modulars defined on spaces of scalar-valued measurable functions. From this perspective, the reader who is not interested in the general theory may skip this section.

Let S denote a Banach space equipped with a norm $\|\cdot\|_S$. A function $f : \Omega \to S$ is called measurable if there exists a sequence of $\mathscr{P}$-measurable simple functions $\{s_n\}$ such that $s_n(\omega) \to f(\omega)$ for every $\omega \in \Omega$. It is well known that a function f is measurable if and only if it is separably valued in S and for every open $B \subset S$, $f^{-1}(B) \in \Sigma$. Thus, if f is measurable then $\|f(\omega)\|_S$ is a scalar-valued measurable function of ω. It follows from the properties of scalar measurable functions that there exists a sequence of $\mathscr{P}$-measurable simple functions $\{s_n\}$ such that $\|s_n(\omega)\|_S \uparrow \|f(\omega))\|_S$ for every $\omega \in \Omega$. Let us denote by $\mathscr{M}(\Omega, S)$ the vector space of all measurable functions acting from Ω to S.

Let us recall the definitions of a subadditive and a σ-subadditive measures.

Definition 3.15. A set function $\mu : \Sigma \to [0, \infty)$ is called a subadditive measure if

1. $\mu(\emptyset) = 0$
2. $\mu(A \cup B) \leq \mu(A) + \mu(B)$ for $A, B \in \Sigma$ such that $A \cap B = \emptyset$
3. $\mu(A) \leq \mu(B)$ if $A, B \in \Sigma$ with $A \subset B$.

If, moreover,

$$\mu\left(\bigcup_{n=1}^{\infty} E_n\right) \leq \sum_{n=1}^{\infty} \mu(E_n)$$

for every sequence of sets $E_n \in \Sigma$, then μ is called a σ-subadditive measure.

We are now ready to give a general definition of a function modular.

Definition 3.16. Let $\rho : \mathscr{E} \times \Sigma \to [0, \infty)$ satisfy the following conditions:

1. $\rho(0, E) = 0$ for every $E \in \Sigma$
2. $\rho(f, E) \leq \rho(g, E)$ whenever $\|f(\omega)\|_S \leq \|g(\omega)\|_S$ for all $\omega \in E$ and all $f \in \mathscr{E}$, $E \in \Sigma$
3. $\rho(f, \cdot) : \Sigma \to 0, \infty)$ is a σ-subadditive measure for every $f \in \mathscr{E}$
4. $\rho(\alpha, A) \to 0$ as $\alpha \downarrow 0$ for every $A \in \mathscr{P}$
5. If $\rho(\alpha, A) = 0$ for an $\alpha > 0$ then $\sup_{\beta > 0} \rho(\beta, A) = 0$
6. For every $\alpha > 0$, $\rho(\alpha, \cdot)$ is order continuous on $\mathscr{P}$.

The definition of ρ is then extended to all $f \in \mathscr{M}(\Omega, S)$ and all sets $E \in \Sigma$ by defining that

$$\rho(f, E) = \sup\{\rho(g, E) : g \in \mathscr{E}, \|g(\omega)\|_S \leq \|f(\omega)\|_S, \ \omega \in E\}.$$

As a convention, we set $\rho(f) = \rho(f, \Omega)$. As already mentioned in Sect. 3.1, it is a straightforward exercise to prove that if $S = \mathbb{R}$ and if in addition ρ is convex in its first argument and it has the Fatou property, that is $\rho(f_n) \uparrow \rho(f)$ whenever $\|f_n(\omega)\|_S \uparrow \|f(\omega)\|_S$ for every $\omega \in \Omega$, then ρ is a regular convex function modular in the sense of Definition 3.2.

This general definition allows us to investigate a wide range of special cases and applications. To the end of this section we introduce few examples of such applications and refer the reader to the corresponding literature. Since these constructs and results will not be used in this book, we limit our discussion to giving some basic facts without proofs.

The wide range of applications of the general theory of modular function spaces suggests that extending the fixed point theory to this more general framework will produce results of high interest to researchers from many areas of mathematics and beyond. Since there exist only very few fix point results for nonconvex function modulars or for spaces of functions with values in Banach spaces, this looks like a promising new research direction.

3.6.2 Nonlinear Operator Valued Measures

The relationship between the theory of modular function spaces and the theory of nonlinear operator valued measures, discussed below, shows how close the results presented in this book are to the theory of abstract integration and representation of nonlinear operators. From this point of view, it is not surprising that several applications of fixed point theory in modular function spaces point to problems of nonlinear operator theory and related theory of integral and differential equations.

Let F be a Banach space. By $N(S,F)$, we denote the space of all mappings $U : S \to F$ such that $U(0) = 0$ and U are uniformly continuous on bounded subsets of S. A set function $\mu : \mathscr{P} \to N(S,F)$ is called an operator valued measure if μ satisfies the following conditions:

1. $\mu(0) = 0$
2. μ is countably additive in the pointwise sense
3. $\lim_{\delta \to 0} sv_\delta(\mu, \alpha, E) = 0$ for every $E \in \mathscr{P}$ and $\alpha > 0$
4. $\widetilde{\mu}(\cdot)r$ is order continuous for every $r \in S$.

Remember that

$$sv_\delta(\mu, \alpha, E) = \sup \left\{ \left\| \sum_{i=1}^{n} (\mu(E_i)r_i - \mu(E_i)r_i') \right\|_F \right\}$$

where the supremum is taken over all finite sequences of sets E_i, and over elements r_i, r_i' in S such that

$$\bigcup_{i=1}^{n} E_i \subset E, E_i \in \mathscr{P}, \|r\|_S, \|r'\|_S \le \alpha, \|r - r'\|_S \le \delta, 1 \le i \le n, n \in \mathbb{N}.$$

Recall also the following definition:

$$\widetilde{\mu}(E)r = \sup\{\|\mu(A)r\|_F : A \subset E, A \in \mathscr{P}\}.$$

For a $\mathscr{P}$-simple function, $f = \sum_{i=1}^{n} r_i 1_{E_i}$, where $r_i \in S$ and $E_i \in \mathscr{P}$. We define the integral with respect to the operator valued measure μ over a set $E \in \Sigma$ by

$$\int_E f d\mu = \sum_{i=1}^{n} \mu(E \cap E_i)r_i.$$

The domain of the integration is then extended to the space $\mathscr{M}(\mu)$ of all measurable functions f such that there exists a sequence of simple functions f_n converging everywhere to f for which the integrals $\int_{(\cdot)} f_n d\mu$ are uniformly countably additive on Σ. Let us assume also that there exists $\alpha > 0$ such that if $\int_E f d\mu = 0$ for every $f \in \mathscr{E}$ such that $\|f(\omega)\|_S \le \alpha$ in E, then $\int_E g d\mu = 0$ for all $g \in \mathscr{E}$. Define

$$\rho(f, E) = \sup \left\{ \left\| \int_E g d\mu \right\|_F : g \in \mathscr{E}, \|g(\omega)\|_S \le \|f(\omega)\|_S, \omega \in E \right\}.$$

It is not difficult to show that ρ defined by (3.6.2) is a function modular in the sense of Definition 3.16. The integral introduced above may be regarded as a nonlinear operator defined in $\mathscr{M}(\mu)$ with values in Banach space F. It may be proved that $L_\rho^0 \subset \mathscr{M}(\mu)$ and that the integral is continuous in E_ρ, see [131]. Hence the apparatus of the theory of modular function spaces can be applied to spaces of functions integrable with respect to operator valued measures.

The linear operator valued measures were investigated by Dobrakov in a series of papers starting from [60] and [61]. Integration of bounded functions with respect to nonlinear operator valued measures was considered by Batt [15] and Friedman and Tong [75] in connection with the representation of orthogonally additive operators. This integration was extended to the larger class $\mathscr{M}(\mu)$ by Kozlowski and Szczypinski in [145, 146]. For the exposition of this theory we refer the reader to the paper by Szczypinski [189].

Example 3.13. Let S and F be two Banach spaces and let $\varphi : S \to F$ be a locally uniformly continuous function with $\varphi(0) = 0$. Assume that m is a positive finite measure on Σ. Let us define an operator valued measure $\mu(A)r = m(A)\varphi(r)$, for $A \in \Sigma$ and $r \in S$. As shown in [189], if $\varphi \circ f : \Omega \to F$ is integrable in the sense of Bochner, then f is integrable with respect to μ and

$$\int_E f d\mu = \int_E (\varphi \circ f) dm.$$

Example 3.14. Let F be a Banach space, $S = \mathbb{R}$ and let m be a countably additive vector valued measure $m : \Sigma \to F$. Define an operator valued measure $\mu(E)r = rm(E)$ for $r \in \mathbb{R}$ and $E \in \Sigma$. The integral of $f : \Omega \to \mathbb{R}$ with respect to μ is equal to the integral introduced by Dunford and Schwartz [70].

The following two examples show that some nonlinear operators important for the theory of integral and differential equations can be represented as integrals with respect to respective operator valued measures. These examples resonate with examples presented in this book, see Example 3.12.

Example 3.15. Let m be a positive σ-additive measure on Σ. Let $S = \mathbb{R}$ and let $F = L^1(\Omega)$ and the function $k : \Omega \times \Omega \times \mathbb{R} \to \mathbb{R}$ satisfy the following conditions:

1. $k(\omega, \sigma, 0) \equiv 0$
2. $k(\omega, \sigma, r)$ is continuous in r almost everywhere in $\Omega \times \Omega$
3. $k(\omega, \sigma, r)$ is measurable in (ω, σ) for every fixed r
4. To every $\varepsilon > 0$, there exists $\delta > 0$ such that if $m(A) < \delta$ then

$$\int_\Omega \left| \int_A k(\omega, \sigma, r) dm(\sigma) \right| dm(\omega) < \varepsilon$$

uniformly in r. Define $\mu : \Sigma \to N(\mathbb{R}, L^1(\Omega))$ by

$$\mu(E)r = \int_E k(\cdot, \sigma, r) dm(\sigma) \in L^1(\Omega).$$

It is easy to verify that μ is an operator valued measure and that for $f \in \mathcal{M}(\mu)$, we have

$$\left(\int_E f d\mu \right)(\omega) = \int_E k(\omega, \sigma, f(\sigma)) dm(\sigma).$$

Hence the integral of f with respect to μ is actually the Urysohn integral operator.

Example 3.16. Let m be a positive σ-additive measure on Σ. Let $S = \mathbb{R}$ and $F = L^1(\Omega)$ and let $H : \Omega \times \mathbb{R} \to \mathbb{R}$ be such that $H(\cdot, r) \in L^1(\Omega)$ for every $r \in \mathbb{R}$, $H(\omega, \cdot)$ is continuous almost everywhere in ω, and $H(\cdot, 0) \equiv 0$. Define μ as

$$\mu(A)r = H(\cdot, r 1_A)$$

for $A \in \Sigma$. Then μ is an operator valued measure such that for any $f \in \mathcal{M}(\mu)$,

$$\left(\int_A f d\mu \right)(\omega) = H(\omega, f(\omega)),$$

that is taking the integral with respect to μ is equivalent to applying the Nemytskii operator defined by H.

3.6.3 Nonlinear Fourier Transform

It is well known that the linear Fourier transform

$$\mathcal{F}(f)(x) = \int_{\mathbb{R}} e^{-tx} f(t) dm(t) \tag{3.12}$$

is the linear isometry from the Hilbert space L^2 onto itself. The situation is much more complicated with the nonlinear version of the Fourier transform defined as

$$F(f)(x) = \int_{\mathbb{R}} e^{-tx} \varphi(f(t)) dm(t), \tag{3.13}$$

where $\varphi : \mathbb{R} \to \mathbb{R}$ is an increasing, continuous, odd function such that

$$\lim_{u \to +\infty} \varphi(u) = +\infty, \ and \ \lim_{u \to -\infty} \varphi(u) = -\infty.$$

The major question is how to define the proper domain of F, that is, such a subspace E of the space $\mathcal{M}$ of all m-measurable functions that $T : E \to L^2$ and is continuous. This task can be achieved by using the tools and methods of modular function spaces. Indeed, let us define

$$\rho(f,A) = \sup\{\|T(g)\|_{L^2} : g \in \mathcal{E}, \|g(\omega)\|_S \leq 1_A(\omega)\|f(\omega)\|_S\}. \qquad (3.14)$$

It is not difficult to prove that the formula (3.14) defines a function modular as well as that $T : E_\rho \to L^2$ and that T is continuous (see e.g., [132]). We will show that $E_\rho = E^\psi$ where ψ is an Orlicz function defined by $\psi(u) = (\varphi(u))^2$, and

$$E^\psi = \{f \in \mathcal{M} : \rho(\lambda f) < \infty, \text{ for all } \lambda > 0\}.$$

Let $f \in E^\psi$. Let $\alpha > 0$ and $A_k \in \mathcal{P}$ (that is A_k are of finite measure) be such that $A_k \downarrow \emptyset$. To prove that $f \in E_\rho$ we need to show that $\rho(\alpha f, A_k) \to 0$. Since for every constant $\alpha > 0$ and every $g \in \mathcal{E}$ the function $\varphi(\alpha g(\cdot))$ belongs to L^2 and the linear Fourier transform $\mathcal{F} : L^2 \to L^2$ is an isometry, then we get the following sequence of equalities:

$$\rho(\alpha f, A_k) = \sup\left\{\left(\int_{\mathbb{R}} \left|\int_{A_k} e^{-itx} \varphi(\alpha g(t)) dm(t)\right|^2 dm(x)\right)^{\frac{1}{2}} : g \in \mathcal{E}, |g| \leq |f|\right\}$$

$$= \sup\left\{\left(\int_{\mathbb{R}} \left|\mathcal{F}\left(1_{A_k}(\cdot)\varphi(\alpha g(\cdot))\right)(x)\right|^2 dm(x)\right)^{\frac{1}{2}} : g \in \mathcal{E}, |g| \leq |f|\right\}$$

$$= \sup\left\{\left\|\mathcal{F}\left(1_{A_k}(\cdot)\varphi(\alpha g(\cdot))\right)\right\|_{L^2} : g \in \mathcal{E}, |g| \leq |f|\right\}$$

$$= \sup\left\{\left\|1_{A_k}(\cdot)\varphi(\alpha g(\cdot))\right\|_{L^2} : g \in \mathcal{E}, |g| \leq |f|\right\}$$

$$= \sup\left\{\left(\int_{A_k} \varphi^2(\alpha g(t)) dm(t)\right)^{\frac{1}{2}} : g \in \mathcal{E}, |g| \leq |f|\right\}$$

$$\leq \left(\int_{A_k} \varphi^2(\alpha f(t)) dm(t)\right)^{\frac{1}{2}} = \left(\int_{A_k} \psi(\alpha f(t)) dm(t)\right)^{\frac{1}{2}} \to 0.$$

Hence, $E^\psi \subset E_\varphi$. In order to prove the inverse inclusion, let us take $f \in E_\varphi$ and hence note that $\rho(\alpha f, A_k) \to 0$. For every simple function $g_k \in \mathcal{E}$ such that $|g_k| \leq |f 1_{A_k}|$, we get

$$\left(\int_{\mathbb{R}} \left|\int_{A_k} e^{-itx} \varphi(\alpha g_k(t)) dm(t)\right|^2 dm(x)\right)^{\frac{1}{2}} \leq \rho(\alpha f, A_k) \to 0,$$

which means that $\mathcal{F}\left(\varphi(g_k(\cdot))\right) \to 0$. Since $\mathcal{E} \subset E^\varphi$, it follows that the function $\varphi(\alpha g_k(\cdot))$ is a member of the space L^2. The inverse Fourier transform acts continuously from L^2 to L^2 and therefore $\{\varphi(\alpha g_k(\cdot))\}$ converges to zero in L^2, that is,

$$\int_{\mathbb{R}} \varphi^2(\alpha g_k(t)) dm(t) \to 0. \qquad (3.15)$$

We claim now that $f \in E^{\psi}$. Indeed, assume to the contrary that there exists $\alpha > 0$ such that

$$\int_{\mathbb{R}} \varphi^2(\alpha f(t)) dm(t) = \infty.$$

There exist then $\varepsilon > 0$ and a sequence $A_k \downarrow \emptyset$ such that for every k,

$$\int_{A_k} \varphi^2(\alpha f(t)) dm(t) > \varepsilon.$$

Hence we may choose a sequence of simple functions g_k so that $|g_k| \leq |f 1_{A_k}|$ and

$$\int_{\mathbb{R}} \varphi^2(\alpha g_k(t)) dm(t) > \frac{\varepsilon}{2},$$

which contradicts (3.15). The contradiction completes the proof of the equality $E_\rho = E^{\psi}$.

Chapter 4
Geometry of Modular Function Spaces

In this chapter, we discuss the general geometrical methods for consideration of fixed point properties, similarly, in the case of Banach space setting. The most common approach in the Banach space fixed point theory for nonexpansive mappings is to assume the uniform convexity of the norm, which implies the reflexivity, which in turn guarantees the weak compactness of the closed bounded sets. As we will see further, the notion of a uniform convexity of function modulars in conjunction with the property (R) being the modular equivalence of the Banach space reflexivity equips us with powerful tools for proving the fixed point property in modular function spaces. Also, this theory provides a set of powerful techniques for proving existence of common fixed points for commutative families of mappings acting in modular function spaces, and for investigating the topological properties of the set of common fixed points.

4.1 Uniform Convexity in Modular Function Spaces

Let us start by reminding ourselves that the modulus of convexity of a Banach space X is defined as

$$\delta(\varepsilon) = \inf\left\{ 1 - \left\| \frac{x+y}{2} \right\| : x,y \in X,\ \|x\| \leq 1,\ \|y\| \leq 1,\ \|x-y\| \geq \varepsilon \right\}.$$

A Banach space X is called *uniformly convex* if $\delta(\varepsilon) > 0$ for every $\varepsilon > 0$. We will investigate notions of the modular equivalents of uniform convexity of ρ. As demonstrated below, one concept of uniform convexity in normed spaces generates several different types of uniform convexity in modular function spaces. This is due primarily to the fact that, in general, modulars are not homogeneous.

Definition 4.1. Let $\rho \in \Re$. We define the following *uniform convexity type* properties of the function modular ρ:

© Springer International Publishing Switzerland 2015   79
M. A. Khamsi, W. M. Kozlowski, *Fixed Point Theory in Modular Function Spaces*,
DOI 10.1007/978-3-319-14051-3_4

(a) Let $r > 0, \varepsilon > 0$. Define

$$D_1(r,\varepsilon) = \left\{ (f,g) : f,g \in L_\rho, \rho(f) \le r, \rho(g) \le r, \rho(f-g) \ge \varepsilon r \right\}.$$

Let

$$\delta_1(r,\varepsilon) = \inf\left\{ 1 - \frac{1}{r}\rho\left(\frac{f+g}{2}\right) : (f,g) \in D_1(r,\varepsilon) \right\}, \text{ if } D_1(r,\varepsilon) \ne \emptyset,$$

and $\delta_1(r,\varepsilon) = 1$ if $D_1(r,\varepsilon) = \emptyset$. We say that ρ satisfies $(UC1)$ if for every $r > 0, \varepsilon > 0$, we have $\delta_1(r,\varepsilon) > 0$. Note, that for every $r > 0$, $D_1(r,\varepsilon) \ne \emptyset$, for $\varepsilon > 0$ small enough.

(b) We say that ρ satisfies $(UUC1)$ if for every $s \ge 0$, $\varepsilon > 0$, there exists $\eta_1(s,\varepsilon) > 0$ depending on s and ε such that

$$\delta_1(r,\varepsilon) > \eta_1(s,\varepsilon) > 0 \text{ for } r > s.$$

(c) Let $r > 0, \varepsilon > 0$. Define

$$D_2(r,\varepsilon) = \left\{ (f,g) : f,g \in L_\rho, \rho(f) \le r, \rho(g) \le r, \rho\left(\frac{f-g}{2}\right) \ge \varepsilon r \right\}.$$

Let

$$\delta_2(r,\varepsilon) = \inf\left\{ 1 - \frac{1}{r}\rho\left(\frac{f+g}{2}\right) : (f,g) \in D_2(r,\varepsilon) \right\}, \text{ if } D_2(r,\varepsilon) \ne \emptyset,$$

and $\delta_2(r,\varepsilon) = 1$ if $D_2(r,\varepsilon) = \emptyset$. We say that ρ satisfies $(UC2)$ if for every $r > 0, \varepsilon > 0$, we have $\delta_2(r,\varepsilon) > 0$. Note, that for every $r > 0$, $D_2(r,\varepsilon) \ne \emptyset$, for $\varepsilon > 0$ small enough.

(d) We say that ρ satisfies $(UUC2)$ if for every $s \ge 0$, $\varepsilon > 0$ there exists $\eta_2(s,\varepsilon) > 0$ depending on s and ε such that

$$\delta_2(r,\varepsilon) > \eta_2(s,\varepsilon) > 0 \text{ for } r > s.$$

(e) We say that ρ is strictly convex, (SC), if for every $f,g \in L_\rho$ such that $\rho(f) = \rho(g)$ and

$$\rho\left(\frac{f+g}{2}\right) = \frac{\rho(f)+\rho(g)}{2}$$

we have $f = g$.

Remark 4.1. Let us observe that for $i = 1,2$, $\delta_i(r,0) = 0$, and $\delta_i(r,\varepsilon)$ is an increasing function of ε for every fixed r. Note also that

$$\delta_1(r,\varepsilon) = \inf\left\{ \delta'(r,h) : h \in L_\rho, \rho(h) \ge r\varepsilon \right\},$$

$$\delta_2(r,\varepsilon) = \inf\left\{ \delta'(r,h) : h \in L_\rho, \rho\left(\frac{h}{2}\right) \ge r\varepsilon \right\},$$

where

$$\delta'(r,h) = \inf\left\{1 - \frac{1}{r}\rho\left(f + \frac{h}{2}\right) \,:\, f \in L_\rho, \rho(f) \leq r, \rho(f+h) \leq r\right\}.$$

Proposition 4.1. *The following conditions characterize relationship between the above defined notions:*

(a) (UUCi) implies (UCi) for $i = 1,2$;
(b) $\delta_1(r,\varepsilon) \leq \delta_2(r,\varepsilon)$;
(c) (UC1) implies (UC2);
(d) (UC2) implies (SC);
(e) (UUC1) implies (UUC2);

Proof. The proofs of these implications follow immediately from Definition 4.1.
□

Remark 4.2. Observe that, denoting $\rho_\alpha(u) = \alpha\rho(u)$, and the corresponding moduli of convexity by $\delta_{\rho_\alpha,i}$, where $i = 1,2$, we have

$$\delta_{\rho_\alpha,i}(r,\varepsilon) = \delta_{\rho,i}\left(\frac{r}{\alpha},\varepsilon\right),$$

and

$$\delta_{\rho,i}(r,\varepsilon) = \delta_{\rho_\alpha,i}(r\alpha,\varepsilon).$$

Hence, ρ is (UCx), where (UCx) is any of the conditions from Definition 4.1, if and only if there exists $\alpha > 0$ such that ρ_α is (UCx). In particular, taking $\alpha = \frac{1}{r}$, it is enough to prove any of the conditions defining (UCx) with $r = 1$.

Example 4.1. It is known that in Orlicz spaces, the Luxemburg norm is uniformly convex if and only if φ is uniformly convex and Δ_2 property holds; this result can be traced to early papers by Luxemburg [157], Milnes [163], Akimovic [4], and Kaminska [101]. It is also known that, under suitable assumptions, $(UC2)$ in Orlicz spaces is equivalent to the very convexity of the Orlicz function [117, 44]. Remember that the function φ is called *very convex* if or every $\varepsilon > 0$ and any $x_0 > 0$, there exists $\delta > 0$ such that

$$\varphi\left(\frac{1}{2}(x-y)\right) \geq \frac{\varepsilon}{2}\left(\varphi(x) + \varphi(y)\right) \geq \varepsilon\varphi(x_0),$$

implies

$$\varphi\left(\frac{1}{2}(x+y)\right) \leq \frac{1}{2}(1-\delta)\left(\varphi(x) + \varphi(y)\right).$$

Typical examples of Orlicz functions that are are very convex and do not satisfy Δ_2 condition are: $\varphi_1(t) = e^{|t|} - |t| - 1$ and $\varphi_2(t) = e^{t^2} - 1$, [163, 147]. Therefore, these are the examples of Orlicz spaces that are not uniformly convex in the norm sense, and hence the classical Kirk theorem cannot be applied. However, these spaces are uniformly convex in the modular sense, and respective modular fixed point results can be applied.

Let us recall a notion of bounded away sequences of real numbers.

Definition 4.2. A sequence $\{t_n\} \subset (0,1)$ is called *bounded away from* 0 if there exists $0 < a < 1$ such that $t_n \geq a$ for every $n \in \mathbb{N}$. Similarly, $\{t_n\} \subset (0,1)$ is called *bounded away from* 1 if there exists $0 < b < 1$ such that $t_n \leq b$ for every $n \in \mathbb{N}$.

We need to prove a series of technical results that will play an essential role in proving our fixed point theorems. First, let us mention the following result whose proof is elementary. Note that for $t = \frac{1}{2}$, this result follows directly from Definition 4.1.

Lemma 4.1. *Let* $\rho \in \mathfrak{R}$ *be* $(UUC1)$ *and let* $t \in (0,1)$. *Then for every* $s > 0$, $\varepsilon > 0$ *there exists* $\eta_1^t(s,\varepsilon) > 0$ *depending only on s and* ε *such that*

$$\delta_1^t(r,\varepsilon) > \eta_1^t(s,\varepsilon) > 0 \ \text{for any} \ r > s.$$

The next lemma introduces a useful technique which is used extensively for investigating convergence to fixed points in the $(UUC1)$ modular function spaces. It was introduced in [115] for the case $t_n = \frac{1}{2}$ and extended to more general case in [54], see, for example, [188] for the Banach space equivalent.

Lemma 4.2. *Let* $\rho \in \mathfrak{R}$ *be* $(UUC1)$ *and let* $\{t_n\} \subset (0,1)$ *be bounded away from* 0 *and* 1. *If there exists* $R > 0$ *such that*

$$\limsup_{n \to \infty} \rho(f_n) \leq R, \ \limsup_{n \to \infty} \rho(g_n) \leq R,$$

$$\lim_{n \to \infty} \rho(t_n f_n + (1 - t_n)g_n) = R,$$

then,

$$\lim_{n \to \infty} \rho(f_n - g_n) = 0.$$

Proof. Assume to the contrary that this is not the case and fix an arbitrary $\gamma > 0$. Passing to a subsequence if necessary, we may assume that there exists an $\varepsilon > 0$ such that

$$\rho(f_n) \leq R + \gamma, \ \rho(g_n) \leq R + \gamma \tag{4.1}$$

while

$$\rho(f_n - g_n) \geq (R + \gamma)\varepsilon. \tag{4.2}$$

Since $\{t_n\}$ is bounded away from 0 and 1, there exists $0 < a < b < 1$ such that $a \leq t_n \leq b$ for all natural n. Passing to a subsequence if necessary, we can assume that there exists $t_0 \in [a,b]$ with $t_n \to t_0$. For every $t \in [0 1]$ and $f,g \in D_1(R + \gamma, \varepsilon)$, let us define $\lambda_{f,g}(t) = \rho(tf + (1 - t)g)$. Observe that the function $\lambda_{f,g} : [0,1] \to [0, R + \gamma]$ is a convex function. Hence, the function

$$\lambda(t) = \sup\{\lambda_{f,g}(t) : f,g \in D_1(R + \gamma, \varepsilon)\}$$

is also convex on $[0,1]$, and consequently it is a continuous function on $[a,b]$. Noting that

$$\delta_1^t(R+\gamma,\varepsilon) = 1 - \frac{1}{r}\lambda_{f,g}(t),$$

we conclude that $\delta_1^t(R+\gamma,\varepsilon)$ is a continuous function of $t \in [a,b]$. Hence

$$\lim_{n\to\infty} \delta_1^{t_n}(R+\gamma,\varepsilon) = \delta_1^{t_0}(R+\gamma,\varepsilon). \tag{4.3}$$

By (4.1) and (4.2)

$$\delta_1^{t_n}(R+\gamma,\varepsilon) \le 1 - \frac{1}{R+\gamma}\rho(t_n f_n + (1-t_n)g_n). \tag{4.4}$$

From (4.3), we deduce that the left hand side of (4.4) tends to $\delta_1^{t_0}(R+\gamma,\varepsilon)$ as $n \to \infty$ while the right hand side tends to $\frac{\gamma}{R+\gamma}$ in view of (4.2). Hence

$$\delta_1^{t_0}(R+\gamma,\varepsilon) \le \frac{\gamma}{R+\gamma}. \tag{4.5}$$

By $(UUC1)$ and Lemma 4.1, there exists $\eta_1^{t_0}(R,\varepsilon) > 0$ satisfying

$$0 < \eta_1^{t_0}(R,\varepsilon) \le \delta_1^{t_0}(R+\gamma,\varepsilon). \tag{4.6}$$

Combining (4.5) with (4.6), we get

$$0 < \eta_1^{t_0}(R,\varepsilon) \le \frac{\gamma}{R+\gamma}.$$

Letting $\gamma \to 0$, we get a contradiction that completes the proof. $\quad\square$

Lemma 4.3. *Let ρ satisfies $(UC2)$. Let $f,g \in L_\rho$ and $r > 0$ be such that $f \ne g$, $\rho(f) \le r$ and $\rho(g) \le r$. Then,*

$$\rho\left(\frac{f+g}{2}\right) < r.$$

Proof. Set $h = f - g$. Since $f \ne g$, then $\rho\left(\frac{h}{2}\right) > 0$. By the definition of δ_2, it follows from

$$\rho\left(\frac{f-g}{2}\right) = \rho\left(\frac{h}{2}\right) = \left[\frac{1}{r}\rho\left(\frac{h}{2}\right)\right]r$$

that

$$\rho\left(\frac{f+g}{2}\right) \le r\left[1 - \delta_2\left(r, \frac{1}{r}\rho\left(\frac{h}{2}\right)\right)\right].$$

Note that

$$\delta_2\left(r, \frac{1}{r}\rho\left(\frac{h}{2}\right)\right) > 0$$

because ρ satisfies $(UC2)$ and $\dfrac{1}{r}\rho\left(\dfrac{h}{2}\right) > 0$. Hence

$$\rho\left(\frac{f+g}{2}\right) < r,$$

as claimed. $\square$

Lemma 4.4. *Let ρ satisfies $(UC2)$. Then, for every $r > 0$, $\delta_2(r, 1) = 1$.*

Proof. Fix $r > 0$. Let $(x, y) \in D_2(r, 1)$, i.e., $\rho(x) \leq r, \rho(y) \leq r$, and

$$\rho\left(\frac{x-y}{2}\right) \geq r.$$

Observe that

$$r \leq \rho\left(\frac{x-y}{2}\right) \leq \frac{\rho(x) + \rho(y)}{2} \leq r.$$

Hence

$$\rho\left(\frac{x-y}{2}\right) = r.$$

Setting $z = -y$, we get

$$r = \rho\left(\frac{x-z}{2}\right) = \rho\left(\frac{x+y}{2}\right) \leq r.$$

By applying Lemma 4.3 to $f = x$ and $g = z$, we obtain that $x = z$, i.e., $x = -y$, which implies that

$$\rho\left(\frac{x+y}{2}\right) = 0.$$

Hence, by the definition of δ_2, we conclude that $\delta_2(r, 1) = 1$ as claimed. $\square$

In the next theorem, we investigate relationship between the uniform convexity of function modulars and the *unique best approximant property* which generalizes well known properties of Hilbert and uniformly convex Banach spaces. For other results on best approximation in modular function spaces see [142, 143, 118, 119].

Theorem 4.1. *Assume $\rho \in \Re$ is $(UUC2)$. Let $C \subset L_\rho$ be nonempty, convex, and ρ-closed. Let $f \in L_\rho$ be such that $d = d_\rho(f, C) < \infty$. There exists then a unique best ρ-approximant of f in C, that is, a unique $g_0 \in C$ such that*

$$\rho(f - g_0) = d_\rho(f, C).$$

Proof. The uniqueness follows from the strict convexity of ρ; remember that $(UUC2)$ implies (SC). Let us prove the existence of the ρ-approximant. Since C is ρ-closed, we may assume without loss of any generality that $d := d_\rho(f, C) > 0$.

Clearly there exists a sequence $\{f_n\} \in C$ such that

$$\rho(f - f_n) \le d \left(1 + \frac{1}{n}\right).$$

We claim that $\left\{\frac{1}{2}f_n\right\}$ is ρ-Cauchy. Assume to the contrary that this is not the case. There exist an $\varepsilon_0 > 0$ and a subsequence $\{f_{n_k}\}$ of $\{f_n\}$ such that

$$\rho\left(\frac{f_{n_k} - f_{n_p}}{2}\right) \ge \varepsilon_0,$$

for any $p, k \ge 1$. Since ρ is (UUC2) then ρ is (UC2). Hence

$$\rho\left(f - \frac{f_{n_k} + f_{n_p}}{2}\right) \le \left(1 - \delta_2\left(d(k,p), \frac{\varepsilon_0}{d(k,p)}\right)\right) d(k,p),$$

where $d(k,p) = \left(1 + \frac{1}{\min(n_p, n_k)}\right) d$. For $p, k \ge 1$, we have $d(k,p) \le 2d$. Hence

$$\delta_2\left(d(k,p), \frac{\varepsilon_0}{d(k,p)}\right) \ge \delta_2\left(d(k,p), \frac{\varepsilon_0}{2d}\right).$$

Since ρ is (UUC2) then there exists $\eta > 0$ such that

$$\delta_2\left(r, \frac{\varepsilon_0}{2d}\right) \ge \eta$$

for any $r > \frac{d}{3}$. Since $d(k,p) \ge d > \frac{d}{3}$, we get

$$\rho\left(f - \frac{f_{n_k} + f_{n_p}}{2}\right) \le (1 - \eta) d(k,p),$$

for any $k, p \ge 1$. By the convexity of C, $\frac{f_{n_k} + f_{n_p}}{2} \in C$. Using the definition of d, we get

$$d \le \rho\left(f - \frac{f_{n_k} + f_{n_p}}{2}\right) \le (1 - \eta) d(k,p),$$

for any $k, p \ge 1$. If we let k, p go to infinity, we get $d \le (1 - \eta)d$, which is impossible. Hence $\left\{\frac{1}{2}f_n\right\}$ is ρ-Cauchy. By the completeness of L_ρ, $\left\{\frac{1}{2}f_n\right\}$ ρ-converges to a $g \in L_\rho$. Fix $m \ge 1$. Since $\left\{\frac{f_m + f_n}{2}\right\} \in C$ and ρ-converges to $\frac{f_m}{2} + g$ and C is ρ-closed, then we have $\frac{f_m}{2} + g \in C$. Letting $m \to \infty$, we get $2g \in C$. By

Propositions 3.4 and 3.5, passing to a subsequence if necessary, we get

$$\rho(f-2g) \le \liminf_{n\to\infty} \rho\left(f-g-\frac{f_n}{2}\right) \le \liminf_{n\to\infty} \liminf_{m\to\infty} \rho\left(f-\frac{f_n+f_m}{2}\right).$$

Since ρ is convex, we get

$$\liminf_{n\to\infty} \liminf_{m\to\infty} \rho\left(f-\frac{f_n+f_m}{2}\right) \le \liminf_{n\to\infty} \liminf_{m\to\infty} \frac{\rho(f-f_n)+\rho(f-f_m)}{2} \le d.$$

Hence, $\rho(f-2g) \le d$. Since $2g \in C$, we get $d \le \rho(f-2g)$. Therefore, $\rho(f-2g) = d$. In other words, $g_0 = 2g$ is the ρ-approximant of f in C. $\quad\square$

We will use Theorem 4.1 in the proof of our next result to establish a relationship between the modular uniform convexity and the property (R), which is a modular equivalent of the Milman–Pettis theorem, which states that the uniform convexity of a Banach space implies its reflexivity.

Theorem 4.2. *Let $\rho \in \mathfrak{R}$ be (UUC2), then L_ρ has property (R).*

Proof. Let $\{C_n\}$ be a nonincreasing sequence of nonempty, ρ-bounded, ρ-closed, convex subsets of L_ρ. According to Definition 3.9 we need to demonstrate that $\{C_n\}$ has nonempty intersection. Fix any $f \in C_1$. By the ρ-boundedness of C_1, there exists a finite constant $M > 0$ such that for any $n \ge 1$, $\rho(f-g) < M$ for any $g \in C_n \subset C_1$. Hence,

$$\sup_{n\ge 1} d_\rho(f,C_n) < \infty.$$

Using the proximinality of ρ-closed convex subsets of L_ρ (Theorem 4.1), for every $n \ge 1$ there exists $f_n \in C_n$ such that $\rho(f-f_n) = d_\rho(f,C_n)$. It is easy to show that $\{d_\rho(f,C_n)\}$ is nondecreasing and bounded. Hence $\lim_{n\to\infty} d_\rho(f,C_n) = d$ exists. Assume that $d = 0$, then $d_\rho(f,C_n) = 0$, for any $n \ge 1$. Since $\{C_n\}$ are ρ-closed, we get $f \in C_n$ for any $n \ge 1$, which implies $\bigcap_{n\ge 1} C_n \ne \emptyset$. Therefore, we can assume $d > 0$. In this case we claim that $\left\{\frac{1}{2}f_n\right\}$ is ρ-Cauchy. Indeed if we assume the contrary, then there exists an $\varepsilon_0 > 0$ and a subsequence $\{f_{n_k}\}$ of $\{f_n\}$ such that

$$\rho\left(\frac{f_{n_k}-f_{n_p}}{2}\right) \ge \varepsilon_0,$$

for any $p,k \ge 1$. Since ρ is (UUC2) it follows that ρ is (UC2). Hence

$$\rho\left(f-\frac{f_{n_k}+f_{n_p}}{2}\right) \le \left(1-\delta_2\left(d,\frac{\varepsilon_0}{d}\right)\right)d,$$

for any $p, k \geq 1$. So

$$d_\rho\left(f, C_{\max(n_p, n_k)}\right) \leq \rho\left(f - \frac{f_{n_k} + f_{n_p}}{2}\right) \leq \left(1 - \delta_2\left(d, \frac{\varepsilon_0}{d}\right)\right) d,$$

for any $p, k \geq 1$. If we let $p, k \to \infty$, we will get

$$d \leq \left(1 - \delta_2\left(d, \frac{\varepsilon_0}{d}\right)\right) d,$$

which is a contradiction because $\delta_2\left(d, \frac{\varepsilon_0}{d}\right) > 0$ by $(UC2)$. Hence the sequence $\left\{\frac{1}{2} f_n\right\}$ is ρ-Cauchy, and therefore it ρ-converges to some $g \in L_\rho$. Let us prove that $2g \in C_n$, for any $n \geq 1$. Indeed, we have $\frac{f_k + f_p}{2} \in C_n$, for any $p, k \geq n$. Fix any $k \geq n$. Since $\left\{\frac{f_k + f_p}{2}\right\}$ ρ-converges to $\frac{f_k}{2} + g$ as $p \to \infty$, and C_n is ρ-closed, then $\frac{f_k}{2} + g \in C_n$, for any $k \geq n$. If we let $k \to \infty$, we get $2g \in C_n$, for any $n \geq 1$. Hence $\bigcap_{n \geq 1} C_n \neq \emptyset$, which completes the proof. $\square$

In fact, the conclusion of Theorem 4.2 may be improved to have an intersection property for any family of subsets.

Theorem 4.3. *Let $\rho \in \Re$ be $(UC2)$. Let $\{C_\alpha\}_{\alpha \in \Gamma}$ be a decreasing family of nonempty, convex, ρ-closed subsets of L_ρ, where $(\Gamma, \prec)$ is upward directed. Assume that there exists $f \in L_\rho$ such that $\sup_{\alpha \in \Gamma} d_\rho(f, C_\alpha) < \infty$. Then, $\bigcap_{\alpha \in \Gamma} C_\alpha \neq \emptyset$.*

Proof. Set $d = \sup_{\alpha \in \Gamma} d_\rho(f, C_\alpha)$. Without loss of generality, we may assume $d > 0$. For Any $n \geq 1$, there exists $\alpha_n \in \Gamma$ such that

$$d\left(1 - \frac{1}{n}\right) < d_\rho(f, C_{\alpha_n}) \leq d.$$

Since $(\Gamma, \prec)$ is upward directed, we may assume $\alpha_n \prec \alpha_{n+1}$. In particular we have $C_{\alpha_{n+1}} \subset C_{\alpha_n}$ for any $n \geq 1$. Since ρ is $(UC2)$, we get $C_0 = \bigcap_{n \geq 1} C_{\alpha_n} \neq \emptyset$. Clearly C_0 is ρ-closed and

$$d_\rho(f, C_0) = \sup_{n \geq 1} d_\rho(f, C_{\alpha_n}) = d.$$

Again using the property $(UC2)$ satisfied by ρ, there exists unique $g_0 \in C_0$ such that $d_\rho(f, C_0) = \rho(f - g_0)$. Let us prove that $g_0 \in C_\alpha$ for any $\alpha \in \Gamma$. Fix $\alpha \in \Gamma$. If for some $n \geq 1$ we have $\alpha \prec \alpha_n$, then obviously we have $g_0 \in C_{\alpha_n} \subset C_\alpha$. Therefore, let us assume that $\alpha \not\prec \alpha_n$, for any $n \geq 1$. Since Γ is upward directed, there exists $\beta_n \in \Gamma$ such that $\alpha_n \prec \beta_n$ and $\alpha \prec \beta_n$ for any $n \geq 1$. We can also assume that $\beta_n \prec \beta_{n+1}$ for any $n \geq 1$. Again we have $C_1 = \bigcap\limits_{n \geq 1} C_{\beta_n} \neq \emptyset$. Since $C_{\beta_n} \subset C_{\alpha_n}$, for any $n \geq 1$, we get $C_1 \subset C_0$. Moreover we have

$$d = d_\rho(f, C_0) \leq d_\rho(f, C_1) = \sup_{n \geq 1} d_\rho(f, C_{\beta_n}) \leq d.$$

Hence, $d_\rho(f, C_1) = d$ which implies the existence of a unique point $g_1 \in C_1$ such that $d_\rho(f, C_1) = \rho(f - g_1) = d$. Since ρ is uniformly convex, it must be (SC). Hence, $g_0 = g_1$. In particular, we have $g_0 \in C_{\beta_n}$, for any $n \geq 1$. Since $\alpha \prec \beta_n$, we conclude that $C_{\beta_n} \subset C_\alpha$, for any $n \geq 1$, which implies $g_0 \in C_\alpha$. Since α was taking arbitrary in Γ, we get $g_0 \in \bigcap\limits_{\alpha \in \Gamma} C_\alpha$, which implies $\bigcap\limits_{\alpha \in \Gamma} C_\alpha \neq \emptyset$. $\square$

Since ρ is convex, ρ-closed balls are convex. Theorem 4.3 implies the following.

Corollary 4.1. *Let $\rho \in \Re$ and $C \subset L_\rho$ be nonempty, convex, ρ-closed, and ρ-bounded. Assume ρ is (UC2). Then $\mathscr{A}(C)$ is compact.*

4.2 Parallelogram Inequality and Minimizing Sequence Property

We will establish now a modular version of the *parallelogram inequality* for uniformly convex modular function spaces . See [201] and [18] for the norm version of this inequality.

Lemma 4.5. *For each $0 < s < r$ and $\varepsilon > 0$, set*

$$\Psi(r, s, \varepsilon) = \inf\left\{ \frac{1}{2}\rho^2(f) + \frac{1}{2}\rho^2(g) - \rho^2\left(\frac{f+g}{2}\right) \right\},$$

where the infimum is taken over all $f, g \in L_\rho$ such that $\rho(f) \leq r$, $\rho(g) \leq r$, $\max(\rho(f), \rho(g)) \geq s$, and $\rho(f - g) \geq r\varepsilon$. If $\rho \in \Re$ is (UUC1), then $\Psi(r, s, \varepsilon) > 0$ for any $0 < s < r$ and $\varepsilon > 0$. Moreover, for a fixed $r, s > 0$, we have

(a) $\Psi(r, s, 0) = 0$;
(b) $\Psi(r, s, \varepsilon)$ *is a nondecreasing function of ε;*
(c) *if $\lim\limits_{n \to \infty} \Psi(r, s, t_n) = 0$, then $\lim\limits_{n \to \infty} t_n = 0$.*

Proof. It is easy to see that $\Psi(r, s, \varepsilon) \geq 0$ for any $0 < s < r$ and $\varepsilon > 0$. We want to show that $\Psi(r, s, \varepsilon) > 0$. Assume to the contrary that there exist $0 < s < r$ and $\varepsilon > 0$ such that $\Psi(r, s, \varepsilon) = 0$. Then there exist $\{f_n\}$ and $\{g_n\}$ such that

$$\lim_{n \to \infty} \frac{1}{2}\rho^2(f_n) + \frac{1}{2}\rho^2(g_n) - \rho^2\left(\frac{f_n + g_n}{2}\right) = 0 \tag{4.7}$$

and $\rho(f_n) \leq r$, $\rho(g_n) \leq r$, $\max\left(\rho(f_n), \rho(g_n)\right) \geq s$, and $\rho(f_n - g_n) \geq r\varepsilon$. Since ρ is convex, we get

$$\rho^2\left(\frac{f_n + g_n}{2}\right) \leq \left(\frac{\rho(f_n) + \rho(g_n)}{2}\right)^2 \leq \frac{\rho^2(f_n) + \rho^2(g_n)}{2},$$

where we used the inequality $2ab \leq a^2 + b^2$, for any $a, b \in \mathbb{R}$. Hence

$$\left(\frac{\rho(f_n) - \rho(g_n)}{2}\right)^2 \leq \frac{1}{2}\rho^2(f_n) + \frac{1}{2}\rho^2(g_n) - \rho^2\left(\frac{f_n + g_n}{2}\right),$$

which implies $\lim\limits_{n \to \infty}\left(\rho(f_n) - \rho(g_n)\right) = 0$. Without loss of any generality, we may assume that $\lim\limits_{n \to \infty}\rho(f_n) = R$ exists. This also implies that $\lim\limits_{n \to \infty}\rho(g_n) = R$. By (4.7) we get then

$$\lim\limits_{n \to \infty}\rho(g_n) = \lim\limits_{n \to \infty}\rho^2\left(\frac{f_n + g_n}{2}\right) = R.$$

Observe that

$$R = \lim\limits_{n \to \infty}\max\left(\rho(f_n), \rho(g_n)\right) \geq s > 0.$$

By Lemma 4.2 then $\rho(f_n - g_n) \to 0$ contradicting the fact that $\rho(f_n - g_n) \geq r\varepsilon > 0$. The proofs of (a), (b), and (c) are straightforward. $\quad\square$

Let us introduce a notion of a ρ-*type*, a powerful technical tool that will be used in the proofs of our fixed point results.

Definition 4.3. Let $C \subset L_\rho$ be convex and ρ-bounded. A function $\tau : C \to [0, \infty]$ is called a ρ-*type* (or shortly a type) if there exists a sequence $\{g_k\}$ of elements of C such that for any $f \in C$ there holds

$$\tau(f) = \limsup\limits_{k \to \infty}\rho(g_k - f).$$

The following lemma establishes a crucial *minimizing sequence property* of uniformly convex modular function spaces. It will be used in conjunction with the parallelogram property in the proof of the main fixed point result for nonexpansive mappings acting in modular function spaces.

Lemma 4.6. *Assume that $\rho \in \mathfrak{R}$ is (UUC1). Let C be a ρ-closed ρ-bounded convex nonempty subset of L_ρ. Let τ be a ρ-type defined on C. Then, any minimizing sequence of τ is ρ-convergent. Its ρ-limit is independent of the minimizing sequence.*

Proof. Let $\{f_n\} \subset C$ be such that $\tau(f) = \limsup\limits_{n \to \infty}\rho(f_n - f)$, $\tau_0 = \inf\{\tau(h) : h \in C\}$. Let $\{g_k\}$ be a minimizing sequence of τ. Since C is ρ-bounded, there exists $R > 0$ such that $\rho(f - g) \leq R$ for any $f, g \in C$. The rest of the proof is split into two cases.
Case 1: Assume that $\tau_0 > 0$. Let us choose a $\sigma > 0$ such that $\tau_0 - \sigma > 0$. Let us fix

g_m *and* g_k and select a subsequence $\{h_n\}$ of $\{f_n\}$ such that

$$0 < \tau_0 \leq \tau\left(\frac{g_m + g_k}{2}\right) = \lim_{n\to\infty} \rho\left(\frac{g_m + g_k}{2} - h_n\right).$$

Then, for n sufficiently large, we have

$$0 < \tau_0 - \sigma \leq \rho\left(\frac{g_m + g_k}{2} - h_n\right) \leq \max\left(\rho(g_m - h_n), \rho(g_k - h_n)\right).$$

Using (4.7) from Lemma 4.5, we have

$$\rho^2\left(\frac{g_m + g_k}{2} - h_n\right) \leq \frac{1}{2}\rho^2(g_m - h_n) + \frac{1}{2}\rho^2(g_k - h_n) - \Psi\left(R, \tau_0 - \sigma, \frac{1}{R}\rho(g_m - g_k)\right),$$

and after passing with n to infinity, we get

$$\tau^2\left(\frac{g_k + g_m}{2}\right) \leq \frac{1}{2}\tau^2(g_k) + \frac{1}{2}\tau^2(g_m) - \Psi\left(R, \tau_0 - \sigma, \frac{1}{R}\rho(g_k - g_m)\right).$$

Hence,

$$\tau_0^2 \leq \frac{1}{2}\tau^2(g_k) + \frac{1}{2}\tau^2(g_m) - \Psi\left(R, \tau_0 - \sigma, \frac{1}{R}\rho(g_k - g_m)\right),$$

for any $k, m \geq 1$, which implies that

$$\Psi\left(R, \tau_0 - \sigma, \frac{1}{R}\rho(g_k - g_m)\right) \leq \frac{1}{2}\tau^2(g_k) + \frac{1}{2}\tau^2(g_m) - \tau_0^2.$$

Hence, $\lim\limits_{k,m\to\infty} \Psi\left(R, \tau_0 - \sigma, \frac{1}{R}\rho(g_k - g_m)\right) = 0$. The properties of Ψ imply that $\{g_k\}$ is ρ-Cauchy. Since L_ρ is ρ-complete and C is ρ-closed, then $\{g_k\}$ is ρ-convergent to a $g \in C$. Let us prove that any other minimizing sequence also ρ-converges to g. Indeed, let $\{u_n\} \in C$ be any minimizing sequence of τ. Using the same argument as previously, we have

$$\tau_0^2 \leq \tau^2\left(\frac{g_n + u_n}{2}\right) \leq \frac{1}{2}\tau^2(g_n) + \frac{1}{2}\tau^2(u_n) - \Psi\left(R, \tau_0 - \sigma, \frac{1}{R}\rho(g_n - u_n)\right),$$

for some $\sigma > 0$ such that $\tau_0 - \sigma > 0$ and

$$\Psi\left(R, \tau_0 - \sigma, \frac{1}{R}\rho(g_n - u_n)\right) \leq \frac{1}{2}\tau^2(g_n) + \frac{1}{2}\tau^2(u_n) - \tau_0^2,$$

for any $n \geq 1$. As before, we get $\lim\limits_{n\to\infty} \rho(g_n - u_n) = 0$. Since ρ is convex, we have

$$\rho\left(\frac{u - g}{3}\right) \leq \frac{1}{3}\rho(u - u_n) + \frac{1}{3}\rho(u_n - g_n) + \frac{1}{3}\rho(g_n - g),$$

where u is the ρ-limit of $\{u_n\}$. Clearly, our assumptions imply that $\rho\left(\dfrac{u-g}{3}\right) = 0$, hence $u = g$. This completes the proof of the lemma for Case 1.

Case 2: Assume that $\tau_0 = 0$. Let

$$K = \bigcap_{n \geq 1} \overline{\mathrm{conv}}_\rho\left(\{f_k; k \geq n\}\right),$$

which is nonempty in view of property (R): recall $(UUC1)$ implies (R) by Theorem 4.2. Let $f_\infty \in K$, $h \in C$, $\varepsilon > 0$. By definition of τ, there exists $n_0 > 0$ such that for every $n > n_0$

$$\rho(f_n - h) \leq \tau(h) + \varepsilon.$$

Therefore, $f_n \in B_\rho(h, \tau(h) + \varepsilon)$ for $n > n_0$. This fact implies

$$K \subset \overline{\mathrm{conv}}_\rho\left(\{f_n : n \geq n_0\}\right) \subset B_\rho(h, \tau(h) + \varepsilon).$$

Hence, $f_\infty \in B_\rho(h, \tau(h) + \varepsilon)$. Since this is true for every $\varepsilon > 0$, $f_\infty \in B_\rho(h, \tau(h))$, that is,

$$\rho(f_\infty - h) \leq \tau(h). \tag{4.8}$$

Let $\{g_k\}$ be a minimizing sequence of τ. Using (4.8) with $h = g_k$, we get

$$\rho(f_\infty - g_k) \leq \tau(g_k) \to \tau_0 = 0 \text{ as } k \to \infty,$$

which means that $\{g_k\}$ is ρ-convergent to f_∞. Since this limit is independent of the sequence $\{g_k\}$, the proof of Case 2 and Lemma 4.6 is complete. $\quad\square$

4.3 Uniform Noncompact Convexity in Modular Function Spaces

Compactness plays an essential role in many fundamental theorems. This is the case with Schauder fixed point theorem for example. However, there are major problems when dealing with mappings, which are not compact. In order to investigate non-compactness, Kuratowski [150] was the first one to introduce the concept of "measure on noncompactness" in connection with certain problems in General Topology. In this section, we extend this concept to modular function spaces.

Definition 4.4. Let $\rho \in \mathcal{R}$. Let A be a nonempty ρ-bounded subset of L_ρ. Define the *Kuratowski measure of noncompactness* of A by

$$\alpha(A) = \inf\left\{ \begin{array}{l} \varepsilon > 0 : A \text{ can be covered by a finite number} \\ \text{of sets with } \rho - \text{diameter smaller than } \varepsilon \end{array} \right\},$$

and the *Hausdorff measure of noncompactness* of A by

$$\chi(A) = \inf \left\{ \begin{array}{c} \varepsilon > 0 : A \ can \ be \ covered \ by \ a \ finite \ number \\ of \ \rho - balls \ of \ radius \ smaller \ than \ \varepsilon \end{array} \right\}.$$

for any subset A of L_ρ. We will make the obvious convention that the infimum over an empty set is infinite.

One can easily notice that $\chi(A) \leq \alpha(A) \leq \delta_\rho(A)$, for every ρ-bounded $A \subset L_\rho$.

Proposition 4.2. *The following properties hold.*

(1) If $A \subset B$, then $\chi(A) \leq \chi(B)$ and $\alpha(A) \leq \alpha(B)$.
(2) $\chi(\bar{A}) = \chi(A)$ and $\alpha(\bar{A}) = \alpha(A)$.
(3) If $\alpha(A) = 0$, then $\bar{A}$ is ρ-sequentially compact.
(4) Let $\{A_n\}$ be a decreasing sequence of nonempty ρ-closed ρ-bounded subset of L_ρ. Assume that $\lim\limits_{n \to \infty} \alpha(A_n) = 0$, then $\bigcap\limits_{n \geq 1} A_n$ is a nonempty ρ-compact set.
(5) Let $\{A_n\}$ be a decreasing sequence of nonempty ρ-closed ρ-bounded convex subset of L_ρ. Assume that $\lim\limits_{n \to \infty} \chi(A_n) = 0$. Then $\bigcap\limits_{n \geq 1} A_n$ is nonempty and

$$\chi\left(\bigcap_{n \geq 1} A_n \right) = 0.$$

Recall that $\bar{A}$ is the smallest ρ-closed subset of L_ρ which contains A.

Proof. The proof of (1) is obvious. In order to prove (2), let us first notice that $\chi(A) \leq \chi(\bar{A})$. Let $\varepsilon > 0$, then there exists $B_\rho(x_1, r), ..., B_\rho(x_n, r)$ such that

$$A \subset B_\rho(x_1, r) \cup .. \cup B_\rho(x_n, r),$$

where $r \leq \chi(A) + \varepsilon$. The Fatou property implies that ρ-balls are ρ-closed and therefore

$$\bar{A} \subset B_\rho(x_1, r) \cup .. \cup B_\rho(x_n, r),$$

which implies $\chi(\bar{A}) \leq r \leq \chi(A) + \varepsilon$. Since ε was taken arbitrarily positive, we get $\chi(\bar{A}) \leq \chi(A)$. In order to prove the second claim of (2), we will only need to show that $\delta_\rho(A) = \delta_\rho(\bar{A})$. Indeed set $\delta_\rho(A) = d$. Then for any $f \in A$, we have $A \subset B_\rho(f, d)$. Since ρ-balls are ρ-closed, we get $\bar{A} \subset B_\rho(f, d)$. So for any $g \in \bar{A}$, we have $\rho(f - g) \leq d$. Hence $A \subset B_\rho(g, d)$, for all $g \in \bar{A}$. Again using the fact that ρ-balls are ρ-closed, we get $\bar{A} \subset B_\rho(g, d)$. Hence $\rho(f - g) \leq d$, for any $f, g \in \bar{A}$, i.e. $\delta_\rho(\bar{A}) \leq d = \delta_\rho(A)$. Since $\delta_\rho(A) \leq \delta_\rho(\bar{A})$ is obvious, we deduce $\delta_\rho(A) = \delta_\rho(\bar{A})$. Let us prove (3). Assume that $\alpha(\bar{A}) = 0$. Let $\{f_n\}$ be any sequence of elements in $\bar{A}$. Let $\varepsilon > 0$. Using the definition of α, we can find a subsequence $\{f_{n'}\}$ such that $\rho(f_{n'} - f_{m'}) \leq \varepsilon$, for all $n', m' \in N$. The diagonal argument will show that $\{f_n\}$ has a ρ-Cauchy subsequence which is ρ-convergent since L_ρ is ρ-complete. In order to prove (4), let $\{A_n\}$ be a decreasing sequence of ρ-closed ρ-bounded nonempty subsets of L_ρ such that $\lim\limits_{n\to\infty} \alpha(A_n) = 0$. Choose $f_n \in A_n$ and let $\varepsilon > 0$. Then, there exists $n_0 \geq 1$ such that $\alpha(A_n) < \varepsilon$, for $n \geq n_0$. Therefore, there exist $D_1, .., D_k$ such that

$$A_{n_0} \subset \bigcup_{1 \leq i \leq k} D_i,$$

with $\delta_\rho(D_i) \leq \varepsilon$, for $i = 1, \cdots, k$. So $f_n \in \bigcup_{1 \leq i \leq k} D_i$, for all $n \geq n_0$. Hence one D_i contains infinitely many f_n. So there exists a subsequence $\{f_{n'}\}$ of $\{f_n\}$ such that $\rho(f_{n'} - f_{m'}) \leq \varepsilon$, for all n', m'. Again the diagonal argument will imply that $\{f_n\}$ has a ρ-Cauchy subsequence which is therefore ρ-convergent to $f \in L_\rho$. Since $\{A_n\}$ are ρ-closed and decreasing, we deduce that $f \in A_n$ for all $n \geq 1$. Hence $\bigcap\limits_{n \geq 1} A_n$ is not empty and $\alpha(\cap A_n) = 0$ holds from (1). Using (3), we conclude that $\bigcap\limits_{n \geq 1} A_n$ is ρ-compact.

Finally, let us prove (5). Let $\{A_n\}$ be a decreasing sequence of ρ-closed ρ-bounded nonempty convex subsets of L_ρ such that $\lim\limits_{n\to\infty} \chi(A_n) = 0$. Let $\varepsilon > 0$, then there exists $n_0 \in N$ such that $\chi(A_{n_0}) < \varepsilon$. Let $f_n \in A_n$, for any $n \geq 1$. Then there exist infinitely many elements from $\{f_n\}$ which belong to a ρ-ball $B_\rho(f, \varepsilon)$, for some $f \in L_\rho$. This implies that there exists a subsequence $\{f_{n'}\}$ such that

$$\rho\left(\frac{1}{2}(f_{n'} - f_{m'})\right) \leq \rho(f_{n'} - f) + \rho(f - f_{m'}) \leq 2\,\varepsilon,$$

for every n', m'. As noted before, we secure the existence of a subsequence $\{f_{n'}\}$ such that $\left\{\frac{1}{2}f_{n'}\right\}$ is ρ-Cauchy. Let $h \in L_\rho$ be its ρ-limit. Fix $n' \geq 1$. Then $\left\{\frac{1}{2}(f_{n'} + f_{m'})\right\}_{m' \geq n'}$ is ρ-convergent to $\frac{1}{2}f_{n'} + h$. Since $A_{n'}$ is ρ-closed and convex, we deduce that $\frac{1}{2}f_{n'} + h \in A_{n'}$. Since (A_n) is decreasing, we get $\frac{1}{2}f_{m'} + h \in A_{n'}$, for any $m' \geq n'$. Again since $(\frac{1}{2}f_{m'} + h)_{m' \geq n'}$ is ρ-convergent to $h + h = 2h$, we obtain that $2h \in A'_n$. This clearly implies that $2h \in \cap A_n$. Using (3) we get $\chi(\cap A_n) = 0$. The proof of Proposition 4.2 is complete. $\square$

Remark 4.3. Since in general ρ is not subadditive, there is no reason to have $\alpha(A) = 0$ whenever A is ρ-compact. In fact, it can be shown that ρ satisfies the Δ_2-condition if and only if $\alpha(A) = 0$ whenever A is ρ-compact. For more on α and χ, one can consult [17].

Similarly as Goebel and Sekowski [83] did for Banach spaces, we can define a new scaling for modular spaces using the measures of noncompactness α and χ.

Definition 4.5. The ρ-*modulus of noncompact convexity* Δ_χ (resp. Δ_α) is defined as

$$\Delta_\chi(r,\varepsilon) = \sup \left\{ \begin{array}{c} c > 0: \; for \; any \; \rho - bounded \; convex \; A \subset L_\rho \\ \rho - bounded \; and \; f \in L_\rho \; such \; that \; A \subset B_\rho(f,r) \\ with \; \chi(A) \geq r\varepsilon, \; then \; d_\rho(f,A) \leq r(1-c) \end{array} \right\}, \quad \text{(II)}$$

for every $r > 0$ and $\varepsilon > 0$. For the definition of Δ_α, one will replace $\chi(A)$ by $\alpha(A)$ in (II). Define the *characteristic of noncompact convexity* of L_ρ by

$$\varepsilon_\chi(r,L_\rho) = \sup \left\{ \varepsilon > 0: \; \Delta_\chi(r,\varepsilon) = 0 \right\},$$

and

$$\varepsilon_\alpha(r,L_\rho) = \sup \left\{ \varepsilon > 0: \; \Delta_\alpha(r,\varepsilon) = 0 \right\},$$

for every $r > 0$.

Since $\chi \leq \alpha$, we get $\Delta_\chi \leq \Delta_\alpha$ and $\varepsilon_\alpha \leq \varepsilon_\chi$. In any Banach space X, it can be easily proved that $\varepsilon_\chi(X) = 0$ if and only if $\varepsilon_\alpha(X) = 0$. In modular spaces, it is not the case in general.

Definition 4.6. The modular function space L_ρ is said to be α *(resp. χ)-uniformly ρ-noncompact convex* if and only if $\varepsilon_\alpha(r,L_\rho) = 0$ (resp. $\varepsilon_\chi(r,L_\rho) = 0$), for every $r > 0$.

Clearly if L_ρ is χ-uniformly ρ-noncompact convex, then L_ρ is α-uniformly ρ-noncompact convex.

Let $\rho \in \mathcal{R}$. In Definition 4.1, we introduced the ρ-modulus of uniform convexity of L_ρ as

$$\delta_1(r,\varepsilon) = \inf \left\{ 1 - \frac{1}{r}\rho(f+h) \right\},$$

for any $r > 0$ and $\varepsilon > 0$, where the infimum is taken over all $f, h \in L_\rho$ such that

$$\rho(f) \leq r, \;\; \rho(f+h) \leq r, \;\; and \;\; \rho(h) \geq r\varepsilon.$$

Let us discuss the connection between ρ-uniform convexity and α-uniform ρ-noncompact convexity. Let $A \subset L_\rho$ be nonempty ρ-bounded and convex such that $A \subset B_\rho(f,r)$ with $\alpha(A) \geq r\varepsilon$ for some $f \in L_\rho, r > 0$ and $\varepsilon > 0$. Let $\zeta < 1$ then, by definition of α, one can find $h_1, h_2 \in A$ such that $\rho(h_1 - h_2) \geq r\zeta\varepsilon$. Therefore, we have

$$\rho\left(\frac{(f-h_1) + (f-h_2)}{2} \right) \leq r(1 - \delta_1(r,\zeta\varepsilon)).$$

Since $\dfrac{h_1 + h_2}{2} \in A$, we deduce that

$$d_\rho(f,A) \leq r(1 - \delta_1(r,\zeta\varepsilon)).$$

Which clearly implies that

$$\delta_1(r,\zeta\varepsilon)) \leq \Delta_\alpha(r,\varepsilon),$$

for every $r > 0$, $\varepsilon > 0$ and $\zeta < 1$.

Goebel and Sekowski [83] proved that whenever the characteristic of uniform non-compact convexity of any Banach space is less than 1, then the space is reflexive. In what follows, we investigate the validity of this result in modular function spaces. First we investigate the link between proximinality and the ρ-modulus of noncompact convexity. More exactly, given $f \in L_\rho$, we consider the minimization problem of finding $h \in C$ such that

$$\rho(f-h) = \inf\{\rho(f-g) : g \in C\},$$

where $C \subset L_\rho$ is not empty. Such a h is called a best approximant. Problems of finding best approximants are important in approximation theory [168, 118] and probability theory [52].

Theorem 4.4. *Let $\rho \in \Re$. Assume that L_ρ is α-uniformly ρ-noncompact convex. Then for any nonempty C ρ-bounded ρ-closed convex subset of L_ρ and $f \in L_\rho$ such that $d_\rho(f,C) < \infty$, the set*

$$P_\rho(f,C) = \{g \in C : d_\rho(f,C) = \rho(f-g)\}$$

is nonempty ρ-compact and convex.

Proof. We assume without any loss of generality that $d = d_\rho(f,C) > 0$. Consider the sets $C_n = C \cap B_\rho(f, d + \frac{1}{n})$ for $n \geq 1$. Clearly $\{C_n\}$ is a decreasing sequence of ρ-closed, ρ-bounded, nonempty, convex subsets of C. Assume that $\inf_{n\geq 1} \alpha(C_n) = \lim_{n\to\infty} \alpha(C_n) > 0$. Then, since L_ρ is α-uniformly ρ-noncompact convex, there exists $\Delta > 0$ such that

$$d_\rho(f,C_n) \leq (1-\Delta)\left(d + \frac{1}{n}\right),$$

for every $n \geq 1$. Since $d_\rho(f,C) \leq d_\rho(f,C_n)$, we get

$$d \leq (1-\Delta)\left(d + \frac{1}{n}\right),$$

for every $n \geq 1$. This contradicts the fact that $d > 0$. Hence $\inf_{n\geq 1} \alpha(C_n) = \lim_{n\to\infty} \alpha(C_n) = 0$, and by Proposition 4.2, we deduce that $\bigcap C_n$ is a nonempty ρ-compact convex subset of C. Clearly

$$P_\rho(f,C) = \bigcap_{n\geq 1} C_n.$$

The proof of Theorem 4.4 is hence complete. $\square$

It is not clear whether the conclusion of Theorem 4.4 remains true if we assume that $\varepsilon_\alpha(L_\rho) < 1$. Note also how Theorem 4.4 relates to Theorem 4.1. In a similar manner, our next result is parallel to Theorem 4.2.

Theorem 4.5. *Let $\rho \in \mathfrak{R}$. Assume L_ρ is α-uniformly ρ-noncompact convex. Then L_ρ has the Property (R).*

Proof. Let $(C_n)_{n\geq 0}$ be a decreasing sequence of ρ-bounded ρ-closed nonempty convex subsets of L_ρ. Fix $f \in C_0$, then

$$r = \sup_{n\geq 1} d_\rho(f, C_n) \leq \delta_\rho(C_0) < \infty.$$

Define $K_n = C_n \cap B_\rho(f, r)$. Clearly $\{K_n\}$ is a decreasing sequence of nonempty ρ-closed convex subsets of L_ρ. Using the same argument as in the proof of Theorem 4.4, one can show that $\lim_{n\to\infty} \alpha(K_n) = 0$. So, $\cap K_n$ is a nonempty ρ-compact convex subset of L_ρ. Therefore $\cap C_n$ is nonempty, which completes the proof of Theorem 4.5. $\square$

One may ask if under the assumptions of Theorem 4.5, we have the intersection property for any family of subsets. The answer to this question is in the affirmative.

Theorem 4.6. *Let $\rho \in \mathfrak{R}$. Assume L_ρ is α-uniformly ρ-noncompact convex. Then for any decreasing family $(A_\beta)_{\beta\in\Gamma}$ of ρ-bounded, ρ-closed nonempty convex subsets of L_ρ and any directed set $(\Gamma, \preceq)$, we have $\bigcap_{\beta\in\Gamma} A_\beta$ is nonempty.*

Proof. Let $(A_\beta)_{\beta\in\Gamma}$ be a decreasing family of ρ-bounded, ρ-closed convex subsets of L_ρ. We can assume without any loss of generality that Γ has a minimum, say β_0. Let $f \in A_{\beta_0}$, then

$$r = \sup\{d_\rho(f, A_\beta) : \beta \in \Gamma\} \leq \delta_\rho(A_{\beta_0}) < \infty.$$

Set $K_\beta = A_\beta \cap B_\rho(f, r)$, for any $\beta \in \Gamma$. Then $(K_\beta)_{\beta\in\Gamma}$ is a decreasing family of nonempty ρ-closed convex subsets of L_ρ. Using the same argument as in the proof of Theorem 4.5, we get $\inf\{\alpha(K_\beta) : \beta \in \Gamma\} = 0$. Therefore for every $n \geq 1$, there exists $\beta_n \in \Gamma$ such that

$$\alpha(K_{\beta_n}) \leq \frac{1}{n}.$$

Clearly one can choose $\{\beta_n\} \subset \Gamma$ such that $\{K_{\beta_n}\}$ is decreasing. Using Proposition 4.2, we obtain that $K_\infty = \cap K_{\beta_n}$ is a nonempty ρ-compact subset. We can assume that there exists $\beta' \in \Gamma$ such that $\beta_n \leq \beta'$, for every $n \geq 1$. Otherwise, we would have

$$K_\infty = \bigcap_{n\geq 1} K_{\beta_n} = \bigcap_{\beta\in\Gamma} K_\beta.$$

Since (K_β) is decreasing, we deduce that $K_\beta \subset K_{\beta'} \subset K_\infty$, for every $\beta \geq \beta'$. Put $\delta = \delta_\rho(K_{\beta'})$. As $\alpha(K_{\beta'}) = 0$, one can find $\{A_i\}_{1 \leq i \leq k}$ such that

$$K_{\beta'} \subset \bigcup_{1 \leq i \leq k} A_i \ \text{ with } \ \delta_\rho(A_i) \leq \frac{\delta}{2}.$$

For any $\beta \geq \beta'$ choose $f_\beta \in K_\beta$. It is not hard to show that there exists $i_0 \in \{1, 2, .., k\}$ such that for every $\beta \geq \beta'$ there exists $\gamma \geq \beta$ such that $f_\gamma \in A_{i_0}$. Indeed, assume to the contrary that for every $i \in \{1, .., k\}$ there exists $\beta_i \geq \beta'$ such that f_γ does not belong to A_i, for all $\gamma \geq \beta_i$. Since Γ is a directed set, there exists $\beta \in \Gamma$ such that $\beta \geq \beta_i$ for every $i \in \{1, .., k\}$. So f_β does not belong to any A_i. Contradicting the fact that $f_\beta \in K_\beta \subset K_{\beta'} \subset A_1 \cup .. \cup A_k$. Therefore, there exists such $i_0 \in \{1, .., k\}$. Set

$$C_\beta = \{f_\gamma; \ \gamma \geq \beta \ \text{ and } \ f_\gamma \in C_{i_0}\}.$$

Then for every $\beta \geq \beta'$, C_β is nonempty. Moreover, we have for every $\zeta \geq \beta$, there exists $\gamma \geq \zeta$ such that $f_\gamma \in A_\beta$. Since $C_\beta \subset A_i$, we get $\delta_\rho(C_\beta) \leq \frac{\delta}{2}$. Let $L_\beta = \overline{conv}_\rho(C_\beta)$, for all $\beta \geq \beta'$, where $\overline{conv}_\rho(A)$ is the smallest ρ-closed convex subset, which contains A. Then $\delta_\rho(L_\beta) \leq \frac{\delta}{2}$ and $\{L_\beta\}_{\beta \geq \beta'}$ is a decreasing family of nonempty ρ-closed convex sets with $L_\beta \subset K_\beta$, for every $\beta \geq \beta'$. Using $\alpha(K_\beta) = 0$ we deduce that $\alpha(L_\beta) = 0$, for every $\beta \geq \beta'$. We may repeat the same construction to get another decreasing family (L_β^2) of ρ-closed convex subsets with $L_\beta^2 \subset L_\beta$ and $\delta_\rho(L_\beta^2) \leq \frac{\delta}{4}$, since $\alpha(L_{\beta'}) = 0$, for all $\beta \geq \beta'$. So by induction, we obtain a decreasing family $\{L_\beta^n\}_{\beta \geq \beta'}$ of ρ-closed convex nonempty subsets with

$$L_\beta^n \subset L_\beta^{n-1} \ \text{ and } \ \delta_\rho(L_\beta^n) \leq \frac{\delta}{2^n},$$

for all $\beta \geq \beta'$ and all $n \geq 2$. Since L_ρ is ρ-complete, we get

$$\bigcap_{n \geq 1} L_\beta^n \neq \emptyset, \ \text{ for any } \beta \geq \beta'.$$

Clearly $\cap L_\beta^n$ is reduced to one point, say h_β for all $\beta \geq \beta'$. Since $L_\beta^n \subset L_\gamma^n$ for any $\beta \geq \gamma \geq \beta'$ and any $n \geq 1$, we obtain $h_\beta = h_\gamma$ for every $\beta \geq \gamma \geq \beta'$. Therefore, $h_{\beta'} \in L_\beta^n \subset K_\beta$ for all $n \geq 1$ and all $\beta \geq \beta'$, which clearly implies

$$\emptyset \neq \bigcap_{\beta \geq \beta'} K_\beta \subset \bigcap_{\beta \in \Gamma} A_\beta.$$

This completes the proof of Theorem 4.6. $\square$

4.4 Opial and Kadec–Klee Properties

The original definitions of Opial and Kadec–Klee properties are closely linked to
the weak topology in Banach spaces, as presented in Sect. 2.5.4. In recent years,
many authors tried to extend these definitions to other topologies. The investigation
of these two properties in modular function spaces finds its root in [109].

Definition 4.7. Let $\rho \in \Re$.

1. We will say that L_ρ satisfies the ρ-*a.e.-Opial property* if for every $\{f_n\} \subset L_\rho$
   ρ-a.e. convergent to 0, such that there exists $\beta > 1$ for which

$$\sup_n \rho(\beta f_n) = M < \infty,$$

   then we have

$$\liminf_{n \to \infty} \rho(f_n) < \liminf_{n \to \infty} \rho(f_n + f),$$

   for every $f \in E_\rho$ not equal to 0.

2. We will say that L_ρ satisfies the ρ-*a.e.-uniform Opial property* if for every $\varepsilon > 0$
   there exists $\eta > 0$ such that for every $\{f_n\} \subset L_\rho$ ρ-a.e. convergent to 0 such that
   there exists $\beta > 1$ for which

$$\sup_n \rho(\beta f_n) = M < \infty$$

   then, for every $f \in E_\rho$ such that $\rho(f) \geq \varepsilon$, we have

$$\liminf_{n \to \infty} \rho(f_n) + \eta \leq \liminf_{n \to \infty} \rho(f_n + f).$$

These definitions can be strengthen to the following one, which will play an important role in this book.

Definition 4.8. We say that L_ρ satisfies the ρ-*a.e. strong Opial property* or shortly
strong Opial property if for every $\{f_n\} \in L_\rho$ which is ρ-a.e. convergent to 0 such
that there exists a $\beta > 1$ for which

$$\sup_n \rho(\beta f_n) < \infty,$$

the following equality holds for any $g \in E_\rho$

$$\liminf_{n \to \infty} \rho(f_n + g) = \liminf_{n \to \infty} \rho(f_n) + \rho(g). \tag{4.9}$$

Clearly, the ρ-a.e.-uniform Opial property implies ρ-a.e.-Opial property. Also, it is not difficult to prove that the ρ-a.e. strong Opial property implies ρ-a.e.-uniform Opial property.

Remark 4.4. If we consider the Luxemburg norm on L_ρ, then we can define a variant of the Opial properties in this case. Indeed, we will say that $(L_\rho, \|.\|_\rho)$ satisfies the ρ-a.e.-uniform Opial property if

$$\inf \left\{ \liminf_{n \to \infty} \|f_n + f\|_\rho - 1 \right\} > 0,$$

where the infimum is taken over all $\{f_n\} \subset L_\rho$ ρ-a.e. convergent to 0 such that $\liminf_{n \to \infty} \|f_n\|_\rho \geq 1$, and $\|f\|_\rho \geq c$, for any $c > 0$.

Let us give the definition of the Kadec–Klee property in modular function spaces.

Definition 4.9. We will say that L_ρ satisfies ρ-*a.e.-Kadec-Klee property* if for some $\varepsilon \in (0, 2]$ and every $r > 0$, there exists $\eta > 0$ such that $\rho(f) \leq r(1 - \eta)$ for every $\{f_n\} \subset E_\rho$ ρ-a.e. convergent to $f \in E_\rho$ for which

(i) there exists $\beta > 1$ such that $\sup_n \rho \left(\beta [f_n - f] \right) = M < \infty$,

(ii) $\rho(f_n) \leq r$, for every $n \geq 1$,

(iii) $sep \left\{ \dfrac{1}{2} f_n \right\} = \inf \left\{ \rho \left(\dfrac{f_n - f_m}{2} \right) : n \neq m \right\} > r \varepsilon.$

We will say that L_ρ satisfies ρ-*a.e.-uniform Kadec–Klee property* if the above still holds for every $\varepsilon \in (0, 2]$.

Remark 4.5. If we consider the Luxemburg norm on L_ρ, then we can define a variant of the Kadec–Klee property in this case. Indeed, we will say that $(L_\rho, \| \cdot \|_\rho)$ satisfies the ρ-a.e.-uniform Kadec–Klee property if for any $\varepsilon \in (0, 2]$, we have

$$\inf \left\{ 1 - \|f\|_\rho \right\} > 0,$$

where the infimum is taken over all $\{f_n\} \subset L_\rho$ ρ-a.e. convergent to f, such that $\|f_n\|_\rho \leq 1$, for any $n \geq 1$, and $sep\{f_n\} = \inf \{ \|f_n - f_m\|_\rho : n \neq m \} > \varepsilon.$

In the next theorem, we show that a large class of modular function spaces enjoy the ρ-a.e strong Opial property. Recall that a function modular ρ is said to be *orthogonally additive* if $\rho(f, A \cup B) = \rho(f, A) + \rho(f, B)$ whenever $A \cap B = \emptyset$. Note that all Orlicz and Musielak–Orlicz modulars are orthogonally additive.

Let us begin with the following technical result.

Lemma 4.7. *[27] Let $\rho \in \Re$. Let $\varepsilon > 0$ and $k > 1$ be such that $k\,\varepsilon < 1$. Then for every $f, g \in L_\rho$ such that $\rho(kf) < \infty$ and $\rho(1/\varepsilon\,(k-1)g) < \infty$, we have*

$$|\rho(f+g) - \rho(f)| \leq \varepsilon\,|\rho(kf) - k\rho(f)| + 2\,\rho(C_\varepsilon\,g),$$

where $C_\varepsilon = 1/\varepsilon\,(k-1)$.

Proof. Put

$$\alpha = 1 - k\varepsilon, \quad \beta = \varepsilon, \quad \gamma = (k-1)\varepsilon.$$

Then, we have $\alpha + \beta + \gamma = 1$ and

$$f + g = \alpha\,f + \beta k f + \gamma C_\varepsilon g.$$

Since ρ is convex, we get

$$\rho(f+g) \leq \alpha\,\rho(f) + \beta\rho(kf) + \gamma\rho(C_\varepsilon g).$$

Hence

$$\rho(f+g) - \rho(f) \leq \varepsilon\,(\rho(kf) - k\rho(f)) + (k-1)\varepsilon\rho(C_\varepsilon g)$$

which implies

$$\rho(f+g) - \rho(f) \leq \varepsilon\,(\rho(kf) - k\rho(f)) + \rho(C_\varepsilon g).$$

On the other hand, if we put

$$\alpha = \frac{1}{1+k\varepsilon}, \quad \beta = \frac{\varepsilon}{1+k\varepsilon}, \quad \gamma = \frac{\varepsilon(k-1)}{1+k\varepsilon},$$

then $f = \alpha(f+g) + \beta k f + \gamma(-C_\varepsilon g)$. Hence

$$\rho(f) - \rho(f+g) \leq \varepsilon\,[\rho(kf) - k\rho(f)] + \varepsilon(k-1)\rho(-C_\varepsilon g).$$

This will imply

$$\rho(f) - \rho(f+g) \leq \varepsilon\,[\rho(kf) - k\rho(f)] + \rho(C_\varepsilon g).$$

The proof is therefore complete.   □

Theorem 4.7. *[109] Let $\rho \in \Re$. Assume that ρ is orthogonally additive. Let $\{f_n\} \subset L_\rho$ be ρ-a.e. convergent to 0. Assume there exists $\beta > 1$ such that*

$$\sup_n \rho(\beta f_n) = M < \infty.$$

Let $g \in E_\rho$. Then

$$\lim_{n \to \infty} \left(\rho(f_n + g) - \rho(f_n) - \rho(g) \right) = 0,$$

which implies

$$\liminf_{n \to \infty} \rho(f_n + g) = \liminf_{n \to \infty} \rho(f_n) + \rho(g),$$

i.e., L_ρ posses the ρ-a.e. strong Opial property.

Proof. Since $\{f_n\}$ converges ρ-a.e. to 0, then by Egoroff's Theorem, there exists an increasing sequence of sets $H_k \in \mathscr{P}$ such that $\Omega = \bigcup H_k$ and $\{f_n\}$ converges uniformly to f on every H_k. On the other hand we have

$$|\rho(f_n + g) - \rho(f_n) - \rho(g)| \leq |\rho(f_n + g, H_m) - \rho(f_n, H_m) - \rho(g, H_m)|$$

$$+ |\rho(f_n + g, H_m^c) - \rho(f_n, H_m^c) - \rho(g, H_m^c)|,$$

where A^c denotes the complement of the subset A. Using Lemma 4.7, we get

$$|\rho(f_n + g, H_m) - \rho(g, H_m)| \leq \varepsilon \left(\rho(kg, H_m) - k\rho(g, H_m) \right) + 2\rho(C_\varepsilon f_n, H_m),$$

for every $\varepsilon > 0$ such that $\varepsilon k < 1$. Since $\{f_n\}$ converges uniformly to 0 on every H_m we have

$$\limsup_{n \to \infty} |\rho(f_n + g, H_m) - \rho(f_n, H_m) - \rho(g, H_m)| \leq \varepsilon \rho(kg).$$

Using the same ideas, we get

$$\limsup_{n \to \infty} |\rho(f_n + g, H_m^c) - \rho(f_n, H_m^c) - \rho(g, H_m^c)| \leq \varepsilon \limsup_{n \to \infty} \rho(kf_n) + 2\rho(C_\varepsilon g, H_m^c)$$

$$+ \rho(g, H_m^c).$$

Hence

$$\limsup_{n \to \infty} |\rho(f_n + g, H_m^c) - \rho(f_n, H_m^c) - \rho(g, H_m^c)| \leq \varepsilon \sup_{n} \rho(kf_n) + 2\rho(C_\varepsilon g, H_m^c)$$

$$+ \rho(g, H_m^c).$$

Therefore

$$\limsup_{n \to \infty} |\rho(f_n + g) - \rho(f_n) - \rho(g)| \leq \varepsilon \rho(kg) + \varepsilon \sup_{n} \rho(kf_n) + 2\rho(C_\varepsilon g, H_m^c) + \rho(g, H_m^c).$$

Let m go to infinity. Using the fact that $g \in E_\rho$, we get

$$\limsup_{n \to \infty} |\rho(f_n + g) - \rho(f_n) - \rho(g)| \leq \varepsilon \rho(kg) + \varepsilon \sup_{n} \rho(kf_n).$$

Finally, we let ε go to 0 to get

$$\limsup_{n \to \infty} |\rho(f_n + g) - \rho(f_n) - \rho(g)| \leq 0.$$

The proof is therefore complete. $\square$

Remark 4.6. Observe that the Opial property in the norm sense, [173], does not necessarily hold for several classical Banach function spaces. For instance, the norm Opial property does not hold for L^p spaces for $1 \leq p \neq 2$ while the modular strong Opial property holds in L^p for all $p \geq 1$.

Note also that if ρ satisfies the Δ_2-condition, then one may assume $f \in L_\rho$ in the definition of the strong Opial property. As an implication of Theorem 4.7, one can get the following result in L^p spaces.

Corollary 4.2. *Let $\rho \in \mathfrak{R}$. Let $p \geq 1$ and $\{f_n\}$ be a sequence of L^p-uniformly bounded functions on a measure space. Assume that $\{f_n\}$ converges almost everywhere to $f \in L^p$. Then*

$$\liminf_{n\to\infty} ||f_n||^p = \liminf_{n\to\infty} ||f_n - f||^p + ||f||^p.$$

When $p = 1$, the conclusion of Corollary 4.2 gives the main result of [25, 152]. Let us add that when $p < 1$ the conclusion of Corollary 4.2 is still true. This will not be a simple deduction from Theorem 4.7 since the function $\varphi(t) = t^p$ is not convex. A technical assumption [27] can be added to get a more general result (see also [151]).

Theorem 4.8. *Let $\rho \in \mathfrak{R}$. Assume that ρ has the ρ-a.e. strong Opial property. Let C be a convex, strongly ρ-bounded and ρ-a.e. compact nonempty subset of the modular function space L_ρ. Let $\{f_n\}_n$ be a sequence of elements of C. Let $\Phi : C \to [0, +\infty)$ be a ρ-type defined by $\Phi(g) = \limsup_{n\to\infty} \rho(f_n - g)$. Then, for any sequence $\{g_m\}$ in C which ρ-a.e. converges to $g \in C$ we have*

$$\Phi(g) \leq \liminf_{m\to\infty} \Phi(g_m),$$

i.e., ρ-types are lower semicontinuous in the sense of convergence ρ-a.e. Moreover, Φ attains its minimum in C.

Proof. Since C is ρ-a.e. compact, there exists a subsequence $\{f_{\phi(n)}\}$ of $\{f_n\}$ such that $f_{\phi(n)} \overset{\rho-a.e}{\longrightarrow} f \in C$ and $\lim_{n\to\infty} \rho(f_{\phi(n)} - g) = \limsup_{n\to\infty} \rho(f_n - g)$. Hence

$$\Phi(g_m) = \limsup_{n\to\infty} \rho(f_n - g_m) \geq \limsup_{n\to\infty} \rho(f_{\phi(n)} - g_m) \geq \liminf_{n\to\infty} \rho(f_{\phi(n)} - g_m).$$

Equality (4.9) of the strong Opial property implies that

$$\liminf_{n\to\infty} \rho(f_{\phi(n)} - g_m) = \liminf_{n\to\infty} \rho(f_{\phi(n)} - f) + \rho(f - g_m).$$

Thus, $\Phi(g_m) \geq \liminf_{n\to\infty} \rho(f_{\phi(n)} - f) + \rho(f - g_m)$, for any $m \leq 1$. Hence

$$\liminf_{m\to\infty} \Phi(g_m) \geq \liminf_{n\to\infty} \rho(f_{\phi(n)} - f) + \liminf_{m\to\infty} \rho(f - g_m).$$

Using again (4.9), we get

$$\liminf_{m\to\infty} \rho(f - g_m) = \liminf_{m\to\infty} \rho(g_m - g) + \rho(g - f),$$

which implies

$$\liminf_{m\to\infty} \Phi(g_m) \geq \liminf_{n\to\infty} \rho(f_{\phi(n)} - f) + \liminf_{m\to\infty} \rho(g_m - g) + \rho(g - f). \quad (4.10)$$

On the other hand,

$$\Phi(g) = \limsup_{n\to\infty} \rho(f_n - g) = \lim_{n\to\infty} \rho(f_{\phi(n)} - g) = \liminf_{n\to\infty} \rho(f_{\phi(n)} - g),$$

which implies

$$\Phi(g) = \liminf_{n\to\infty} \rho(f_{\phi(n)} - f) + \rho(f - g). \quad (4.11)$$

From (4.10) and (4.11), it is clear that

$$\Phi(g) \leq \liminf_{m\to\infty} \Phi(g_m).$$

To prove that Φ attains its minimum in C, let $\Phi_0 = \inf\{\Phi(f) : f \in C\}$ which is finite due to the ρ-boundedness of C. There exists then a sequence of $g_n \in C$ such that $\Phi_0 = \lim_{n\to\infty} \Phi(g_n)$. By the ρ-a.e. compactness of C, passing to a subsequence if necessary, there exists $g \in C$ such that $g_n \to g$ ρ-a.e. Using the ρ-a.e. lower semi-continuity of Φ we have

$$\Phi_0 \leq \Phi(g) \leq \lim_{n\to\infty} \Phi(g_n) = \Phi_0,$$

which completes the proof of Theorem 4.8.

$\square$

An easy consequence of Theorem 4.7 and Theorem 4.8 is the ρ-a.e. lower semicontinuity of the ρ-types for orthogonally additive function modulars for C satisfying hypothesis of Theorem 4.7.

In the next theorem, we discuss the Kadec–Klee property in modular function spaces.

Theorem 4.9. *Let $\rho \in \mathfrak{R}$. Assume that ρ is orthogonally additive. Let $\{f_n\} \subset L_\rho$ be ρ-a.e. convergent to 0. Assume there exists $\beta > 1$ such that*

$$\sup_n \rho(\beta f_n) = M < \infty.$$

Then the modular function space L_ρ has the ρ-a.e.-uniform Kadec–Klee property.

Proof. Let $\varepsilon > 0$, $r > 0$ and $\{f_n\}$ be as in Definition 4.9. Assume $\{f_n\}$ is ρ-a.e. convergent to f. Theorem 4.7 implies that ρ has the strong Opial property. Hence

$$\liminf_{n\to\infty} \rho(f_n - f) + \rho(f) = \liminf_{n\to\infty} \rho(f_n).$$

Our assumption on $\{f_n\}$ implies that

$$\liminf_{n\to\infty} \rho(f - f_n) \ge r\,\frac{\varepsilon}{2}.$$

Therefore

$$\rho(f) \le r\left(1 - \frac{\varepsilon}{2}\right).$$

The proof is complete.

$\square$

We will now discuss the Opial and Kadec–Klee properties in modular function spaces seen as a Banach space endowed with the Luxemburg norm. Most of the results obtained here are found in [100].

Lemma 4.8. *[100] Let $\rho \in \Re$. Assume that ρ is an orthogonally additive modular function with the Δ_2-type condition. Let $\{f_n\}$ be a sequence in L_ρ which ρ-a.e. converges to 0. Let $f \in L_\rho$. Then to every $\varepsilon > 0$, there exist a subsequence $\{f_{n_k}\}$, a sequence $\{g_k\}$ in L_ρ and a function $g \in L_\rho$ such that*

(i) $\lim\limits_{k\to\infty} \|f_{n_k} - g_k\| = 0,$
(ii) supp $g_k \cap$ supp $g = \emptyset$, for any $k \ge 1,$
(iii) $\|f - g\| < \varepsilon.$

Proof. Using the Egoroff theorem , we are assured of the existence of a sequence of sets $\Omega_k \in \mathscr{P}$ such that $\bigcup \Omega_k = \Omega$, $\Omega_k \cap \Omega_l = \emptyset$, whenever $k \ne l$, and $\{f_n\}$ converges uniformly to the null function on every Ω_k. Using the orthogonal additivity of ρ and the Δ_2-condition, we get $\rho(f) < \infty$, and

$$\rho(f) = \sum_{k\ge 1} \rho(f, \Omega_k).$$

for any $f \in L_\rho$. Let $\varepsilon > 0$, and consider $\delta > 0$ given by Lemma 3.2. Let us fix $f \in L_\rho$. Then there exists $k_0 \ge 1$ such that

$$\sum_{k>k_0} \rho(f, \Omega_k) < \delta.$$

Let $\{\varepsilon_n\}$ be a real sequence in $(0,1)$ such that $\lim\limits_{n\to\infty} \varepsilon_n = 0$. For every ε_n, consider the corresponding δ_n given by Lemma 3.2. Using the properties of the function modular ρ, we know that $\rho\left(\alpha, \bigcup\limits_{k=1}^{k=k_0} \Omega_k\right) \to 0$, as α decreases to 0. Thus there exists $\alpha_0 > 0$ such that

$$\rho\left(\alpha_0, \bigcup_{k=1}^{k=k_0} \Omega_k\right) < \frac{\delta_1}{2}.$$

Since $\{f_n\}$ converges uniformly to 0 on $\bigcup_{k=1}^{k=k_0} \Omega_k$, there exists $n_1 \geq 1$ such that for

any $n \geq n_1$, we have $|f_n(x)| \leq \alpha_0$, for any $x \in \bigcup_{k=1}^{k=k_0} \Omega_k$. If we consider the function

f_{n_1}, then there exists $k_1 > k_0$ such that

$$\sum_{k>k_1} \rho(f_{n_1}, \Omega_k) < \frac{\delta_1}{2}.$$

Define the function $g_1 = f_{n_1} 1_{A_{k_0,k_1}}$, where $A_{k_0,k_1} = \bigcup_{k_0 < k \leq k_1} \Omega_k$. So $g_1 \in L_\rho$ and

$$\rho(f_{n_1} - g_1) = \rho\left(f_{n_1}, \bigcup_{k=1}^{k_0} \Omega_k \cup \bigcup_{k>k_1} \Omega_k\right) = \rho\left(f_{n_1}, \bigcup_{k=1}^{k_0} \Omega_k\right) + \rho\left(f_{n_1}, \bigcup_{k>k_1} \Omega_k\right)$$

which implies

$$\rho(f_{n_1} - g_1) \leq \rho\left(\alpha_0, \bigcup_{k=1}^{k_0} \Omega_k\right) + \sum_{k>k_1} \rho(f_{n_1}, \Omega_k) \leq \frac{\delta_1}{2} + \frac{\delta_1}{2} = \delta_1.$$

Hence $\|f_{n_1} - g_1\| \leq \varepsilon_1$. Suppose that we have constructed $k_0 < k_1 < \cdots < k_l$, $n_1 < n_2 < \cdots < n_l$ such that $\rho(f_{n_i} - g_i) \leq \delta_i$, where

$$g_i = f_{n_i} 1_{A_{k_0,k_1}},$$

for $i \in [1, l]$. Using the properties of ρ, there exists $\alpha_{l+1} > 0$ such that

$$\rho\left(\alpha_{l+1}, \bigcup_{k=1}^{k=k_l} \Omega_k\right) < \frac{\delta_{l+1}}{2}.$$

Since $\{f_n\}$ converges uniformly to 0 on $\bigcup_{k=1}^{k=k_l} \Omega_k$, we can find $n_{l+1} > n_l$ such that

$|f_{n_{l+1}}(x)| \leq \alpha_{l+1}$, for any $x \in \bigcup_{k=1}^{k=k_l} \Omega_k$. Also, we can find $k_{l+1} > k_l$ such that

$$\sum_{k>k_{l+1}} \rho(f_{n_{l+1}}, \Omega_k) < \frac{\delta_{l+1}}{2}.$$

Set

$$g_{l+1} = f_{n_{l+1}} 1_{\bigcup_{k_l < k \leq k_{l+1}} \Omega_k}.$$

Then, we have $\rho(f_{n_{l+1}} - g_{l+1}) \leq \delta_{l+1}$. Therefore, we have $\|f_{n_{l+1}} - g_{l+1}\| \leq \varepsilon_{l+1}$. By induction, we constructed a subsequence $\{f_{n_k}\}$ and a sequence $\{g_k\}$ such that $\lim_{k \to \infty} \|f_{n_k} - g_k\| = 0$. Next, we define the function

$$g = f \, 1 \bigcup_{1 \leq k \leq k_0} \Omega_k.$$

It is obvious that supp $g_k \cap$ supp $g = \emptyset$, for any $k \geq 1$. Moreover, we have

$$\rho(f - g) \leq \sum_{k > k_0} \rho(f, \Omega_k) < \delta,$$

which implies $\|f - g\|_\rho < \varepsilon$. $\square$

As a consequence to the above technical results, we obtain the following theorem.

Theorem 4.10. *Let $\rho \in \mathfrak{R}$. Assume that ρ is a orthogonally additive function modular satisfying the Δ_2-type condition. Let $\varepsilon > 0$. Let $\{f_n\}$ be in L_ρ such that*

(1) $\|f_n\|_\rho \leq 1$, for any $n \geq 1$,
(2) $sep(f_n) = \inf\{\|f_n - f_m\|_\rho : n \neq m\} > \varepsilon$,
(3) $\{f_n\}$ ρ-a.e. converges to $f \in L_\rho$.

Then

$$\|f\|_\rho \leq \cfrac{1}{\omega_\rho^{-1}\left(\cfrac{1}{1 - 1/2\omega_\rho(1/\varepsilon)}\right)} < 1.$$

In other words, $(L_\rho, \|\cdot\|_\rho)$ has the uniform Kadec–Klee property with respect to the ρ-a.e. convergence.

Proof. Fix $\varepsilon > 0$. Let $\{f_n\}$ be a sequence such that (1), (2), and (3) hold. Let $\gamma > 0$ and define

$$g_n = \frac{f_n}{\gamma + 1}, \quad g = \frac{f}{\gamma + 1}, \quad \text{and } \varepsilon' = \frac{\varepsilon}{\gamma + 1}.$$

Then $\|g_n\|_\rho < 1$, for any $n \geq 1$. Hence $\rho(g_n) \leq \|g_n\|_\rho < 1$. Since $\|g_n - g_m\|_\rho > \varepsilon'$, we get $\rho\big((g_n - g_m)/\varepsilon'\big) > 1$, for any $n \neq m$. Using the definition of the growth function, we have

$$\rho\left(\frac{g_n - g_m}{\varepsilon'}\right) \leq \omega_\rho\left(\frac{1}{\varepsilon'}\right)\rho(g_n - g_m),$$

which implies

$$1 < \omega_\rho\left(\frac{1}{\varepsilon'}\right)\rho(g_n - g_m),$$

for any $n \neq m$. Hence

$$\frac{1}{\omega_\rho\left(\dfrac{1}{\varepsilon'}\right)} < \rho(g_n - g_m),$$

for any $n \neq m$. Fix $m \geq 1$. Using Theorem 4.7, we get

$$\frac{1}{\omega_\rho\left(\frac{1}{\varepsilon'}\right)} \leq \liminf_{n\to\infty}\rho(g_n - g_m) = \liminf_{n\to\infty}\rho(g_n - g) + \rho(g - g_m).$$

If we let $m \to \infty$, we get

$$\frac{1}{\omega_\rho\left(\frac{1}{\varepsilon'}\right)} \leq 2\liminf_{n\to\infty}\rho(g_n - g).$$

On the other hand, we also have

$$\liminf_{n\to\infty}\rho(g_n) = \liminf_{n\to\infty}\rho(g_n - g) + \rho(g).$$

Hence

$$0 \leq \rho(g) = \liminf_{n\to\infty}\rho(g_n) - \liminf_{n\to\infty}\rho(g_n - g) \leq 1 - \frac{1}{2\omega_\rho\left(\frac{1}{\varepsilon'}\right)}.$$

Using property (v) of Lemma 3.1 and the monotonicity of ω_ρ^{-1}, we get

$$\|f\|_\rho \leq (1+\gamma)\frac{1}{\omega_\rho^{-1}\left(\frac{1}{1-1/2\omega_\rho((1+\gamma)/\varepsilon)}\right)}.$$

Since γ is arbitrary and both functions ω_ρ and ω_ρ^{-1} are continuous, we deduce

$$\|f\|_\rho \leq \frac{1}{\omega_\rho^{-1}\left(\frac{1}{1-1/2\omega_\rho(1/\varepsilon)}\right)} = \eta(\varepsilon).$$

In order to finish the proof of Theorem 4.10, we need to check that $\eta(\varepsilon) < 1$. Since $\omega_\rho(1) = 1$, then $\eta(\varepsilon) < 1$ if and only if $\omega_\rho\left(\frac{1}{\varepsilon}\right) > 0$. Since ρ satisfies the Δ_2-type condition, then there exists $K > 0$ such that $\rho(2f) \leq K\rho(f)$, for any $f \in L_\rho$. This will imply that $\omega_\rho\left(\frac{1}{2}\right) \geq \frac{1}{K}$. Since $\varepsilon \in (0,2]$, then we have

$$\omega_\rho\left(\frac{1}{\varepsilon}\right) \geq \omega_\rho\left(\frac{1}{2}\right) \geq \frac{1}{K} > 0.$$

$\square$

The last result of this section deals with the Opial property.

Theorem 4.11. *Let $\rho \in \mathfrak{R}$. Assume that ρ is an orthogonally additive modular function satisfying the Δ_2-type condition. Let $c > 0$. Let $\{f_n\}$ be in L_ρ such that $\{f_n\}$ ρ-a.e. converges to 0, and $\liminf\limits_{n\to\infty} \|f_n\|_\rho \geq 1$. Then*

$$\liminf_{n\to\infty} \|f_n + f\|_\rho \geq \omega_\rho^{-1} \left(1 + \frac{1}{\omega_\rho(1/c)} \right) > 1,$$

for any $f \in L_\rho$ such that $\|f\|_\rho \geq c$. In other words, $(L_\rho, \|\cdot\|_\rho)$ has the uniform Opial property with respect to the ρ-a.e. convergence.

Proof. Fix $c > 0$. Let $\{f_n\}$ be in L_ρ such that $\{f_n\}$ ρ-a.e. converges to 0, and $\liminf\limits_{n\to\infty} \|f_n\|_\rho \geq 1$. Let $f \in L_\rho$ be such that $\|f\|_\rho \geq c$. Let $\varepsilon \in (0, c)$. Lemma 4.8 will force the existence of a subsequence $\{f_{n_k}\}$ of $\{f_n\}$, a sequence $\{g_n\}$ and a function $g \in L_\rho$ such that $\lim\limits_{k\to\infty} \|f_{n_k} - g_k\|_\rho = 0$, $\|f - g\|_\rho < \varepsilon$, and supp $g_k \cap$ supp $g = \emptyset$, for any $k \geq 1$. We may also assume that

$$\liminf_{n\to\infty} \|f_n + f\|_\rho = \liminf_{k\to\infty} \|f_{n_k} + f\|_\rho.$$

Hence

$$\liminf_{k\to\infty} \|g_k\|_\rho \geq 1, \quad \text{and} \quad \|g\|_\rho \geq c - \varepsilon.$$

Fix $t \in (0,1)$. Define the functions $h = \dfrac{1}{t} g$ and $h_k = \dfrac{1}{t} g_k$, for any $k \geq 1$. Then we have

$$\|h\|_\rho = \frac{\|g\|_\rho}{t} > \|g\|_\rho \geq c - \varepsilon, \quad \text{and} \quad \liminf_{k\to\infty} \|h_k\|_\rho \geq \frac{1}{t} > 1.$$

So, for k large enough, we may assume $\|h_k\|_\rho > 1$. Set

$$\delta_\varepsilon = \omega_\rho^{-1} \left(1 + \frac{1}{\omega_\rho(1/(c - \varepsilon))} \right).$$

Since h_k and h have disjoint support, the additivity of ρ implies

$$\rho\left(\frac{h_k + h}{\delta_\varepsilon} \right) = \rho\left(\frac{h_k}{\delta_\varepsilon} \right) + \rho\left(\frac{h}{\delta_\varepsilon} \right)$$

Using the properties of ω_ρ, we get

$$\rho\left(\frac{h_k}{\delta_\varepsilon} \right) + \rho\left(\frac{h}{\delta_\varepsilon} \right) \geq \frac{1}{\omega_\rho(\delta_\varepsilon)} \rho(h_k) + \frac{1}{\omega_\rho(\delta_\varepsilon/(c - \varepsilon))} \rho\left(\frac{h}{c - \varepsilon} \right)$$

$$> \frac{1}{\omega_\rho(\delta_\varepsilon)} + \frac{1}{\omega_\rho(\delta_\varepsilon)\omega_\rho(1/(c - \varepsilon))}$$

$$= \frac{1}{\omega_\rho(\delta_\varepsilon)} \left[1 + \frac{1}{\omega_\rho(1/(c - \varepsilon))} \right] = 1$$

Hence $\liminf\limits_{k\to\infty}\|g_k+g\|_\rho \geq t\delta_\varepsilon$, which implies $\liminf\limits_{k\to\infty}\|g_k+g\|_\rho \geq \delta_\varepsilon$, by letting $t\to 1$. Therefore, we have

$$
\begin{aligned}
\liminf_{n\to\infty}\|f_n+f\|_\rho &= \lim_{k\to\infty}\|f_{n_k}+f\|_\rho \\
&\geq \liminf_{k\to\infty}\|g_k+g\|_\rho - \lim_{k\to\infty}\|f_{n_k}-g_k\|_\rho - \|f-g\|_\rho \\
&\geq \delta_\varepsilon - \varepsilon.
\end{aligned}
$$

Since ε was arbitrary and the functions ω_ρ and ω_ρ^{-1} are continuous, we get

$$
\liminf_{n\to\infty}\|f_n+f\|_\rho \geq \omega_\rho^{-1}\left(1+\frac{1}{\omega_\rho(1/c)}\right).
$$

In order to complete the proof of Theorem 4.11, we need to show that

$$
\omega_\rho^{-1}\left(1+\frac{1}{\omega_\rho(1/c)}\right) > 1.
$$

Clearly this will happen if and only if $\omega_\rho(1/c) > 0$. Using the same ideas as those applied at the end of the proof of Theorem 4.10, we can show that $\omega_\rho(1/c) > 0$. $\qquad\square$

Remark 4.7. The lower bound found in Theorem 4.11 is optimal. Indeed, assume $L_\rho = L_p([0,1])$, $1\leq p<\infty$, and $\rho(f)=\int_{[0,1]}|f(x)|^p dx$. Then the ρ-a.e. is the classical a.e. convergence, which is equivalent to *convergence locally in measure* . In this case, the best lower bound is $(c^p+1)^{1/p}$ which is the exact value that we obtained in Theorem 4.11 since $\omega_\rho(t)=t^p$.

Chapter 5
Fixed Point Existence Theorems in Modular Function Spaces

This chapter presents a series of fixed point existence theorems for nonlinear mappings acting in modular functions spaces. We cover a range of different types of mappings including ρ-contractions and their pointwise asymptotic versions and ρ-nonexpansive mappings and pointwise asymptotic ρ-nonexpansive mappings, under various assumptions on the function modular ρ and on the sets these mappings are defined in. In addition, we will discuss some examples and applications to the theory of differential equations.

5.1 Definitions

Let us introduce definitions of modular contractions, pointwise contractions, asymptotic pointwise mappings, and associated notions. Frequently, they are collectively referenced in the literature as *generalized contractions* and *generalized nonexpansive mappings*, respectively.

Definition 5.1. Let $\rho \in \Re$ and $C \subset L_\rho$ be nonempty, and ρ-closed. A mapping $T : C \to C$ is called a ρ-*contraction* or shortly *contraction* if there exists a number $c \in [0,1)$ such that

$$\rho(T(f) - T(g)) \leq c\rho(f - g), \quad \text{for all } f, g \in C.$$

Definition 5.2. Let $\rho \in \Re$ and $C \subset L_\rho$ be nonempty, and ρ-closed. A mapping $T : C \to C$ is called a *pointwise ρ-contraction* or shortly *pointwise contraction* if there exists $\alpha : C \to [0,1)$ such that

$$\rho(T(f) - T(g)) \leq \alpha(f)\rho(f - g), \quad \text{for all } f, g \in C.$$

Definition 5.3. Let $\rho \in \Re$ and $C \subset L_\rho$ be nonempty, and ρ-closed. A mapping $T : C \to C$ is called an *asymptotic pointwise mapping* if there exists a sequence

© Springer International Publishing Switzerland 2015

M. A. Khamsi, W. M. Kozlowski, *Fixed Point Theory in Modular Function Spaces*,
DOI 10.1007/978-3-319-14051-3_5

of mappings $\alpha_n : C \to [0, \infty)$ such that

$$\rho(T^n(f) - T^n(g)) \leq \alpha_n(f)\rho(f - g), \quad \text{for all } f, g \in L_\rho.$$

(a) If $\alpha_n(f) = 1$ for every $f \in L_\rho$ and $n \in \mathbb{N}$, then T is called ρ-*nonexpansive* or shortly nonexpansive.
(b) If $\{\alpha_n\}$ converges pointwise to $\alpha : C \to [0, 1)$, then T is called *asymptotic pointwise ρ-contraction* or shortly *asymptotic pointwise contraction*.
(c) If $\limsup\limits_{n \to \infty} \alpha_n(f) \leq 1$ for any $f \in L_\rho$, then T is called *asymptotic pointwise ρ-nonexpansive* or shortly *asymptotic pointwise nonexpansive*.
(d) If α_n is a constant function for every n, and $\limsup\limits_{n \to \infty} \alpha_n \leq 1$ for any $f \in L_\rho$, then T is called *asymptotically ρ-nonexpansive* or shortly *asymptotically nonexpansive*.

Remark 5.1. It follows immediately from the definition that every ρ-contraction is a pointwise ρ-contraction, and every pointwise ρ-contraction is an asymptotic pointwise ρ-contraction.

Remark 5.2. It follows immediately from the definition that every ρ-nonexpansive mapping is asymptotically ρ-nonexpansive, and every asymptotically ρ-nonexpansive mapping is an asymptotic pointwise ρ-nonexpansive mapping.

Let us look at some examples which illustrate the role of the above defined notions. We start with a typical application of the methods of modular function spaces to the theory of nonlinear integral equations.

Example 5.1. As suggested in the introduction, an operator itself may be used for the construction of a function modular and hence a space in which this operator has required properties like ρ-nonexpansiveness or ρ-contraction. Let us consider for instance the following, important for the theory of integral equations, Urysohn operator:

$$T(f)(x) = \int_0^1 k(x, y, |f(y)|)dy + f_0(x),$$

where f_0 is a fixed function and $f : [0, 1] \to \mathbb{R}$ is Lebesgue measurable. For the kernel k we assume that

(a) $k : [0, 1] \times [0, 1] \times \mathbb{R}_+ \to \mathbb{R}_+$ is Lebesgue measurable
(b) $k(x, y, 0) = 0$
(c) $k(x, y, \cdot)$ is continuous, convex and increasing to $+\infty$
(d) $\int_0^1 k(x, y, t)dx > 0$ for $t > 0$ and $y \in (0, 1)$.

Assume, in addition, that for almost all $t \in [0, 1]$ and all measurable functions f, g there exists a constant $K > 0$ such that

$$\int_0^1 \left\{ \int_0^1 k(t, u, |k(u, v, |f(v)|) - k(u, v, |g(v)|)|)dv \right\} du \leq K \int_0^1 k(t, u, |f(u) - g(u)|)du.$$

Setting $\rho(f) = \int_0^1 \left\{ \int_0^1 k(x,y,|f(y)|)dy \right\} dx$ and using Jensen's inequality it is easy to show that ρ is a function modular and that $\rho(T(f) - T(g)) \leq K\rho(f - g)$ on L_ρ, that is, T is K-Lipschitzian with respect to ρ. Let us summarize what we have done: given an integral operator we have constructed a modular function space in which this operator is Lipschitzian with the constant K. Obviously, if $K < 1$ then T becomes a ρ-contraction, or if $K \leq 1$, ρ-nonexpansive mapping. As we will see, applying relevant fixed point theorems one can solve the corresponding Urysohn integral equation.

The next example, taken from [79], illustrates the role of the asymptotic assumptions. Please remember that the Hilbert space l^2 is a modular function space too.

Example 5.2. Let C denote the unit ball in l^2 and let T be defined as follows:

$$T : (x_1, x_2, x_3, \ldots) \mapsto (0, x_1^2, A_2 x_2, A_3 x_3, \ldots)$$

where $A_i \in (0,1)$ are numbers such that $\prod_{i=2}^{\infty} A_i = \frac{1}{2}$. Then T is Lipschitzian with constant 2 but it is not nonexpansive. Moreover,

$$\|T^i(x) - T^i(y)\| \leq 2 \prod_{j=2}^{i} A_j \|x - y\|$$

for $i = 2, 3, \ldots$, which implies that

$$\lim_{i \to \infty} k_i = 2 \lim_{i \to \infty} \prod_{j=2}^{i} A_j = 1,$$

that is, T is asymptotically nonexpansive.

The following simple finite-dimensional example gives another view on asymptotic pointwise contractions and asymptotic pointwise nonexpansive mappings.

Example 5.3. Let a be any real number from $(0,1)$ and $C = [a,1]$. Let $T : [a,1] \to \mathbb{R}$ be defined simply as $T(x) = \sqrt{x}$. Note that if $x, y \in C$ with $x \leq y$ then

$$|\sqrt{x} - \sqrt{y}| \leq \frac{1}{2\sqrt{x}} |x - y|,$$

which makes T pointwise nonexpansive (and actually nonexpansive) but only if $a \geq \frac{1}{4}$. We know that T has an obvious fixed point at $x = 1$. However, if we started with $a < \frac{1}{4}$ we would not be able to use fixed point results like Banach contraction principle or Kirk's theorem to prove its existence. On the other hand, since in the same situation

$$|T^n(x) - T^n(y)| \leq \frac{1}{2^n} x^{\frac{1}{2^n} - 1} |x - y|,$$

it is easy to see that T is an asymptotic pointwise contraction in $C = [a,1]$ for any $a \in (0,1)$ and hence, by Theorem 5.3, it will have a unique fixed point in C which can be found by taking the limit of orbits. An easy calculation will show that this limit is indeed equal to 1.

5.2 Contractions in Modular Function Spaces

5.2.1 Banach Contraction Principle in Modular Function Spaces

As we already discussed in Sect. 2.1 the Banach contraction principle (Theorem 2.1) is an extremely important tool used by many to guarantee the existence and unique-ness of solutions seen as fixed point of self-mappings defined on metric spaces. It also provides a constructive method to find those fixed points. It was natural to try to extend such celebrated theorem to the context of modular function spaces. That is exactly what we will do in our next result.

Theorem 5.1. Let $\rho \in \Re$. Let $C \subset L_\rho$ be nonempty, ρ-closed and ρ-bounded. Let $T : C \to C$ be a ρ-contraction. Then, T has a unique fixed point $\bar{f} \in C$. Moreover, for any $f \in C$, $\rho \left(T^n(f) - \bar{f} \right) \to 0$ as $n \to \infty$, where T^n is the n-th iterate of T.

Proof. Since T is ρ-contraction, there exists $\alpha < 1$ such that

$$\rho(T(f) - T(g)) \leq \alpha \, \rho(f - g), \quad \text{for all } f, g \in C.$$

Let us fix $f_0 \in C$. Since C is ρ-bounded, hence

$$\delta_\rho(C) = \sup\{\rho(f - g) : f, g \in C\} < \infty.$$

Observe that

$$\rho \left(T^{n+k}(f_0) - T^n(f_0) \right) \leq \alpha \, \rho \left(T^{n+k-1}(f_0) - T^{n-1}(f_0) \right)$$
$$\leq \alpha^n \, \rho \left(T^k(f_0) - f_0 \right)$$
$$\leq \alpha^n \, \delta_\rho(C) \to 0,$$

for any $n, k \geq 1$. Since $\alpha < 1$ and $\delta_\rho(C) < \infty$, we conclude that $\{T^n(f_0)\}$ is ρ-Cauchy. The ρ-completeness of L^ρ (Theorem 3.2) implies the existence of $\bar{f} \in L^\rho$ such that $\lim_{n \to \infty} \rho \left(T^n(f_0) - \bar{f} \right) = 0$. Because C is ρ-closed, we get $\bar{f} \in C$. Since

$$\rho \left(\frac{\bar{f} - T(\bar{f})}{2} \right) \leq \rho \left(\bar{f} - T^n(f_0) \right) + \rho \left(T^n(f_0) - T(\bar{f}) \right)$$
$$\leq \rho \left(\bar{f} - T^n(f_0) \right) + \alpha \, \rho \left(T^{n-1}(f_0) - \bar{f} \right) \to 0, \quad \text{as } n \to \infty.$$

Therefore, $T(\bar{f}) = \bar{f}$, which means that $\bar{f}$ is a fixed point of T. To prove the unique-ness part observe that if $T(f_1) = f_1$ and $T(f_2) = f_2$, then

$$\rho(f_1 - f_2) = \rho \left(T(f_1) - T(f_2) \right) \leq \alpha \, \rho(f_1 - f_2). \tag{5.1}$$

Since $\alpha < 1$ and the right-hand side is finite, equality (5.1) can hold only if $f_1 = f_2$.
$\square$

Next, we discuss the extension of Theorem 5.1 to the case of pointwise ρ-contractions or asymptotic pointwise ρ-contractions.

Theorem 5.2. *Let us assume that $\rho \in \mathfrak{R}$. Let $C \subset L_\rho$ be nonempty, ρ-closed, and ρ-bounded. Let $T : C \to C$ be a pointwise ρ-contraction or asymptotic pointwise ρ-contraction. Then T has at most one fixed point $f_0 \in C$. Moreover, if f_0 is a fixed point of T, then the orbit $\{T^n(f)\}$ ρ-converges to f_0 for any $f \in C$.*

Proof. Since every pointwise ρ-contraction is an asymptotic pointwise ρ-contraction, we can assume that T is an asymptotic pointwise ρ-contraction, i.e., there exists a sequence of mappings $\alpha_n : C \to [0, \infty)$ such that

$$\rho(T^n(f) - T^n(g)) \leq \alpha_n(f)\,\rho(f - g), \quad \text{for all } f, g \in C.$$

where $\{\alpha_n\}$ converges pointwise to $\alpha : C \to [0, 1)$. Let $f_1, f_2 \in C$ be two fixed points of T. Then we have

$$\rho(f_1 - f_2) = \rho\left(T^n(f_1) - T^n(f_2)\right) \leq \alpha_n(f_1)\,\rho(f_1 - f_2),$$

for any $n \geq 1$. If we let $n \to \infty$, we will get

$$\rho(f_1 - f_2) \leq \alpha(f_1)\,\rho(f_1 - f_2).$$

Since $\alpha(f_1) < 1$ and C is ρ-bounded, we conclude that $\rho(f_1 - f_2) = 0$, i.e. $f_1 = f_2$. This proves that T has at most one fixed point. To prove the convergence, assume that f_0 is the fixed point of T. Fix an arbitrary $f \in C$. Let us prove that $\{T^n(f)\}$ ρ-converges to f_0. Indeed we have

$$\rho\left(T^{n+m}(f) - f_0\right) = \rho\left(T^{n+m}(f) - T^n(f_0)\right) \leq \alpha_n(f_0)\,\rho\left(T^m(f) - f_0\right),$$

for any $n, m \geq 1$. Hence

$$\limsup_{m \to \infty} \rho\left(T^{n+m}(f) - f_0\right) \leq \limsup_{m \to \infty} \alpha_n(f_0)\,\rho\left(T^m(f) - f_0\right).$$

Since $\limsup_{m \to \infty} \rho\left(T^{n+m}(f) - f_0\right) = \limsup_{m \to \infty} \rho\left(T^m(f) - f_0\right)$, we get

$$\limsup_{m \to \infty} \rho\left(T^m(f) - f_0\right) \leq \alpha_n(f_0)\,\limsup_{m \to \infty} \rho\left(T^m(f) - f_0\right),$$

for any $n \geq 1$. If we let $n \to \infty$, we obtain

$$\limsup_{m \to \infty} \rho\left(T^m(f) - f_0\right) \leq \alpha(f_0)\,\limsup_{m \to \infty} \rho\left(T^m(f) - f_0\right).$$

Since $\alpha(x_0) < 1$, we get $\limsup\limits_{n \to \infty} \rho\left(T^n(f) - f_0\right) = 0$, which implies the desired conclusion

$$\lim_{n \to \infty} \rho\left(T^n(f) - f_0\right) = 0.$$

$\square$

5.2.2 Case of Uniformly Continuous Function Modulars

We will start our discussion with the existence of fixed point results in the case of uniformly continuous function modulars.

Definition 5.4. Let $\rho \in \mathfrak{R}$. We will say that a function modular ρ *is uniformly continuous* if for every $\varepsilon > 0$ and $L > 0$, there exists $\delta > 0$ such that

$$|\rho(g) - \rho(h + g)| \leq \varepsilon, \text{ if } \rho(h) \leq \delta \text{ and } \rho(g) \leq L, \ f, g \in L_\rho.$$

Remark 5.3. Let us mention that uniform continuity holds for a large class of function modulars. For instance, it can be proved that in Orlicz spaces over a finite atomless measure [44] or in sequence Orlicz spaces [101] the uniform continuity of the Orlicz modular is equivalent to the Δ_2 property.

Lemma 5.1. *Let $\rho \in \mathfrak{R}$ be uniformly continuous. Let $K \subset L_\rho$ be nonempty, convex, ρ-closed, and ρ-bounded. Then, any ρ-type $\tau : K \to [0, \infty)$ is ρ-lower semicontinuous in K.*

Proof. Let τ be a ρ-type defined on K. Then there exists $\{f_n\}$ in K such that

$$\tau(f) = \limsup_{n \to \infty} \rho(f - f_n).$$

Since K is ρ-bounded, then $\tau(f) \leq \delta_\rho(K) < \infty$. Let us fix arbitrarily $\alpha > 0$. Define $C_\alpha = \{f \in K : \tau(x) \leq \alpha\}$. We need to prove that C_α is ρ-closed. Without loss of generality we can assume that C_α is nonempty. Let $\{g_k\}$ be a sequence in C_α such that $\lim\limits_{k \to \infty} \rho(g_0 - g_k) = 0$, with $g_0 \in K$. We need to prove that $g_0 \in C_\alpha$. Since K is ρ-bounded, then $L = \sup\limits_{n \geq 1} \rho(g_0 - f_n) < \infty$. Let us fix an arbitrary $\varepsilon > 0$. By uniform continuity of ρ there exists $\delta > 0$ such that

$$|\rho(g) - \rho(h + g)| \leq \varepsilon \ \text{ if } \ \rho(h) \leq \delta \ \text{ and } \ \rho(g) \leq L. \tag{5.2}$$

Since $\lim\limits_{k \to \infty} \rho(g_0 - g_k) = 0$, there exists $p \geq 1$ such that $\rho(g_0 - g_k) \leq \delta$, for $k \geq p$. Using (5.2) with $h = g_0 - g_p$ and $g = f_n - g_0$ we have

$$|\rho(f_n - g_0) - \rho(f_n - g_p)| \leq \varepsilon, \ \text{for every } \ n \geq 1.$$

By the definition of τ, we have then that $\tau(g_0) \leq \alpha + 2\varepsilon$ because $g_p \in C_\alpha$. Since ε was chosen arbitrarily we have finally $\tau(g_0) \leq \alpha$, i.e., $g_0 \in C_\alpha$, as claimed.
□

We will utilize Lemma 5.1 to prove the following fixed point theorem.

Theorem 5.3. *Assume that $\rho \in \mathfrak{R}$ is uniformly continuous and has property (R). Let $K \subset L_\rho$ be nonempty, convex, ρ-closed, and ρ-bounded. Let $T : K \to K$ be an asymptotic pointwise ρ-contraction or asymptotic pointwise ρ-contraction. Then, T has a unique fixed point $f_0 \in K$. Moreover, the orbit $\{T^n(f)\}$ ρ-converges to f_0 for any $f \in K$.*

Proof. Since every pointwise ρ-contraction is an asymptotic pointwise ρ-contraction we can assume that T is an asymptotic pointwise ρ-contraction. In view of Theorem 5.2, it is enough to show that T has a fixed point. Let us fix $f \in K$ and define the ρ-type

$$\tau(g) = \limsup_{n \to \infty} \rho(T^n(f) - g),$$

for $g \in K$. By Lemma 5.1 the ρ-type τ is ρ-lower semicontinuous in K. Since ρ has property (R), Lemma 3.3 ensures us the existence of $f_0 \in K$ such that

$$\tau(f_0) = \inf\{\tau(g) : g \in K\}.$$

Let us prove that $\tau(f_0) = 0$. Indeed, for any $n, m \geq 1$ we have

$$\rho(T^{n+m}(f) - T^m(f_0)) \leq \alpha_m(f_0)\, \rho(T^n(f) - f_0).$$

If we let n go to infinity, we get

$$\tau(T^m(f_0)) \leq \alpha_m(f_0)\, \tau(f_0),$$

which implies

$$\tau(f_0) = \inf\{\tau(g) : g \in K\} \leq \tau(T^m(f_0)) \leq \alpha_m(f_0)\, \tau(f_0).$$

Passing with m to infinity, we get $\tau(f_0) \leq \alpha(f_0)\, \tau(f_0)$, which forces $\tau(f_0) = 0$, as $\alpha(f_0) < 1$. Hence, $\limsup_{n \to \infty} \rho(T^n(f) - f_0) = 0$ holds, i.e.,

$$\lim_{n \to \infty} \rho(T^n(f) - f_0) = 0.$$

The ρ-continuity of T implies f_0 is a fixed point of T.
□

Remark 5.4. The conclusions of Theorem 5.3 were originally proved in [114] using slightly different techniques.

5.2.3 Case of Modular Function Spaces with Strong Opial Property

A typical method of proof for the fixed point theorems is to construct a fixed point by finding an element on which a specific type function attains its minimum. As a matter of fact, that is exactly how we proved our Theorem 5.3. To be able to proceed with this method, one has to know that such an element indeed exists. In case of Theorem 5.3 the assumption of uniform continuity of ρ and property (R) ensured such an existence because in this case ρ-types are ρ-lower semicontinuous which was proved in Lemma 5.1. In the case of ρ with the strong Opial property, we cannot guarantee anymore that the ρ-type is ρ-lower semicontinuous. However, in view of Theorem 4.8, ρ-types are ρ-a.e. lower semicontinuous, provided C are strongly ρ-bounded, which ensures that they attain their minimum in C.

Theorem 5.4. *Let $\rho \in \Re$. Assume that L_ρ has the ρ-a.e. strong Opial property. Let $K \subset E_\rho$ be a nonempty, ρ-a.e. compact, strongly ρ-bounded, convex set. Then, any $T : K \to K$ asymptotic pointwise ρ-contraction or pointwise ρ-contraction has a unique fixed point $f_0 \in K$. Moreover, the orbit $\{T^n(f)\}$ ρ-converges to f_0, for any $f \in K$.*

Proof. Since every pointwise ρ-contraction is an asymptotic pointwise ρ-contraction we can assume that T is an asymptotic pointwise ρ-contraction. In view of Theorem 5.2 it is enough to show that T has a fixed point. Let us fix then $f \in K$ and define the ρ-type by

$$\tau(h) = \limsup_{n \to \infty} \rho(T^n(f) - h),$$

for $h \in K$. By Theorem 4.8 there exists $f_0 \in K$ such that

$$\tau(f_0) = \inf\{\tau(h) : h \in K\}.$$

Using the same argument as in the proof of Theorem 5.3, we will get $\tau(f_0) = 0$. Hence, $\lim_{n \to \infty} \rho(T^n(f) - f_0)) = 0$. The ρ-continuity of T forces f_0 to be a fixed point of T.
□

Remark 5.5. The conclusions of Theorem 5.4 were proven by Khamsi and Kozlowski in [114] using different methods.

5.2.4 Quasi-Contraction Mappings in Modular Function Spaces

As a generalization to Banach contraction principle, Ćirić [49] introduced the concept of quasi-contraction mappings. In this section, we investigate this kind of mappings in modular function spaces.

Similarly to Ćirić definition, we introduce the concept of quasi-contractions in modular spaces.

Definition 5.5. Let $\rho \in \mathscr{R}$. Let C be a nonempty subset of L_ρ. A mapping $T : C \to C$ is said to be a *quasi ρ-contraction* if there exists $k < 1$ such that for any $f, g \in C$

$$\rho(T(f) - T(g)) \leq k \max \Big(\rho(f - g); \rho(f - T(f)); \rho(g - T(g));$$
$$\rho(f - T(g)); \rho(g - T(f)) \Big).$$

In the sequel we prove an existence fixed point theorem for such mappings. First, let T and C be as in the above definition. For any $f \in C$ define the orbit $\mathscr{O}(f) = \{f, T(f), T^2(f), \cdots\}$, and its ρ-diameter by

$$\delta_\rho(f) = \operatorname{diam}(\mathscr{O}(f)) = \sup\{\rho(T^n(f) - T^m(f)) : n, m \in \mathbb{N}\}.$$

Lemma 5.2. *Let $\rho \in \mathfrak{R}$. Let C be a nonempty subset of L^ρ and $T : C \to C$ be a quasi ρ-contraction. Let $f \in C$ be such that $\delta_\rho(f) < \infty$. Then for any $n \geq 1$, we have*

$$\delta_\rho(T^n(f)) \leq k^n \delta_\rho(f),$$

where k is the constant associated with the quasi-contraction definition of T. Moreover, we have

$$\rho(T^n(f) - T^{n+m}(f)) \leq k^n \delta_\rho(f)$$

for any $n, m \in \mathbb{N}$.

Proof. Let $n, m \in \mathbb{N}$. We have

$$\rho(T^n(f) - T^m(g)) \leq k \max \Big(\rho(T^{n-1}(f) - T^{m-1}(g)); \rho(T^{n-1}(f) - T^n(f));$$
$$\rho(T^m(y) - T^{m-1}(g)); \rho(T^{n-1}(f) - T^m(g)); \rho(T^n(f) - T^{m-1}(g)) \Big)$$

for any $f, g \in C$. This obviously implies that

$$\delta_\rho(T^n(f)) \leq k \delta_\rho(T^{n-1}(f))$$

for any $n \geq 1$. Hence for any $n \geq 1$, we have

$$\delta_\rho(T^n(f)) \leq k^n \delta_\rho(f).$$

Moreover, for any $n, m \in \mathbb{N}$, we have

$$\rho(T^n(f) - T^{n+m}(f)) \leq \delta_\rho(T^n(f)) \leq k^n \delta_\rho(f).$$

□

The next lemma will be helpful to prove the main result of this section.

Lemma 5.3. *Let* $\rho \in \Re$. *Let C be a ρ-closed nonempty subset of L_ρ and $T : C \to C$ a quasi ρ-contraction. Let $f \in C$ such that $\delta_\rho(f) < \infty$. Then $\{T^n(f)\}$ ρ-converges to $h \in C$. Moreover, we have*

$$\rho(T^n(f) - h) \le k^n \delta_\rho(f),$$

for any $n \ge 1$.

Proof. Lemma 5.2 implies that $\{T^n(f)\}$ is ρ-Cauchy. Since L_ρ is ρ-complete and C is ρ-closed, then there exists $h \in C$ such that $\{T^n(f)\}$ ρ-converges to h. Since

$$\rho(T^n(f) - T^{n+m}(f)) \le k^n \delta_\rho(f),$$

for any $n, m \in \mathbb{N}$, the Fatou property (once we let $m \to \infty$) will imply

$$\rho(T^n(f) - h) \le k^n \delta_\rho(f) .$$

$\square$

Next we prove that h is in fact a fixed point of T and it is unique provided some extra assumptions.

Theorem 5.5. *Let C, T, and f be as in Lemma 5.3. Assume $\rho(h - T(h)) < \infty$ and $\rho(f - T(h)) < \infty$. Then the ρ-limit h of $\{T^n(f)\}$ is a fixed point of T. Moreover, if g is any fixed point of T in C such that $\rho(h - g) < \infty$, then we have $h = g$.*

Proof. We have

$$\rho(T(f) - T(h)) \le k \, \max \Big(\rho(f - h); \rho(f - T(f)); \rho(T(h) - h);$$
$$\rho(T(f) - h); \rho(f - T(h)) \Big).$$

From the previous results, we get

$$\rho(T(f) - T(h)) \le k \max \Big(\delta_\rho(f); \rho(h - T(h)); \rho(f - T(h)) \Big).$$

Assume that for $n \ge 1$ we have

$$\rho(T^n(f) - T(h)) \le \max \Big(k^n \delta_\rho(f); k\rho(h - T(h)); k^n \rho(f - T(h)) \Big).$$

Then

$$\rho(T^{n+1}(f) - T(h)) \le k \max \Big(\rho(T^n(f) - h); \rho(T^n(f) - T^{n+1}(f));$$
$$\rho(h - T(h)); \rho(T^{n+1}(f) - h); \rho(T^n(f) - T(h)) \Big) .$$

Hence

$$\rho(T^{n+1}(f) - T(h)) \leq k \max\left(k^n \delta_\rho(f); \rho(h - T(h)); \rho(T^n(f) - T(h))\right).$$

Using our previous assumption, we get

$$\rho(T^{n+1}(f) - T(h)) \leq \max\left(k^{n+1} \delta_\rho(f); k\rho(h - T(h)); k^{n+1}\rho(x - T(h))\right).$$

So by induction, we have

$$\rho(T^n(f) - T(h)) \leq \max\left(k^n \delta_\rho(f); k\rho(h - T(h)); k^n\rho(f - T(h))\right),$$

for any $n \geq 1$. Therefore we have

$$\limsup_{n \to \infty} \rho(T^n(f) - T(h)) \leq k\rho(h - T(h)).$$

Using the Fatou property, we get

$$\rho(h - T(h)) \leq \liminf_{n \to \infty} \rho(T^n(f) - T(h)) \leq k\rho(h - T(h)).$$

Since $k < 1$, we get $\rho(h - T(h)) = 0$ implying $T(h) = h$. Let g be another fixed point of T such that $\rho(h - g) < \infty$. Then we have

$$\rho(h - g) = \rho(T(h) - T(g)) \leq k\rho(h - g)$$

which implies $h = g$. This completes the proof of Theorem 5.5.

$\square$

5.3 Nonexpansive and Pointwise Asymptotic Nonexpansive Mappings

5.3.1 Case of Uniformly Convex Function Modulars

Similarly as in Banach and metric spaces, the fixed point existence theorems for the ρ-nonexpansive mappings (and for their pointwise and pointwise asymptotic generalizations) acting in modular function spaces are much harder to obtain than for ρ-contractions. Let us first consider the following question: Are the ρ-nonexpansive mappings really different from the mappings nonexpansive with respect to the Luxemburg norm associated with the modular ρ?

First we will show the following simple result.

Proposition 5.1. *Let* $\rho \in \Re$. *If for every* $\lambda > 0$

$$\rho\left(\lambda\left(T(f) - T(g)\right)\right) \leq \rho\left(\lambda(f - g)\right) \tag{5.3}$$

then, $\|T(f) - T(g)\|_\rho \leq \|f - g\|_\rho$.

Proof. Assume to the contrary that there exist $f, g \in L_\rho$ and $\alpha > 0$ such that

$$\|f - g\|_\rho < \alpha < \|T(f) - T(g)\|_\rho.$$

Then, $\left\|\dfrac{f - g}{\alpha}\right\|_\rho < 1$, which by Proposition 3.7 part (a) implies that $\rho\left(\dfrac{f - g}{\alpha}\right) < 1$. It also implies that

$$1 < \left\|\frac{T(f) - T(g)}{\alpha}\right\|_\rho,$$

which, by Proposition 3.7 part (b), yields $1 < \rho\left(\dfrac{T(f) - T(g)}{\alpha}\right)$. Finally, setting $\lambda = \alpha^{-1}$, we obtain

$$\rho\left(\lambda(f - g)\right) < 1 < \rho\left(\lambda\left(T(f) - T(g)\right)\right).$$

This contradiction completes the proof of Proposition 5.1. □

In view of Proposition 5.1, we need to ask whether the inequality (5.3) needs to hold for every $\lambda > 0$ in order to ensure the norm nonexpansiveness? If we knew that it sufficed to assume it merely for $\lambda = 1$, then there would be no real reason to consider ρ-nonexpansiveness. The answer to this question can be found in the following example of a mapping which is ρ-nonexpansive but it is not $\|.\|_\rho$-nonexpansive.

Example 5.4. Let $X = (0, \infty)$ and Σ be the σ-algebra of all Lebesgue measurable subsets of X. Let $\mathscr{P}$ denote the δ-ring of subsets of finite measure. Define a function modular by

$$\rho(f) = \frac{1}{e^2}\int_0^\infty |f(x)|^{x+1} dm(x).$$

Let B be the set of all measurable functions $f : (0, \infty) \to \mathbb{R}$ such that $0 \leq f(x) \leq 1/2$. Consider the map

$$T(f)(x) = \begin{cases} f(x - 1), & \text{for } x \geq 1, \\ 0, & \text{for } x \in [0, 1]. \end{cases}$$

Clearly, we have $T(B) \subset B$. For every $f, g \in B$ and $\lambda \leq 1$, we have

$$\rho\left(\lambda\left(T(f) - T(g)\right)\right) \leq \lambda\rho\left(\lambda(f - g)\right),$$

which implies that T is ρ-nonexpansive. On the other hand, if we take $f = 1_{[0,1]}$, then

$$\|T(f)\|_\rho > e \geq \|f\|_\rho,$$

which clearly implies that T is not $\|.\|_\rho$-nonexpansive. Note that T is linear.

For historical reasons, let us quote without proof an early example of a fixed point theorem for ρ-nonexpansive mappings acting in modular function spaces, see Theorem 2.13 [116].

Theorem 5.6. *Let $\rho \in \Re$ satisfy the regular growth condition , that is, $\omega_\rho(t) < 1$ for all $t \in [0,1)$. Let $C \subset L_\rho$ be a convex, ρ − bounded and ρ-a.e. compact such that $C - C \subset L_\rho^0$. Assume in addition that for every sequence $f_n \in C$ such that $f_n \to f$ ρ-a.e. with $f \in C$, and for every sequence of sets $G_k \downarrow \emptyset$,*

$$\lim_{k \to \infty} \left(\sup_{n \in \mathbb{N}} \rho(f_n - f, G_k) \right) = 0.$$

If $T : C \to C$ is ρ-nonexpansive, then T has a fixed point.

The above theorem sheds an interesting light on the following example by Alspach, [8].

Example 5.5. Define the operator

$$T(f)(x) = \begin{cases} \min\{2, 2f(x)\}, & \text{for } x \in [0, 1/2], \\ \max\{0, 2f(2x - 1) - 2\}, & \text{for } x \in (1/2, 1] \end{cases}$$

on C, a convex subset of $L^1[0,1]$, defined by

$$C = \left\{ f \in L^1[0,1] : 0 \leq f(x) \leq 2 \text{ a.e. and } \int_0^1 f(x)dx = 1 \right\}.$$

Let $L_\rho = L^1[0,1]$. The operator T is an L^1-isometry on C hence is ρ-nonexpansive, but it does not have any fixed points. It is easy to see that all assumptions of Theorem 5.6 are satisfied except for the ρ-a.e. compactness of C. To see that C is not ρ-a.e. compact take a sequence $f_n = 2g_n$ where $g_n = 1_{A_n}$ and $A_n = [2^{-1} - 2^{-n}, 1 - 2^{-n}]$. Obviously $f_n \in C$ but $f_n \to 0$ ρ-a.e. while the zero function does not belong to C. Note that Theorem 5.6 gives an intrinsic reason why T, defined on $B = \{ f \in L^1[0,1] : 0 \leq f(x) \leq 2 \text{ a.e.} \}$, must have a fixed point while $T : C \to C$ does not have to. Moreover, we do not refer to any geometrical properties of L^1 or its subspaces.

A more advanced fixed point results for ρ-nonexpansive mappings started showing up in the early 2000s, see for example [63, 64, 65]. It was only after a proper modular function space geometry was established that it became possible to obtain an elegant modular fixed point theory for ρ-nonexpansive mappings, summarized below in a generalized version which covers also the asymptotic pointwise case, Theorem 5.7. On a high level, the working of this theory can be summarized as follows:

1. The uniform convexity property implies the unique best approximant property (Theorem 4.1).

2. The uniform convexity property via the unique best approximant property implies the property (R) (Theorem 4.2).
3. The uniform convexity property implies the parallelogram property (Lemma 4.5).
4. The parallelogram property implies the minimizing sequence property for type functions when the minimum is strictly positive (Lemma 4.6, Case 1).
5. The property (R) implies the minimizing sequence property for type functions when the minimum is equal to zero (Lemma 4.6, Case 2).
6. The minimizing sequence property for type functions implies the fixed point property for asymptotic pointwise noenxpansive mappings; the modular limit of a minimizing sequence for a type function defined by an orbit is an obvious candidate for a fixed point. As Theorem 5.7 below shows, this is indeed the case.

Theorem 5.7. *Assume $\rho \in \Re$ is (UUC1). Let C be a ρ-closed ρ-bounded convex nonempty subset of L_ρ. Then, any $T : C \to C$ pointwise asymptotically nonexpansive mapping has a fixed point. Moreover, the set of all fixed points $F(T)$ is ρ-closed and convex.*

Proof. Let $f \in C$. Define the ρ-type

$$\tau(h) = \limsup_{n \to \infty} \rho(T^n(f) - h), \text{ for any } h \in C.$$

Define $\tau_0 = \inf\{\tau(h) : h \in C\}$ and note that $\tau_0 < \infty$ due to the ρ-boundedness of C. Let $\{g_n\} \subset C$ be a minimizing sequence of τ and $g \in C$ its ρ-limit which exists in view of Lemma 4.6. We will prove that g is a fixed point of T. First notice that

$$\tau(T^m(h)) \leq \alpha_m(h)\tau(h),$$

for any $h \in C$ and $m \geq 1$. In particular, we have

$$\tau(T^m(g_n)) \leq \alpha_m(g_n)\tau(g_n),$$

for any $n, m \geq 1$. By induction, we will build an increasing sequence $\{m_k\}$ such that

$$\alpha_{m_k+m}(g_k) \leq 1 + \frac{1}{k}, \tag{5.4}$$

for $k, m \geq 1$. Since T is asymptotic pointwise nonexpansive, we have

$$\limsup_{m \to \infty} \alpha_m(g_1) \leq 1.$$

Therefore, there exists $m_1 \geq 1$ such that for any $m \geq 1$ we have

$$\alpha_{m_1+m}(g_1) \leq 1 + \frac{1}{1}.$$

Since $\limsup\limits_{m\to\infty} \alpha_m(g_2) \leq 1$, there exists $m_2 > m_1$ such that for any $m \geq 1$, we have

$$\alpha_{m_2+m}(g_2) \leq 1+\frac{1}{2}.$$

Assume now that m_k has been built. Since

$$\limsup\limits_{m\to\infty} \alpha_m(g_{k+1}) \leq 1,$$

then there exists $m_{k+1} > m_k$ such that for any $m \geq m_{k+1}$, we have

$$\alpha_m(g_{k+1}) \leq 1+\frac{1}{k+1},$$

which completes our induction claim and proves (5.4). Let $n \geq 1$, $k \geq 1$ and $p \geq 0$ be integers. From (5.4), it follows that

$$\rho\left(T^{n+m_k+p}(f) - T^{m_k+p}(g_k)\right) \leq \left(1+\frac{1}{k}\right)\rho\left(T^n(f)-g_k\right). \tag{5.5}$$

By taking $\limsup\limits_{n\to\infty}$ of both sides of (5.5), we obtain

$$\tau_0 \leq \tau\left(T^{m_k+p}(g_k)\right) \leq \left(1+\frac{1}{k}\right)\tau(g_k). \tag{5.6}$$

Passing with $k \to \infty$ in (5.6) gives us

$$\lim\limits_{k\to\infty} \tau\left(T^{m_k+p}(g_k)\right) = \tau_0,$$

which forces $\{T^{m_k+p}(g_k)\}$ to be a minimizing sequence of τ, for any $p \geq 0$. Lemma 4.6 implies $\{T^{m_k+p}(g_k)\}$ is ρ-convergent to g, for any $p \geq 0$. In particular, taking $p = 1$, we conclude that

$$\rho\left(T^{m_k+1}(g_k) - g\right) \to 0. \tag{5.7}$$

From

$$\rho\left(T^{m_k+1}(g_k) - T(g)\right) \leq \alpha_1(g)\rho\left(T^{m_k}(g_k) - g\right),$$

we get the following

$$\rho\left(T^{m_k+1}(g_k) - T(g)\right) \to 0. \tag{5.8}$$

Since the ρ-limit of any ρ-convergent sequence is unique then (5.7) together with (5.8) imply that $T(g) = g$. To prove that $F(T)$ is ρ-closed, let $f_n \in F(T)$ and $\rho(f_n - f) \to 0$. Observe that

$$\rho\left(\frac{1}{3}(T(f) - f)\right) \leq \rho(T(f) - T(f_n)) + \rho(T(f_n) - f_n) + \rho(f_n - f)$$
$$\leq \alpha_1(f)\rho(f_n - f) + \rho(f_n - f),$$

for any $n \geq 1$. If we let $n \to \infty$, we get

$$\rho\left(\frac{1}{3}(T(f) - f)\right) = 0.$$

Hence $f \in F(T)$ proving $F(T)$ is ρ-closed. It remains to prove that $F(T)$ is convex. Note that to this end it suffices to show that $h = \dfrac{f+g}{2} \in F(T)$, for any $f, g \in F(T)$. Without loss of generality, we can assume that $f \neq g$. Let $k > 0$ be an integer. We have

$$\rho(f - T^k(h)) = \rho(T^k(f) - T^k(h)) \leq \alpha_k(f)\rho(f - h),$$

and

$$\rho(g - T^k(h)) = \rho(T^k(g) - T^k(h)) \leq \alpha_k(g)\rho(g - h).$$

Since $\rho(f - h) = \rho(g - h) = \rho\left(\dfrac{f-g}{2}\right)$, and

$$\rho\left(\frac{f-g}{2}\right) \leq \frac{1}{2}\rho(f - T^k(h)) + \frac{1}{2}\rho(g - T^k(h)),$$

we conclude that

$$\lim_{k\to\infty} \rho(f - T^k(h)) = \lim_{k\to\infty} \rho(g - T^k(h)) = \rho\left(\frac{f-g}{2}\right).$$

Similarly we have

$$\rho\left(f - \frac{h + T^k(h)}{2}\right) \leq \frac{1}{2}\rho(f - h) + \frac{1}{2}\rho(f - T^k(h)),$$

and

$$\rho\left(g - \frac{h + T^k(h)}{2}\right) \leq \frac{1}{2}\rho(g - h) + \frac{1}{2}\rho(g - T^k(h)).$$

Since

$$\rho\left(\frac{f-g}{2}\right) \leq \frac{1}{2}\rho\left(f - \frac{h + T^k(h)}{2}\right) + \frac{1}{2}\rho\left(g - \frac{h + T^k(h)}{2}\right),$$

we conclude that

$$\lim_{k\to\infty} \rho\left(f - \frac{h + T^k(h)}{2}\right) = \lim_{k\to\infty} \rho\left(g - \frac{h + T^k(h)}{2}\right) = \rho\left(\frac{f-g}{2}\right).$$

Therefore we have

$$\lim_{k\to\infty} \rho(f - T^k(h)) = \lim_{k\to\infty} \rho\left(f - \frac{h + T^k(h)}{2}\right) = \rho(f - h).$$

Lemma 4.2 applied to $A_k = f - T^k(h)$ and $B_k = T^k(h) - g$ implies that $\rho(A_k - B_k) \to 0$. Hence

$$\lim_{k \to \infty} \rho(h - T^k(h)) = \lim_{k \to \infty} \rho\left(\frac{A_k - B_k}{2}\right) \leq \lim_{k \to \infty} \rho(A_k - B_k) = 0.$$

Clearly we will get $\lim_{k \to \infty} \rho(h - T^{k+m}(h)) = 0$, for any integer $m \geq 0$. Since

$$\rho(T^m(h) - T^{k+m}(h)) \leq \alpha_m(h)\rho(h - T^k(h))$$

we get $\lim_{k \to \infty} \rho(T^m(h) - T^{k+m}(h)) = 0$. Finally using the inequality

$$\rho\left(\frac{h - T^m(h)}{2}\right) \leq \frac{1}{2}\rho(h - T^{k+m}(h)) + \frac{1}{2}\rho(T^m(h) - T^{k+m}(h)),$$

and by letting $k \to \infty$, we get $T^m(h) = h$, for any integer $m \geq 0$. Taking $m = 1$ we get $h \in F(T)$, which completes the proof.
$\square$

Remark 5.6. Observe that the statement of the above theorem is parallel to the celebrated Browder and Göhde fixed point theorems (see [32, 85]) but formulated purely in terms of function modulars without any reference to norms. Also, note that Theorem 5.7 extends outside nonexpansiveness and assumes merely asymptotic pointwise ρ-nonexpansiveness of the mapping T. Therefore, Theorem 5.7 can be actually understood as parallel not only to the Browder and Göhde fixed point theorems for nonexpansive mappings in Banach spaces but also to the result of Kirk and Xu [128] for asymptotic pointwise nonexpansive mappings in Banach spaces.

In the next section, we will discuss the analogue to Kirk's fixed point theorem [120] in modular function spaces.

5.3.2 Normal Structure Property in Modular Function Spaces

Because of the important role played by the normal structure property during the first 20 years (since 1965) of the fixed point theory, it was clear from the beginning that an analogue property in modular function spaces must be defined and investigated [116].

First, we have to introduce some basic definitions.

Definition 5.6. Let C be a ρ-bounded subset of L_ρ.

(1) The quantity $r_\rho(f, C) = \sup\{\rho(f - g) : g \in C\}$ will be called the ρ-*Chebyshev radius of C with respect to f* .

(2) The ρ-*Chebyshev radius of* C is defined by $R_\rho(C) = \inf \{r_\rho(f,C) : f \in C\}$.

(3) The ρ-*Chebyshev center of* C is defined by

$$\mathscr{C}_\rho(C) = \{f \in C : r_\rho(f,C) = R_\rho(C)\}.$$

Note that $R_\rho(C) \leq r_\rho(f,C) \leq \delta_\rho(C)$, for any $f \in C$ and any ρ-bounded nonempty subset C of L_ρ. Observe that $\mathscr{C}_\rho(C)$ may be empty in general.

Definition 5.7. Let C be a ρ-bounded subset of L_ρ.

(1) We say that A is a ρ-*admissible subset of* C if $A = \bigcap_{i \in I} B_\rho(f_i, r_i) \cap C$, where $f_i \in C$,

$r_i \geq 0$, I is an arbitrary index set and $B_\rho(f,r) = \{g \in L_\rho; \rho(f-g) \leq r\}$ is the ρ-closed ball centered at f with radius r. The family of all admissible subsets of C will be denoted by $\mathscr{A}_\rho(C)$. If D is a subset of C, we write

$$co_C(D) = \bigcap_{f \in D} B_\rho(f, r_\rho(f,D)) \cap C.$$

Note that $co_C(D) \in \mathscr{A}_\rho(C)$ and is the smallest ρ-admissible subset of C which contains D.

(2) $\mathscr{A}_\rho(C)$ is said to be compact if any family $\{A_\alpha\}_{\alpha \in \Gamma}$ of elements of $\mathscr{A}_\rho(C)$, has a nonempty intersection provided $\bigcap_{\alpha \in F} A_\alpha \neq \emptyset$, for any finite subset $F \subset \Gamma$.

(3) We will say that $\mathscr{A}_\rho(C)$ is countably compact or satisfies the property (R) if any sequence $\{A_n\}_{n \geq 1}$ of elements of $\mathscr{A}_\rho(C)$, which are nonempty and decreasing, has a nonempty intersection.

(4) $\mathscr{A}_\rho(C)$ is said to be normal if each ρ-admissible subset A of C, not reduced to a single point, we have $R_\rho(A) < \delta_\rho(A)$.

(5) We will say that $\mathscr{A}_\rho(C)$ is uniformly normal if there exists $c \in (0,1)$ such that for any nonempty $A \in \mathscr{A}_\rho(C)$, not reduced to one point, we have $R_\rho(A) \leq c\, \delta_\rho(A)$. Define the ρ-uniform normal structure coefficient of C as

$$\tilde{N}(C) = \sup \left\{ \frac{R_\rho(A)}{\delta_\rho(A)} : A \in \mathscr{A}_\rho(C),\ A \neq \emptyset \text{ and } \delta_\rho(A) > 0 \right\}.$$

Next we give the modular version of Kirk's fixed point theorem [120].

Theorem 5.8. *Let C be a nonempty ρ-closed ρ-bounded subset of L_ρ. Assume that the family $\mathscr{A}_\rho(C)$ is normal and compact. Let $T : C \to C$ be ρ-nonexpansive. Then T has a fixed point.*

Proof. Since $\mathscr{A}_\rho(C)$ is compact there exists a minimal nonempty $A \in \mathscr{A}_\rho(C)$ such that $T(A) \subset A$. It is easy to check that $co_C(T(A)) = A$. Indeed, we know that $co_C(T(A))$ is a nonempty ρ-admissible subset of C. Since $T(A) \subset A$, then $co_C(T(A)) \subset A$. Hence $T(co_C(T(A))) \subset T(A) \subset co_C(T(A))$. The minimality of A

will force $A = co_C(T(A))$ as claimed. Let us prove that $\delta_\rho(A) = 0$, i.e., A is reduced to one point. Suppose that $\delta_\rho(A) \neq 0$. For any $f \in A$, we have

$$r_\rho(f,A) \leq \delta_\rho(A) \leq \delta_\rho(C) < \infty,$$

and $A \subset B_\rho(f, r_\rho(f,A))$. Hence $T(A) \subset B_\rho\left(T(f), r_\rho(f,A)\right)$, for any $f \in A$, since T is ρ-nonexpansive. Thus

$$co_C(T(A)) \subset B_\rho\left(T(f), r_\rho(f,A)\right).$$

So $r_\rho(T(f),A) \leq r_\rho(f,A)$, for any $f \in A$. Next we fix $f_0 \in A$ and define

$$A_0 = \{f \in A : r_\rho(f,(A) \leq r_\rho(f_0,A)\}.$$

Clearly A_0 is not empty since $f_0 \in A_0$. Moreover, we have

$$A_0 = \bigcap_{f \in A} B_\rho\left(f, r_\rho(f_0,A)\right) \cap A \in \mathscr{A}_\rho(C).$$

Since $r_\rho(T(f),A) \leq r_\rho(f,A)$, for any $f \in A$, we get $T(A_0) \subset A_0$. The minimality of A implies $A_0 = A$. In particular we have $r_\rho(f,A) = r_\rho(f_0,A)$, for any $f \in A$. Hence $\delta_\rho(A) = \sup_{f \in A} r_\rho(f,A) = r_\rho(f_0,A)$. Since f_0 was taken arbitrary in A, we get $R_\rho(A) = \delta_\rho(A)$. This is a contradiction with the assumption $\mathscr{A}_\rho(C)$ is normal. Thus, we must have $\delta_\rho(A) = 0$, i.e., A is reduced to one point which is fixed by T. $\square$

Next, we give a constructive result discovered for Banach spaces by Kirk [124] which relaxes the compactness assumption in the above theorem. The main ingredient in Kirk's constructive proof is a technical lemma due to Gillespie and Williams [77]. The next lemma is the modular version of this technical result.

Lemma 5.4. *Let C be a ρ-bounded subset of L_ρ. Let $T : C \to C$ be a ρ-nonexpansive mapping. Assume that $\mathscr{A}_\rho(C)$ is normal. Let $A \in \mathscr{A}_\rho(C)$ be nonempty and T-invariant, i.e., $T(A) \subset A$. Then there exists a nonempty $A_0 \in \mathscr{A}_\rho(C)$ such that A_0 is T-invariant, $A_0 \subset A$ and*

$$\delta_\rho(A_0) \leq \frac{\delta_\rho(A) + R_\rho(A)}{2}.$$

Proof. Set $r = \frac{1}{2}(\delta_\rho(A) + R_\rho(A))$. We assume that $\delta_\rho(A) > 0$, otherwise we can take the set $A = A_0$. Since $\mathscr{A}_\rho(C)$ is normal, we have $R_\rho(A) < \delta_\rho(A)$. Hence $R_\rho(A) < r$, which implies the existence of $f \in A$ such that $r_\rho(f,A) < r$. Therefore, the set

$$D = \{f \in A : A \subset B_\rho(f,r)\} = \bigcap_{g \in A} B_\rho(g,r) \cap A$$

is a nonempty ρ-admissible subset of C. Note that there is no reason for D to be T-invariant. Consider the family

$$\mathscr{F} = \{M \in \mathscr{A}_\rho(C): \ D \subset M \ \text{and} \ T(M) \subset M\}.$$

Note that $\mathscr{F}$ is nonempty, since $C \in \mathscr{F}$. Set $L = \bigcap_{M \in \mathscr{F}} M$. The set L is a ρ-admissible subset of C which contains D. Using the definition of $\mathscr{F}$, we deduce that L is T-invariant. Consider $B = D \cup T(L)$, and observe that $co_C(B) = L$. Indeed, since $B \subset T(L) \subset L$ and $L \in \mathscr{A}_\rho(C)$, we have $co_C(B) \subset L$. From this we obtain

$$T(co_C(B)) \subset T(L) \subset B \subset co_C(B).$$

Hence $co_C(B) \in \mathscr{F}$, and $L \subset co_C(B)$. This gives the desired equality. Define

$$A_0 = \{f \in L: \ L \subset B_\rho(f,r)\} = \bigcap_{g \in L} B_\rho(g,r) \cap L.$$

We claim that A_0 is the desired set. Observe that A_0 is nonempty since it contains D (by definition of D) and is a ρ-admissible subset of C. On the other hand, it is clear that $\delta_\rho(A_0) \leq r$. To complete the proof, we have to show that A_0 is T-invariant. Let $f \in A_0$. By definition of A_0, we have $L \subset B_\rho(f,r)$. Since T is ρ-nonexpansive, we have

$$T(L) \subset B_\rho(T(f),r).$$

For any $g \in D$, there holds $L \subset B_\rho(g,r)$. But $T(f) \in L$, so $T(f) \in B_\rho(g,r)$, which implies $g \in B_\rho(T(f),r)$. Hence $D \subset B_\rho(T(f),r)$ holds. Since $B = D \cup T(L)$, we get $B \subset B_\rho(T(f),r)$. Therefore, we must have

$$co_C(B) = L \subset B_\rho(T(f),r).$$

By the definition of A_0, it follows that $T(f) \in A_0$. In other words, A_0 is T-invariant. Finally, note that since $L \subset A$, then $A_0 \subset A$.
$\square$

Next we give the modular analogue of Kirk's fixed point theorem [124].

Theorem 5.9. *Let C be a ρ-bounded and ρ-closed nonempty subset of L_ρ. Assume that $\mathscr{A}_\rho(C)$ is normal and countably compact. If $T : C \to C$ is ρ-nonexpansive, then T has a fixed point.*

Proof. Let $\mathscr{F} = \{D \in \mathscr{A}_\rho(C): \ D \neq \emptyset \ \text{and} \ T(D) \subset D\}$. The family $\mathscr{F}$ is not empty since $C \in \mathscr{F}$. Define $\tilde{\delta} : \mathscr{F} \to [0,+\infty)$ by

$$\tilde{\delta}(D) = \inf \{\delta_\rho(B): \ B \in \mathscr{F} \ \text{and} \ B \subset D\}.$$

Set $D_1 = C$. By definition of $\tilde{\delta}(D_1)$, there exists $D_2 \in \mathscr{F}$ such that $D_2 \subset D_1$ and $\delta_\rho(D_2) < \tilde{\delta}(D_1) + 1$. Assume $D_1, D_2, \cdots, D_n$, for $n \geq 1$, are constructed. Again by definition of $\tilde{\delta}(D_n)$, there exists $D_{n+1} \in \mathscr{F}$ such that $\delta_\rho(D_{n+1}) < \tilde{\delta}(D_n) + \frac{1}{n}$ and $D_{n+1} \subset D_n$. Since $\mathscr{A}_\rho(C)$ is countably compact, then $D_\infty = \bigcap_{n \geq 1} D_n$ is not empty. Clearly we have $D_\infty \in \mathscr{F}$. Assume that D_∞ is not reduced to one point. Using Lemma 5.4, there exists $D^* \in \mathscr{F}$ and $D^* \subset D_\infty$ such that

$$\delta_\rho(D^*) \leq \frac{R_\rho(D_\infty) + \delta_\rho(D_\infty)}{2}. \tag{5.9}$$

Since $D^* \subset D_n$, we get

$$\delta_\rho(D^*) \leq \delta_\rho(D_\infty) \leq \delta_\rho(D_{n+1}) \leq \tilde{\delta}(D_n) + \frac{1}{n} \leq \delta_\rho(D^*) + \frac{1}{n},$$

for any $n \geq 1$. If we let $n \to \infty$, we get $\delta_\rho(D^*) = \delta_\rho(D_\infty)$. Then the inequality (5.9) implies $\delta_\rho(D_\infty) \leq R_\rho(D_\infty)$. This is in contradiction with the assumption that $\mathscr{A}_\rho(C)$ is normal. Hence, D_∞ is reduced to one point which is a fixed point of T since D_∞ is T-invariant. $\square$

As a corollary we get the following result.

Corollary 5.1. *Let $\rho \in \mathfrak{R}$. Let C be a ρ-bounded and ρ-closed nonempty subset of L_ρ. Assume that C is ρ-a.e. compact and $\mathscr{A}_\rho(C)$ is normal. If $T : C \to C$ is ρ-nonexpansive, then T has a fixed point.*

Proof. Recall that ρ-balls are ρ-a.e. closed because of Proposition 3.4. Hence, any element of $\mathscr{A}_\rho(C)$ is ρ-a.e. closed subset of C. Since C is ρ-a.e. sequentially compact, we deduce that $\mathscr{A}_\rho(C)$ is countably compact. Therefore, the conclusion of Corollary 5.1 follows from Theorem 5.9. $\square$

In the next result we relax the assumption of ρ-a.e. compactness.

Theorem 5.10. *Let C be a ρ-bounded and ρ-closed nonempty subset of L_ρ. Assume that $\mathscr{A}_\rho(C)$ is uniformly normal. If $T : C \to C$ is ρ-nonexpansive, then T has a fixed point.*

Proof. Without loss of generality we may assume that C is not reduced to one point. Set $C_1 = C$. Using Lemma 5.4, there exists $C_2 \in \mathscr{A}_\rho(C)$ such that $C_2 \subset C_1$, $T(C_2) \subset C_2$ and

$$\delta_\rho(C_2) \leq \frac{\delta_\rho(C_1) + R_\rho(C_1)}{2}.$$

If we proceed by induction we will construct a sequence $\{C_n\}$ of nonempty elements of $\mathscr{A}_\rho(C)$ such that $C_{n+1} \subset C_n$, $T(C_n) \subset C_n$ and

$$\delta_\rho(C_{n+1}) \leq \frac{\delta_\rho(C_n) + R_\rho(C_n)}{2}, \quad n = 1, 2, \cdots$$

Since $\mathscr{A}_\rho(C)$ is uniformly normal, there exists $c \in (0,1)$ such that $R_\rho(A) \leq c\, \delta_\rho(A)$, for any $A \in \mathscr{A}_\rho(C)$. Hence

$$\delta_\rho(C_{n+1}) \leq \frac{1+c}{2}\delta_\rho(C_n), \quad n = 1,2,\cdots$$

Thus

$$\delta_\rho(C_n) \leq \left(\frac{1+c}{2}\right)^n \delta_\rho(C), \quad n = 1,2,\cdots$$

Let $f_n \in C_n$, for any $n \geq 1$. Since

$$\rho(f_{n+h} - f_n) \leq \delta_\rho(C_n) \leq \left(\frac{1+c}{2}\right)^n \delta_\rho(C), \quad n,h = 1,2,\cdots$$

and $1 + c < 2$, we conclude that $\{f_n\}$ is ρ-Cauchy. Since L_ρ is ρ-complete, we deduce that $\{f_n\}$ is ρ-convergent to $f \in L_\rho$. Since C is ρ-closed, we conclude that $f \in C$. In fact, we have $f \in C_n$, for any $n \geq 1$. Hence $C_\infty = \bigcap_{n \geq 1} C_n$ is not empty.

Observe that

$$\delta_\rho(C_\infty) \leq \delta_\rho(C_n) \leq \left(\frac{1+c}{2}\right)^n \delta_\rho(C), \quad n = 1,2,\cdots$$

This will force $\delta_\rho(C_\infty) \leq 0$, i.e., C_∞ is reduced to one point, i.e., $C_\infty = \{f\}$. Since each C_n is T-invariant, for any $n \geq 1$, we conclude that C_∞ is also T-invariant, i.e., $T(f) = f$. This completes the proof of Theorem 5.10.
□

Next, we discuss the normal structure property when L_ρ is uniformly ρ-noncompact convex modular space.

Theorem 5.11. *Let $\rho \in \mathfrak{R}$ has property (R'). Assume that the characteristic of noncompact convexity of L_ρ satisfies $\varepsilon_\alpha(r, L_\rho) < 1$, for some $r > 0$. Then for any nonempty, ρ-closed, ρ-bounded convex $K \subset L_\rho$ such that $\delta_\rho(K) \leq r$, there exists $f \in K$ for which*

$$\sup\{\rho(f - g) : g \in K\} < \delta_\rho(K).$$

Proof. Assume this is not the case. Then there exists a nonempty, ρ-closed, ρ-bounded convex, not reduced to one point, set $K \subset L_\rho$ such that $\delta_\rho(K) \leq r$ and

$$\sup\{\rho(f - g) : g \in K\} = \delta_\rho(K) = d > 0,$$

for all $f \in K$. Fix $f_1 \in K$ and let $f_2 \in K$ such that

$$\rho(f_2 - f_1) \geq d\left(1 - \frac{1}{2^3}\right).$$

Assume that $(f_1, .., f_n)$ have been constructed, then let $f_{n+1} \in K$ be such that

$$\rho \left(f_{n+1} - \frac{1}{n} \sum_{i=1}^{n} f_i \right) \geq d \left(1 - \frac{1}{(n+1)^3} \right).$$

Hence there exists a sequence $\{f_n\} \subset K$ such that

$$d \left(1 - \frac{1}{(n+1)^3} \right) \leq \rho \left(f_{n+1} - \frac{1}{n} \sum_{i=1}^{n} f_i \right) \leq d.$$

Using the convexity of ρ, we can get

$$d \left(1 - \frac{1}{(n+1)^2} \right) \leq \rho(f - f_{n+1}) \leq d, \tag{5.10}$$

for every $f = \sum_{1 \leq i \leq n} \alpha_i f_i$, with $\alpha_i \geq 0$ and $\sum \alpha_i = 1$. Indeed put $\alpha = \max\{\alpha_i\}$, then we have

$$\frac{1}{n} \sum_{1 \leq i \leq n} f_i = \frac{1}{n\alpha} f + \sum_{1 \leq i \leq n} \left(\frac{1}{n} - \frac{\alpha_i}{n\alpha} \right) f_i.$$

Since $\dfrac{1}{n\alpha} + \sum (\dfrac{1}{n} - \dfrac{\alpha_i}{n\alpha}) = 1$, we obtain

$$\frac{1}{n} \sum_{i=1}^{n} f_i - f_{n+1} = \frac{1}{n\alpha}(f - f_{n+1}) + \sum_{1 \leq i \leq n} \left(\frac{1}{n} - \frac{\alpha_i}{n\alpha} \right)(f_i - f_{n+1}).$$

Therefore,

$$d \left(1 - \frac{1}{(n+1)^3} \right) \leq \frac{1}{n\alpha} \rho(f - f_{n+1}) + \sum_{1 \leq i \leq n} \left(\frac{1}{n} - \frac{\alpha_i}{n\alpha} \right) \rho(f_i - f_{n+1}),$$

and then

$$d \left(1 - \frac{1}{(n+1)^3} \right) \leq \frac{1}{n\alpha} \rho(f - f_{n+1}) + \left(1 - \frac{1}{n\alpha} \right) d,$$

which implies

$$d \left(1 - \frac{n\alpha}{(n+1)^3} \right) \leq \rho(f - f_{n+1}).$$

Since $n\alpha \leq n+1$, we obtain the inequality (5.10). In particular, we have

$$\rho(f_n - f_m) \geq d \left(1 - \frac{1}{m^2} \right), \tag{5.11}$$

for every $m > n \geq 1$. Since L_ρ satisfies the property (R'), there exists a subsequence $\{f_{n'}\}$ of $\{f_n\}$ such that

$$\bigcap_{n \geq 1} \overline{conv}_\rho\left(\{f_{i'} : i \geq n\}\right) = \{h\}.$$

Denote $C_n = \overline{conv}_\rho\left(\{f_{i'} : i \geq n\}\right)$. Using (5.11), it is not hard to show that $\alpha(C_n) \geq d$, for every $n \geq 1$. Let $f \in K$, then $C_n \subset B_\rho(f, d)$ since $\delta_\rho(K) = d$. Because $\varepsilon_\alpha(r, L_\rho) < 1$, one can find $\Delta > 0$ such that

$$d_\rho(f, C_n) \leq (1 - \Delta)d.$$

Theorem 4.4 implies that

$$C_n \cap B_\rho(f, (1 - \Delta)d) = K_n,$$

is nonempty, for all $n \geq 1$. Since the property (R') implies the property (R) and (K_n) is a decreasing sequence of ρ-closed ρ-bounded nonempty convex subsets, we obtain

$$\bigcap_n K_n = \bigcap_n C_n \cap B_\rho(f, (1 - \Delta)d)$$

is nonempty. This clearly implies that

$$\rho(f - h) \leq (1 - \Delta)d.$$

Since this is true for any $f \in K$, we get

$$\sup\{\rho(f - h) : f \in K\} \leq (1 - \Delta)d,$$

contradicting our assumption on K. This completes the proof of Theorem 5.11.
$\square$

Remark 5.7. As a consequence of this theorem, whenever L_ρ has property (R'), and the characteristic of noncompact convexity of L_ρ satisfies $\varepsilon_\alpha(r, L_\rho) < 1$, for some $r > 0$, then for any nonempty ρ-closed, ρ-bounded convex subset K of L_ρ, and $\delta_\rho(K) \leq r$, we know that $\mathscr{A}_\rho(K)$ is normal and countably compact. Under these assumptions, Theorem 5.9 will imply that any $T : K \to K$ ρ-nonexpansive has a fixed point.

5.3.3 Case of Uniformly Lipschitzian Mappings

In this section, we study the existence of fixed points for a more general class of mappings: uniformly Lipschitzian mappings. Fixed point theorems for this class of mappings in Banach spaces have been studied in [42, 62] and in metric spaces in [154] (for further information about this subject, see Chapter VIII of [11] and references therein). The main tool used in this section is the coefficient of normal structure property in modular sense.

The following technical lemma will be useful to prove the main theorem of this section.

Lemma 5.5. *Let $\rho \in \Re$ be a function modular satisfying the Δ_2-type condition. Assume that ρ has the ρ-a.e. strong Opial property. Let K be a strongly ρ-bounded and ρ-a.e sequentially compact subset of L_ρ. Let $\{f_n\}_n$ and $\{g_n\}_n$ be two sequences in K. Then, there exists $g \in \Omega_{ae}(\{g_n\}) \cap K$ such that*

$$\limsup_{n\to\infty} \rho(g - f_n) \leq \limsup_{j\to\infty} \limsup_{n\to\infty} \rho(g_j - f_n),$$

where $\Omega_{ae}(\{g_n\})$ is the set of all ρ-a.e. limit of subsequences of $\{g_n\}$.

Proof. Let $\{f_n\}_n$ and $\{g_n\}_n$ be two sequences in K. Define $\theta : K \to [0, +\infty]$ by

$$\theta(h) = \limsup_{n\to\infty} \rho(h - f_n).$$

Since K is ρ-a.e. sequentially compact and ρ-bounded, there exist a subsequence $\{g_{\varphi(n)}\}$ of $\{g_n\}$ and a subsequence $\{f_{\psi(n)}\}$ of $\{f_n\}$ such that

$$\begin{cases} g_{\varphi(n)} \overset{\rho-a.e}{\longrightarrow} g, \\ f_{\psi(n)} \overset{\rho-a.e}{\longrightarrow} f, \\ \lim_{n\to\infty} \rho(f_{\psi(n)} - g) = \limsup_{n\to\infty} \rho(f_n - g). \end{cases}$$

Clearly we have $g \in \Omega_{ae}(\{g_n\}) \cap K$ and $f \in K$. Next we prove that

$$\theta(g) \leq \limsup_{j\to\infty} \theta(g_j).$$

Indeed, the ρ-a.e. strong Opial property implies

$$\theta(g_j) = \limsup_{n\to\infty} \rho(f_n - g_j)$$
$$\geq \liminf_{n\to\infty} \rho(f_{\psi(n)} - g_j)$$
$$= \liminf_{n\to\infty} \rho(f_{\psi(n)} - f) + \rho(f - g_j),$$

for any $j \geq 1$. Thus

$$
\begin{aligned}
\limsup_{j\to\infty} \theta(g_j) &\geq \liminf_{n\to\infty} \rho(f_{\psi(n)} - f) + \limsup_{j\to\infty} \rho(f - g_j) \\
&\geq \liminf_{n\to\infty} \rho(f_{\psi(n)} - f) + \liminf_{j\to\infty} \rho(f - g_{\phi(j)}) \\
&= \liminf_{n\to\infty} \rho(f_{\psi(n)} - f) + \liminf_{j\to\infty} \rho(g_{\phi(j)} - g) + \rho(f - g).
\end{aligned}
$$

On the other hand, we have

$$
\theta(g) = \limsup_{n\to\infty} \rho(f_n - g) = \lim_{n\to\infty} \rho(f_{\psi(n)} - g) = \liminf_{n\to\infty} \rho(f_{\psi(n)} - f) + \rho(f - g).
$$

Therefore, we have $\theta(g) \leq \limsup_{j\to\infty} \theta(g_j)$.

$\square$

The following lemma was inspired by [42] where a similar result is proved in reflexive Banach spaces (see also [154] for a version in metric spaces with additional properties).

Lemma 5.6. *Let $\rho \in \Re$ be a function modular satisfying the Δ_2-type condition. Assume that ρ satisfies the ρ-a.e. strong Opial property. Let C be a strongly ρ-bounded and ρ-a.e compact subset of L_ρ. Let $\{f_n\}$ be a sequence in C and c a constant such that $c > \tilde{N}(C)$. Then there exists $f \in C$ such that*

(i) $\limsup_{j\to\infty} \rho(f - f_j) \leq c\, \delta_\rho(\{f_n\})$,

(ii) $\rho(f - g) \leq \limsup_{n\to\infty} \rho(f_n - g),$ *for any $g \in C$.*

Proof. Let $\{f_n\}$ be a sequence in C. Set

$$
A = \bigcap_{m=1}^{\infty} A_m \subset C,
$$

where $A_m = \overline{co}_C\Big(\{f_j : j \geq m\}\Big) \in \mathscr{A}_\rho(C)$, for any $m \geq 1$. Since ρ satisfies the Fatou property, any ρ-ball is ρ-a.e. closed. Using the ρ-a.e compactness of C, then $A \neq \emptyset$. It is easy to check that $\delta_\rho(A_m) \leq \delta_\rho(\{f_n\})$ and $A_m \subset B_\rho(g, \sup_{j\geq m} \rho(g - f_j))$, for any $m \geq 1$ and $g \in C$. Hence for any $f \in A$ and $g \in C$, we have

$$
\rho(g - f) \leq r_\rho(g, A) \leq r_\rho(g, A_n) \leq \sup_{j\geq n} \rho(g - f_j),
$$

for any $n \geq 1$. In fact we have $r_\rho(g, A_n) = \sup_{j\geq n} \rho(g - f_j)$, for any $n \geq 1$. Therefore, we have

$$
\rho(g - f) \leq \limsup_{n\to\infty} \rho(g - f_n),
$$

i.e., (ii) holds for any $f \in A$. Next we prove that there exists $f \in A$ such that (i) holds. Without loss of generality we may assume that $\delta_\rho(\{f_n\}) > 0$. Choose $\varepsilon > 0$ such that $\tilde{N}(C)\delta_\rho(\{f_n\}) + \varepsilon \leq c\,\delta_\rho(\{f_n\})$. By the definition of $R_\rho(A_m)$, there exists $g_m \in A_m$ such that

$$r_\rho(g_m, A_m) < R_\rho(A_m) + \varepsilon \leq \tilde{N}(C)\delta_\rho(A_m) + \varepsilon \leq \tilde{N}(C)\delta_\rho(\{f_n\}) + \varepsilon \leq c\,\delta_\rho(\{f_n\}),$$

for any $m \geq 1$. Since $\sup_{j \geq m} \rho(g_m - f_j) = r_\rho(g_m, A_m)$, we have

$$\limsup_{j \to \infty} \rho(g_m - f_j) \leq c\,\delta_\rho(\{f_n\}), \tag{5.12}$$

for any $m \geq 1$. Using Lemma 5.5, there exists $f \in \Omega_{ae}(\{g_n\}) \cap C$ such that

$$\limsup_{j \to \infty} \rho(f - f_j) \leq \limsup_{n \to \infty} \limsup_{j \to \infty} \rho(g_n - f_j). \tag{5.13}$$

We will check that $f \in A$. Indeed, for all i, n integers such that $i \geq n$ we have $g_i \in A_i \subset A_n$. Thus, $\Omega_{ae}(\{g_n\}) \cap C \subset A_n$ for any $n \geq 1$, which implies $f \in A$. Using the inequalities (5.12) and (5.13), we get

$$\limsup_{j \to \infty} \rho(f - f_j) \leq c\,\delta_\rho(\{f_n\}),$$

which completes the proof of Lemma 5.6.
$\square$

.

Using the two lemmas above, we are ready to prove the main theorem of this section.

Theorem 5.12. *[63] Let $\rho \in \Re$ be a function modular satisfying the Δ_2-type condition. Assume that ρ satisfies the ρ-a.e. strong Opial property. Let K be a strongly ρ-bounded and ρ-a.e compact convex subset of L_ρ. Suppose that $\tilde{N}(K) < 1$. Let $T : K \to K$ be a k-uniformly Lipschitzian mapping, i.e.,*

$$\rho(T^n(f) - T^n(g)) \leq k\,\rho(f - g),$$

for any $f, g \in K$. Suppose that $k < (\tilde{N}(K))^{-1/2}$. Then T has a fixed point.

Proof. If $k \leq 1$, then T is ρ-nonexpansive, so Theorem 5.10 shows that T has a fixed point. Therefore, we may assume that $k > 1$. Choose a constant $c \in (\tilde{N}(K), 1)$ such that $1 < k < c^{-1/2}$. Fix $f_0 \in K$. By Lemma 5.6, we can inductively construct a sequence $\{f_n\} \subset K$ such that

(1) $\limsup\limits_{n \to \infty} \rho(T^n(f_m) - f_{m+1}) \leq c\,\delta(\{T^n(f_m)\}_n)$,

(2) $\rho(f_{m+1} - g) \leq \limsup\limits_{n \to \infty} \rho(T^n(f_m) - g)$,

for all $g \in K$ and $m \geq 0$. Denote $D_m = \limsup\limits_{n \to \infty} \rho(T^n(f_m) - f_{m+1})$. For $n, h \in \mathbb{N}$, we have

$$
\begin{aligned}
\rho(T^{n+h}(f_m) - T^n(f_m)) &\leq k\, \rho(f_m - T^h(f_m)) \\
&\leq k \limsup\limits_{i \to \infty} \rho(T^i(f_{m-1}) - T^h(f_m)) \\
&\leq k^2 \limsup\limits_{i \to \infty} \rho(T^{i-h}(f_{m-1}) - f_m) \\
&\leq k^2 D_{m-1}.
\end{aligned}
$$

Since $D_m = \limsup\limits_{n \to \infty} \rho(T^n(f_m) - f_{m+1}) \leq c\, \delta(\{T^n(f_m)\}_n)$, we obtain

$$
D_m \leq c\, k^2\, D_{m-1} \leq (c\, k^2)^m D_0, \quad m = 1, 2, \cdots
$$

Thus, for any $j \in \mathbb{N}$, we have

$$
\begin{aligned}
\rho(f_{j+1} - f_j) &\leq \omega_\rho(2)\Big(\rho(f_{j+1} - T^n(f_j)) + \rho(f_j - T^n(f_j))\Big) \\
&\leq \omega_\rho(2)\Big(\rho(f_{j+1} - T^n(f_j)) + \limsup\limits_{m \to \infty} \rho(T^m(f_{j-1}) - T^n(f_j))\Big) \\
&\leq \omega_\rho(2)\Big(\rho(f_{j+1} - T^n(f_j)) + k \limsup\limits_{m \to \infty} \rho(T^{m-n}(f_{j-1}) - f_j)\Big) \\
&\leq \omega_\rho(2)\Big(\rho(f_{j+1} - T^n(f_j)) + k\, D_{j-1}\Big),
\end{aligned}
$$

where $\omega_\rho(t)$ is the growth function associated to the modular ρ (see Definition 3.7). Taking $\limsup$ as $n \to \infty$, we obtain

$$
\begin{aligned}
\rho(f_{j+1} - f_j) &\leq \omega_\rho(2)(D_j + k\, D_{j-1}) \\
&\leq \omega_\rho(2)((c\, k^2)^j + k\, (c\, k^2)^{j-1}) D_0 \\
&\leq \omega_\rho(2)(c\, k^2 + k)(c\, k^2)^{j-1} D_0 \\
&\leq A(c\, k^2)^j,
\end{aligned}
$$

where $A = \omega_\rho(2)\frac{c\, k + 1}{c\, k} D_0$ and $j \in \mathbb{N}$, which implies

$$
\frac{1}{A(c\, k^2)^j} \leq \frac{1}{\rho(f_{j+1} - f_j)}.
$$

Using Lemma 3.1, we obtain

$$
\omega_\rho^{-1}\left(\frac{1}{A}\right)\left(\omega_\rho^{-1}\left(\frac{1}{c\, k^2}\right)\right)^j \leq \omega_\rho^{-1}\left(\frac{1}{A(c\, k^2)^j}\right) \leq \omega_\rho^{-1}\left(\frac{1}{\rho(f_{j+1} - f_j)}\right),
$$

for any $j \in \mathbb{N}$. Therefore, we have

$$||f_{j+1} - f_j||_\rho \le \frac{1}{\omega_\rho^{-1}\left(\frac{1}{\rho(f_{j+1}-f_j)}\right)} \le B \frac{1}{\left(\omega_\rho^{-1}\left(\frac{1}{c\,k^2}\right)\right)^j},$$

for any $j \in \mathbb{N}$. Since $\omega_\rho(1) = 1$ and $ck^2 < 1$, we conclude that $\omega_\rho^{-1}\left(\frac{1}{ck^2}\right) > 1$. Hence $\{f_n\}$ is a Cauchy sequence in $(L_\rho, ||.||_\rho)$. Since $(L_\rho, ||.||_\rho)$ is complete, there exists $f \in L_\rho$ such that $||f_n - f||_\rho \to 0$. Since under Δ_2-condition norm convergence and modular convergence are identical, $\{f_n\}$ is ρ-convergent to f. Thus, Proposition 3.5 implies the existence of a subsequence of $\{f_n\}$ ρ-a.e convergent to f. Note that f belongs to K since K is ρ-a.e closed. Next we prove that f is a fixed point of T. Indeed,

$$\rho(f - T(f)) \le \omega_\rho(3)\Big(\rho(f - f_{j+1}) + \rho(f_{j+1} - T^n(f_j)) + \rho(T^n(f_j) - T(f))\Big)$$

$$\le \omega_\rho(3)\Big(\rho(f - f_{j+1}) + \rho(f_{j+1} - T^n(f_j)) + k\,\rho(T^{n-1}(f_j) - f)\Big)$$

$$\le \omega_\rho(3)\Big(\rho(f - f_{j+1}) + \rho(f_{j+1} - T^n(f_j))$$

$$+ k\,\omega_\rho(2)\Big(\rho(T^{n-1}(f_j) - f_{j+1}) + \rho(f_{j+1} - f)\Big)\Big),$$

for any $n \ge 1$ and $j \in \mathbb{N}$. Taking $\limsup$ as $n \to \infty$, we have

$$\rho(f - Tf) \le \omega_\rho(3)\Big(\rho(f - f_{j+1}) + D_j + k\,\omega_\rho(2)\Big(D_j + \rho(f_{j+1} - f)\Big)\Big),$$

for any $j \in \mathbb{N}$. Taking the limit as $j \to \infty$, we obtain $\rho(f - Tf) = 0$, i.e., $T(f) = f$. $\square$

The following lemma gives a relationship between the coefficient of ρ-normal structure and the ρ-modulus of uniform convexity.

Lemma 5.7. *Let $\rho \in \Re$ staisfy the Δ_2-type condition. Let C be a ρ-bounded nonempty convex subset of L_ρ not reduced to one point. Then,*

$$\tilde{N}(C) \le 1 - \inf\Big\{\delta_2(d, \varepsilon) : d \in (0, \delta_\rho(C)]\Big\},$$

for any $\varepsilon \in \left(0, \dfrac{1}{\omega_\rho(2)}\right)$.

Proof. Let $B \in \mathscr{A}_\rho(C)$ be nonempty and not reduced to one point, i.e., $\delta_\rho(B) > 0$. Since C is convex and the ρ-balls are convex, we deduce that B is convex. Denote $d = \delta_\rho(B)$ and $r = R_\rho(B)$. Let $\varepsilon \in (0,1)$. There exist $f,g \in B$ such that

$\rho(f - g) \geq \varepsilon\delta(B)$. Hence

$$\rho\left(\frac{f-g}{2}\right) \geq \frac{\rho(f-g)}{\omega_\rho(2)} \geq \frac{d\,\varepsilon}{\omega_\rho(2)}.$$

Let $h \in B$. We know that $\rho(h - f) \leq d$, $\rho(h - g) \leq d$ and

$$\rho\left(\frac{(h-f)-(h-g)}{2}\right) \geq \frac{d\varepsilon}{\omega_\rho(2)}.$$

By the definition of $\delta_2\left(d, \dfrac{\varepsilon}{\omega_\rho(2)}\right)$, we have

$$\rho\left(h - \frac{f+g}{2}\right) = \rho\left(\frac{(h-f)+(h-g)}{2}\right) \leq d\left(1 - \delta_2\left(d, \frac{\varepsilon}{\omega(2)}\right)\right),$$

for all $h \in B$. Thus,

$$\frac{R_\rho(B)}{\delta_\rho(B)} = \frac{r}{d} \leq 1 - \delta_2\left(d, \frac{\varepsilon}{\omega(2)}\right).$$

Therefore,

$$\tilde{N}(C) \leq \sup\left\{1 - \delta_2\left(d, \frac{\varepsilon}{\omega(2)}\right) : 0 < d \leq \delta_\rho(C)\right\}$$

$$\leq 1 - \inf\left\{1 - \delta_2\left(d, \frac{\varepsilon}{\omega(2)}\right) : 0 < d \leq \delta_\rho(C)\right\}.$$

□

Remark 5.8. If we use the modulus of uniform convexity

$$\delta_1(r, \varepsilon) = \inf\left\{1 - \frac{1}{r}\rho\left(\frac{f+g}{2}\right) : (f, g) \in D_1(r, \varepsilon)\right\},$$

where $D_1(r, \varepsilon) = \{(f, g) : f, g \in L_\rho, \rho(f) \leq r, \rho(g) \leq r, \rho(f - g) \geq \varepsilon r\}$, then we do not need ρ to satisfy the Δ_2-type condition. The estimate found in Lemma 5.7 becomes

$$\tilde{N}(C) \leq 1 - \inf\left\{\delta_2(d, \varepsilon) : d \in (0, \delta_\rho(C)]\right\},$$

for any $\varepsilon \in (0, 1)$.

Next we give an improvement to the conclusion of Theorem 5.12 in the case of asymptotically regular mappings. First we need the following definition.

Definition 5.8. Let C be a nonempty subset of L_ρ and $T : C \to C$. We denote by $|T|$ the ρ-Lipschitz constant of T, i.e.,

$$|T| = \sup \left\{ \frac{\rho T(f) - T(g)}{\rho(f - g)} : f \neq g, \ f, g \in C \right\}$$

and $s(T) = \liminf_{n \to \infty} |T^n|$. We say that T is *asymptotically regular* if

$$\lim_{n \to \infty} \rho(T^{n+1}(f) - T^n(f)) = 0,$$

for any $f \in C$.

The following lemma will be needed to prove the next theorem.

Lemma 5.8. *Let $\rho \in \mathfrak{R}$ satisfy the Δ_2-type condition. Let $\{f_n\}$ and $\{g_n\}$ be two sequences in L_ρ. Assume $\lim_{n \to \infty} \rho(g_n) = 0$. Then*

$$\limsup_{n \to \infty} \rho(f_n + g_n) = \limsup_{n \to \infty} \rho(f_n).$$

Proof. Using the properties of ρ and ω_ρ, we get

$$\rho(f_n + g_n) \leq \rho \left(\frac{f_n}{1 - \varepsilon} \right) + \rho \left(\frac{g_n}{\varepsilon} \right),$$

for any $\varepsilon \in (0, 1)$. Thus

$$\rho(f_n + g_n) \leq \omega_\rho \left(\frac{1}{1 - \varepsilon} \right) \rho(f_n) + \omega_\rho \left(\frac{1}{\varepsilon} \right) \rho(g_n)$$

which implies

$$\limsup_{n \to \infty} \rho(f_n + g_n) \leq \omega_\rho \left(\frac{1}{1 - \varepsilon} \right) \limsup_{n \to \infty} \rho(f_n).$$

Since ε is arbitrary and

$$\omega_\rho \left(\frac{1}{1 - \varepsilon} \right) \to 1 \quad \text{as} \quad \varepsilon \to 0^+$$

we obtain

$$\limsup_{n \to \infty} \rho(f_n + g_n) \leq \limsup_{n \to \infty} \rho(f_n).$$

Furthermore, the same argument proves

$$\limsup_{n \to \infty} \rho(f_n) = \limsup_{n \to \infty} \rho(f_n + g_n - g_n) \leq \limsup_{n \to \infty} \rho(f_n + g_n).$$

Putting the inequalities together we get $\limsup\limits_{n\to\infty} \rho(f_n + g_n) = \limsup\limits_{n\to\infty} \rho(f_n)$ as claimed.

$\square$

Theorem 5.13. *[64] Let $\rho \in \mathfrak{R}$ be a function modular satisfying the Δ_2-type condition. Assume that ρ has the ρ-a.e. strong Opial property. Let C be a strongly ρ-bounded, ρ-a.e. compact subset of L_ρ. Let $T : C \to C$ be an asymptotically regular mapping such that $s(T) < 2$. Then, T has a fixed point.*

Proof. Using the definition of $s(T)$, there exists a subsequence $\{T^{n_k}\}$ such that $s(T) = \lim\limits_{k\to\infty} |T^{n_k}| = \liminf\limits_{n\to\infty} |T^n|$. Define the function $r : C \to [0, +\infty)$ by

$$r(f) = \inf\left\{r > 0 : \exists g \in C \text{ such that } \liminf_{k\to\infty} \rho(f - T^{n_k}(g)) \le r\right\}.$$

Note that $r(f) \le \delta_\rho(C)$, for any $f \in C$. Set $b = \dfrac{1 + s(T)}{2}$. Then our assumption on $s(T)$ implies $s(T) < b < 2$. Since $\omega_\rho(1) = 1$, there exists $\varepsilon \in (0,1)$ such that $\omega_\rho\left(\dfrac{1}{\varepsilon}\right) < \dfrac{1}{b-1}$. Then choose $\gamma \in (0,1)$ such that $\omega_\rho\left(\dfrac{1}{\varepsilon}\right) < \dfrac{\gamma}{b-1}$. Set

$$\delta = \min\left\{\frac{1}{2}, \frac{1}{2}\left(\gamma + (1-b)\omega_\rho\left(\frac{1}{\varepsilon}\right)\right)\right\}.$$

Note that $0 < \delta < 1$. Now, we choose a number $\mu \in (0,1)$ such that

$$\mu < \min\left\{\frac{\delta}{\omega_\rho\left(\frac{1}{1-\varepsilon}\right)}, \frac{\gamma - \delta + \omega_\rho\left(\frac{1}{\varepsilon}\right)(1-b)}{\omega_\rho\left(\frac{1}{\varepsilon}\right)b}\right\}.$$

Finally denote

$$\alpha = \max\left\{1 + \mu - \frac{\delta}{\omega_\rho\left(\frac{1}{1-\varepsilon}\right)}, b(1+\mu) - \frac{\gamma - \delta}{\omega_\rho\left(\frac{1}{\varepsilon}\right)}\right\}.$$

Then $0 < \alpha < 1$. Let $f \in C$. Assume that $r(f) = 0$. For any $m \ge 1$, there exists $g_m \in C$ such that

$$\liminf_{k\to\infty} \rho(f - T^{n_k}(g_m)) \le \frac{1}{m}.$$

Then we have

$$\rho(f - T(f)) \le \omega_\rho(3)\Big[\rho(f - T^{n_k}(g_m)) + \rho(T^{n_k}(g_m) - T^{n_k+1}(g_m))$$
$$+ \rho(T^{n_k+1}(g_m) - T(f))\Big]$$
$$\le \omega_\rho(3)\Big[(1 + |T|)\,\rho(f - T^{n_k}(g_m)) + \rho(T^{n_k}(g_m) - T^{n_k+1}(g_m))\Big].$$

If we take the lim inf when $n_k \to +\infty$, we get

$$\rho(f - T(f)) \le \omega_\rho(3)\left[(1 + |T|)\frac{1}{m} + 0\right],$$

where we used the asymptotic regularity of T. Now if we let $m \to +\infty$, we get $\rho(f - T(f)) = 0$, i.e., $T(f) = f$. Next assume that $r(f) > 0$. Since $\gamma < 1$, we can find a positive integer $k_0 \ge 1$ such that $|T^{n_{k_0}}| < b$ and $\rho(f - T^{n_{k_0}}(f)) > \gamma\, r(f)$. Otherwise, we will have

$$\liminf_{n \to \infty} \rho(f - T^n(f)) \le \liminf_{k \to \infty} \rho(f - T^{n_k}(f)) \le \gamma\, r(f).$$

Hence $r(f) \le \gamma\, r(f) < r(f)$, which is a contradiction since $r(f) > 0$. Therefore there exists $k_0 \ge 1$ such that

$$|T^{n_{k_0}}| < b \quad \text{and} \quad \rho(f - T^{n_{k_0}}(f)) > \gamma r(f).$$

Since $\mu > 0$, we can also find $g \in C$ such that

$$\liminf_{k \to \infty} \rho(f - T^{n_k}(g)) \le r(f)(1 + \mu).$$

Consider a subsequence $\{n_{k'}\}$ of $\{n_k\}$ such that $\{T^{n_{k'}}(g)\}$ is ρ-a.e convergent in C, say to h, and

$$\lim_{k' \to \infty} \rho(f - T^{n_{k'}}(g)) = \liminf_{k \to \infty} \rho(f - T^{n_k}(g)).$$

Using Lemma 5.8 and the asymptotic regularity of T, we get

$$\limsup_{k' \to \infty} \rho(T^{n_{k_0}}(f) - T^{n_{k'}}(g)) \le |T^{n_{k_0}}| \limsup_{k' \to \infty} \rho(f - T^{n_{k'} - n_{k_0}}(g))$$

$$= |T^{n_{k_0}}| \limsup_{k' \to \infty} \rho(f - T^{n_{k'}}(g) + T^{n_{k'}}(g) - T^{n_{k'} - n_{k_0}}(g))$$

$$= |T^{n_{k_0}}| \limsup_{k' \to \infty} \rho(f - T^{n_{k'}}(g)).$$

We claim that

$$\liminf_{k' \to \infty} \rho(T^{n_{k'}}(g) - h) \le \alpha\, r(f). \tag{5.14}$$

Indeed if

$$\rho(f - h) \ge \frac{\delta}{\omega_\rho\left(\frac{1}{1-\varepsilon}\right)} r(f),$$

then using the ρ-a.e. strong Opial property, we get

$$\liminf_{k'\to\infty} \rho(T^{n_{k'}}(g) - h) = \liminf_{k'\to\infty} \rho(T^{n_{k'}}(g) - f) - \rho(f - h)$$

$$\leq (1+\mu)r(f) - \frac{\delta}{\omega_\rho\left(\frac{1}{1-\varepsilon}\right)} r(f)$$

$$= \left[(1+\mu) - \frac{\delta}{\omega_\rho\left(\frac{1}{1-\varepsilon}\right)}\right] r(f)$$

$$\leq \alpha\, r(f).$$

Otherwise, assume that

$$\rho(f - h) < \frac{\delta}{\omega_\rho\left(\frac{1}{1-\varepsilon}\right)} r(f).$$

Using again the ρ-a.e. strong Opial property, we get

$$\liminf_{k'\to\infty} \rho(T^{n_{k'}}(g) - h) = \liminf_{k'\to\infty} \rho(T^{n_{k'}}(g) - T^{n_{k_0}}(f)) - \rho(T^{n_{k_0}}(f) - h).$$

The properties of the modular ρ and the growth function ω_ρ give

$$\rho(T^{n_{k_0}}(f) - f) = \rho\left(\frac{1}{\varepsilon}\varepsilon(T^{n_{k_0}}(f) - h) + \frac{1}{1-\varepsilon}(1-\varepsilon)(f - h)\right)$$

$$\leq \omega_\rho\left(\frac{1}{\varepsilon}\right)\rho(T^{n_{k_0}}(f) - h) + \omega_\rho\left(\frac{1}{1-\varepsilon}\right)\rho(h - f).$$

Hence

$$\rho(T^{n_{k_0}}(f) - h) \geq \frac{1}{\omega_\rho\left(\frac{1}{\varepsilon}\right)}\rho(T^{n_{k_0}}(f) - f) - \frac{\omega_\rho\left(\frac{1}{1-\varepsilon}\right)}{\omega_\rho\left(\frac{1}{\varepsilon}\right)}\rho(h - f).$$

Since

$$\liminf_{k'\to\infty} \rho(T^{n_{k'}}(g) - h) = \liminf_{k'\to\infty} \rho(T^{n_{k'}}(g) - T^{n_{k_0}}(f)) - \rho(T^{n_{k_0}}(f) - h)$$

$$\leq |T^{n_{k_0}}| \liminf_{k'\to\infty} \rho(T^{n_{k'} - n_{k_0}}(g) - f) - \rho(T^{n_{k_0}}(f) - h)$$

$$\leq b(1+\mu)r(f) - \rho(T^{n_{k_0}}(f) - h),$$

we get

$$\liminf_{k'\to\infty} \rho(T^{n_{k'}}(g)-h) \leq b(1+\mu)r(f)-\frac{1}{\omega_\rho\left(\frac{1}{\varepsilon}\right)}\rho(T^{n_{k_0}}f-f)+\frac{\omega_\rho\left(\frac{1}{1-\varepsilon}\right)}{\omega_\rho\left(\frac{1}{\varepsilon}\right)}\rho(h-f)$$

$$\leq b(1+\mu)r(f)-\frac{\gamma}{\omega_\rho\left(\frac{1}{\varepsilon}\right)}r(f)+\frac{\delta}{\omega_\rho\left(\frac{1}{\varepsilon}\right)}r(f)$$

$$\leq \alpha\, r(f),$$

which proves the inequality (5.14). The latter implies, by the definition of the function r, that $r(h) \leq \alpha r(f)$. Moreover

$$\rho(h-f) = \rho\left(2\left(\frac{h-f}{2}\right)\right) \leq \omega_\rho(2)\rho\left(\frac{h-f}{2}\right)$$

$$\leq \omega_\rho(2)\left(\liminf_{k'\to\infty}\rho(h-T^{n_{k'}}(g))+\liminf_{k'\to\infty}\rho(T^{n_{k'}}(g)-f)\right)$$

$$\leq \omega_\rho(2)\left(\alpha r(f)+(1+\mu)r(f)\right) = Ar(f),$$

where $A = \omega_\rho(2)(\alpha+1+\mu)$. Set $h_0 = f$ and $h_1 = h$. Assume that $h_0, h_1, \cdots, h_{n-1}$ are defined, we consider h_n as the corresponding element to h_{n-1} in the above construction. Thus,

$$r(h_n) \leq \alpha r(h_{n-1}) \leq \cdots \leq \alpha^n r(h_0),$$

and $\rho(h_{n+1}-h_n) \leq Ar(h_n)$. Hence, we have constructed $\{h_n\}$ such that

$$\rho(h_{n+1}-h_n) \leq K\alpha^n \leq \beta^n,$$

where $K = Ar(f)$, which implies

$$\frac{1}{K\,\alpha^n} \leq \frac{1}{\rho(h_{n+1}-h_n)}.$$

Using the properties of ω_ρ, we get

$$\omega^{-1}\left(\frac{1}{K}\right)\left(\omega^{-1}\left(\frac{1}{\alpha}\right)\right)^n \leq \omega^{-1}\left(\frac{1}{K\,\alpha^n}\right) \leq \omega^{-1}\left(\frac{1}{\rho(h_{n+1}-h_n)}\right).$$

Since

$$\|h_{n+1}-h_n\|_\rho \leq \frac{1}{\omega^{-1}\left(\frac{1}{\rho(h_{n+1}-h_n)}\right)} \leq \frac{1}{\omega^{-1}\left(\frac{1}{K}\right)\left(\omega^{-1}\left(\frac{1}{\alpha}\right)\right)^n},$$

and $\alpha < 1$, we conclude that $\{h_n\}$ is a Cauchy sequence in $(L_\rho, \|.\|_\rho)$. Therefore there exists $h \in L_\rho$ such that $\lim_{n\to\infty}\|h_n-h\|_\rho = 0$, because $(L_\rho, \|.\|_\rho)$ is complete. Hence, $\{h_n\}$ is ρ-convergent to h which implies the existence of a subsequence $\{h_{\varphi(n)}\}_n$ of $\{h_{\varphi(n)}\}_n$ which converges to h ρ-a.e. Since C is ρ-a.e. closed, we get $h \in C$. We finish the proof of Theorem 5.13 by showing that $r(h) = 0$. Indeed, let

$\varepsilon > 0$. We choose n large enough such that

$$r(h_n) < \frac{\varepsilon}{2}, \text{ and } \rho(h_n - h) < \frac{\varepsilon}{2}.$$

There exists $g \in C$ such that $\liminf\limits_{k\to\infty} \rho(T^{n_k}(g) - h_n) \leq \frac{\varepsilon}{2}$. Hence

$$\liminf_{k\to\infty} \rho\left(\frac{T^{n_k}(g) - h}{2}\right) \leq \liminf_{k\to\infty} \rho(T^{n_k}(g) - h_n) + \rho(h_n - h) \leq \varepsilon,$$

which implies

$$\liminf_{k\to\infty} \rho(T^{n_k}(g) - h) \leq \omega_\rho(2)\varepsilon.$$

Therefore we have $r(h) \leq \omega_\rho(2)\varepsilon$. Since ε was taken arbitrarily positive, we get $r(h) = 0$. This implies, as we proved earlier, that $T(h) = h$, i.e., T has a fixed point.
□

5.3.4 Common Fixed Point Theorems in Modular Function Spaces

The purpose of this section is to give an outline of a common fixed point theory for nonexpansive mappings defined on some subsets of modular function spaces. One frequently used approach to prove the existence of a common fixed point is the investigation of the fixed point set of one mapping. In this section, we lay down the work done in [7] extending the ideas developed in [108] in metric spaces to modular function spaces.

Definition 5.9. Let $\rho \in \Re$ and $C \subset L_\rho$ be nonempty. A nonempty subset D of C is said to be a *one-local retract of* C if for every family $\{B_i : i \in I\}$ of ρ-balls centered in D such that, if

$$C \cap \left(\bigcap_{i\in I} B_i\right) \neq \emptyset$$

holds then the following is true as well

$$D \cap \left(\bigcap_{i\in I} B_i\right) \neq \emptyset.$$

Recall that $D \subset C$ is a ρ-*nonexpansive retract of* C if there exists a ρ-nonexpansive map $R : C \to D$ such that $R(f) = f$, for every $f \in D$. It is immediate that each ρ-nonexpansive retract of L_ρ is a one-local retract (but not conversely). The following result will shed some light on the interest generated around this concept.

Theorem 5.14. *Let $\rho \in \Re$ and $C \subset L_\rho$ be nonempty, ρ-closed, and ρ-bounded. Assume that $\mathscr{A}(C)$ is compact and normal. Then for any ρ-nonexpansive mapping $T : C \to C$, the fixed point set $F(T)$ is a nonempty one-local retract of C.*

Proof. Theorem 5.8 shows that $F(T)$ is nonempty. Let us complete the proof by showing it is a one-local retract of C. Let $\{B_\rho(f_i, r_i)\}_{i \in I}$ be any family of ρ-closed balls such that $f_i \in F(T)$, for any $i \in I$, and

$$C_0 = C \cap \left(\bigcap_{i \in I} B_\rho(f_i, r_i) \right) \neq \emptyset.$$

Let us prove that $F(T) \cap (\cap_{i \in I} B_\rho(f_i, r_i)) \neq \emptyset$. Since $\{f_i\}_{i \in I} \subset F(T)$, and T is ρ-nonexpansive, then $T(C_0) \subset C_0$. Clearly, $C_0 \in \mathscr{A}(C)$ and is nonempty. Then we have $\mathscr{A}(C_0) \subset \mathscr{A}(C)$. Therefore, $\mathscr{A}(C_0)$ is compact and ρ-normal. Theorem 5.8 implies that T has a fixed point in C_0 which will give us

$$F(T) \cap \left(\bigcap_{i \in I} B_\rho(f_i, r_i) \right) \neq \emptyset.$$

$\square$

Next we discuss some properties of one-local retract subsets.

Theorem 5.15. *Let $\rho \in \mathfrak{R}$ and $C \subset L_\rho$ be nonempty. Let D be a nonempty subset of C. The following are equivalent.*

(i) D is a one-local retract of C.
(ii) D is a ρ-nonexpansive retract of $D \cup \{f\}$, for every $f \in C$.

Proof. Let us prove $(i) \Rightarrow (ii)$. Let $f \in C$. We may assume that $f \notin D$. In order to construct a ρ-nonexpansive retract $R : D \cup \{f\} \to D$, we only need to find $R(f) \in D$ such that

$$\rho(R(f) - g) \leq \rho(f - g), \quad \text{for every } g \in D.$$

Since $f \in \bigcap_{g \in D} B_\rho(g, \rho(f - g))$ and $f \in C$, then

$$C \cap \left(\bigcap_{g \in D} B_\rho(g, \rho(f - g)) \right) \neq \emptyset.$$

Since D is a one-local retract of C, we get

$$D_0 = D \cap \left(\bigcap_{g \in D} B_\rho(g, \rho(f - g)) \right) \neq \emptyset.$$

Any point in D_0 will work as $R(f)$. Next, we prove that $(ii) \Rightarrow (i)$. In order to prove that D is a one-local retract of C, let $\{B_\rho(f_i, r_i)\}_{i \in I}$ be any family of ρ-closed balls such that $f_i \in D$, for any $i \in I$, and

$$C_0 = C \cap \left(\bigcap_{i \in I} B_\rho(f_i, r_i) \right) \neq \emptyset.$$

Let us prove that $D \cap (\cap_{i \in I} B_\rho(f_i, r_i)) \neq \emptyset$. Let $f \in C_0$. If $f \in D$, we have nothing to prove. Assume otherwise that $f \notin D$. Property (ii) implies the existence of a ρ-nonexpansive retract $R : D \cup \{f\} \to C$. It is easy to check that

$$R(f) \in D \cap (\cap_{i \in I} B_\rho(f_i, r_i)) \neq \emptyset,$$

which completes the proof of Theorem 5.15.

$\square$

The following technical lemma will be useful for the next results.

Lemma 5.9. *Let* $\rho \in \mathfrak{R}$ *and* $C \subset L_\rho$ *be nonempty, and* ρ-*bounded. Let* D *be a nonempty one-local retract of* C*. Set*

$$co_C(D) = C \cap \bigcap \{A : A \in \mathscr{A}(C) \text{ and } D \subset A\}.$$

Then

  (i) $r_\rho(f, D) = r_\rho(f, co_C(D))$, *for any* $f \in C$
  (ii) $R_\rho(co_C(D)) = R_\rho(D)$
  (iii) $\delta_\rho(co_C(D)) = \delta_\rho(D)$

Proof. Let us first prove (i). Fix $f \in C$. Since $D \subset co_C(D)$, we get

$$r_\rho(f, D) \leq r_\rho(f, co_C(D)).$$

Set $r = r_\rho(f, D)$. We have $D \subset B_\rho(f, r) \in \mathscr{A}(C)$. From the definition of $co_C(D)$ it follows that $co_C(D) \subset B_\rho(f, r)$. Hence $r_\rho(f, co_C(D)) \leq r = r_\rho(f, D)$, which implies $r_\rho(f, D) = r_\rho(f, co_C(D))$.

Next, let us prove (ii). Let $f \in D$. We have $f \in co_C(D)$. Using (i), we get

$$r_\rho(f, D) = r_\rho(f, co_C(D)) \geq R_\rho(co_C(D)).$$

Hence, $R_\rho(D) \geq R_\rho(co_C(D))$. Let $f \in co_C(D)$ and set $r = r_\rho(f, co_C(D))$. Since we have $D \subset co_C(D) \subset B_\rho(f, r)$ then

$$f \in \bigcap_{g \in D} B_\rho(g, r),$$

which implies that

$$C \cap (\bigcap_{g \in D} B_\rho(g, r)) \neq \emptyset.$$

Since D is a one-local retract of C, we get

$$D_0 = D \cap (\cap_{g \in D} B_\rho(g, r)) \neq \emptyset.$$

Let $g \in D_0$. Observe that $r_\rho(g, D) \leq r$. Hence, $R_\rho(D) \leq r$. Since f was arbitrary chosen in $co_C(D)$, we get $R_\rho(D) \leq R_\rho(co_C(D))$, which implies $R_\rho(D) = R_\rho(co_C(D))$.

Finally, let us prove (*iii*). Since $D \subset co_C(D)$, we get $\delta_\rho(D) \leq \delta_\rho(co_C(D))$. Next set $d = \delta_\rho(D)$. Then, for any $f \in D$, we have $D \subset B_\rho(f, d)$. Hence $co_C(D) \subset B_\rho(f, d)$. This implies that

$$ f \in \bigcap_{g \in co_C(D))} B_\rho(g, d). $$

Since f was arbitrarily chosen in D, we get

$$ D \subset \bigcap_{g \in co_C(D))} B_\rho(g, d). $$

The definition of $co_C(D)$ implies

$$ co_C(D) \subset \bigcap_{g \in co_C(D))} B_\rho(g, d). $$

Therefore, for any $f, g \in co_C(D)$, we have $\rho(f - g) \leq d$. Hence $\delta_\rho(co_C(D)) \leq d = \delta_\rho(D)$, which implies $\delta_\rho(D) = \delta_\rho(co_C(D))$. $\square$

As an application of this lemma we get the following result.

Theorem 5.16. *Let $\rho \in \mathfrak{R}$ and $C \subset L_\rho$ be nonempty, ρ-closed, and ρ-bounded. Assume that $\mathscr{A}(C)$ is compact and ρ-normal. If D is a nonempty one-local retract of C, then $\mathscr{A}(D)$ is compact and ρ-normal.*

Proof. Using the definition of one-local retract, it is easy to see that $\mathscr{A}(D)$ is compact. Let us show that $\mathscr{A}(D)$ is ρ-normal. Let $A_0 \in \mathscr{A}(D)$ nonempty and not reduced to one point. Set

$$ co_C(A_0) = C \cap \left(\bigcap \{A : A \in \mathscr{A}(C) \text{ and } A_0 \subset A\} \right). $$

From Lemma 5.9 we have

$$ R_\rho(co_C(A_0)) = R_\rho(A_0), \text{ and } \delta_\rho(co_C(A_0)) = \delta_\rho(A_0). $$

Since $co_C(A_0) \in \mathscr{A}(C)$, then we must have $R_\rho(co_C(A_0)) < \delta_\rho(co_C(A_0))$ because $\mathscr{A}(C)$ is ρ-normal. Therefore, $R_\rho(A_0) < \delta_\rho(A_0)$, which completes the proof of our claim. $\square$

The next result is quite amazing. Its metric counterpart has found many applications. As we will see toward the end this section, we already have few quite important applications of Theorem 5.17 (Theorems 5.18, 5.19, 5.20, and Corollary 5.2) and the future research will for sure extend this list. Most of the ideas in the proof below actually go back to the pioneering work of Baillon [13].

Theorem 5.17. *Let* $\rho \in \mathfrak{R}$ *and* $C \subset L_\rho$ *be nonempty,* ρ-*closed, and* ρ-*bounded. Assume that* $\mathscr{A}(C)$ *is compact and* ρ-*normal. Let* $(C_\beta)_{\beta \in \Gamma}$ *be a decreasing family of one-local retracts of* C, *where* $(\Gamma, \prec)$ *is totally ordered. Then*

$$\bigcap_{\beta \in \Gamma} C_\beta \neq \emptyset$$

and is a one-local retract of C.

Proof. First, let us prove that $\bigcap_{\beta \in \Gamma} C_\beta$ is not empty. Consider the family

$$\mathscr{F} = \left\{ \prod_{\beta \in \Gamma} A_\beta \; : A_\beta \in \mathscr{A}(C_\beta) \text{ and } (A_\beta) \text{ is decreasing} \right\}.$$

$\mathscr{F}$ is not empty since $\prod_{\beta \in \Gamma} C_\beta \in \mathscr{F}$. $\mathscr{F}$ will be ordered by inclusion, i.e., $\prod_{\beta \in \Gamma} A_\beta \subset \prod_{\beta \in \Gamma} B_\beta$ if and only if $A_\beta \subset B_\beta$ for any $\beta \in \Gamma$. From Theorem 5.16, we know that $\mathscr{A}(C_\beta)$ is compact, for every $\beta \in \Gamma$. Therefore, $\mathscr{F}$ satisfies the hypothesis of Zorn's lemma. Hence for every $D \in \mathscr{F}$, there exists a minimal element $A \in \mathscr{F}$ such that $A \subset D$. We claim that if $A = \prod_{\beta \in \Gamma} A_\beta$ is minimal, then there exists $\beta_0 \in \Gamma$ such that $\delta(A_\beta) = 0$ for every $\beta \succ \beta_0$. Assume not, i.e., $\delta(A_\beta) > 0$ for every $\beta \in \Gamma$. Fix $\beta \in \Gamma$. For every $K \subset C$, set

$$co_\beta(K) = \bigcap_{f \in C_\beta} B_\rho(f, r_\rho(f, K)).$$

Consider $A' = \prod_{\alpha \in \Gamma} A'_\alpha$ where

$$\begin{cases} A'_\alpha = co_\beta(A_\beta) \cap A_\alpha \text{ if } \alpha \leq \beta \\ \\ A'_\alpha = A_\alpha \qquad\qquad \text{ if } \alpha \geq \beta. \end{cases}$$

The family $(A'_{\alpha \geq \beta})$ is decreasing since $A \in \mathscr{F}$. Let $\alpha \leq \gamma \leq \beta$. Then $A'_\gamma \subset A'_\alpha$ since $A_\gamma \subset A_\alpha$ and $A_\beta = co_\beta(A_\beta) \cap A_\beta$. Hence the family (A'_α) is decreasing. On the other hand if $\alpha \prec \beta$, then $co_\beta(A_\beta) \cap A_\alpha \in \mathscr{A}(C_\alpha)$ since $C_\beta \subset C_\alpha$. Hence $A'_\alpha \in \mathscr{A}(C_\alpha)$. Therefore, we have $A' \in \mathscr{F}$. Since A is minimal, then $A = A'$. Hence

$$A_\alpha = co_\beta(A_\beta) \bigcap A_\alpha, \text{ for every } \alpha \prec \beta .$$

Let $f \in C_\beta$ and $\alpha \prec \beta$. Since $A_\beta \subset A_\alpha$, then $r_\rho(f, A_\beta) \leq r_\rho(f, A_\alpha)$. Because

$$co_\beta(A_\beta) = \bigcap_{g \in C_\beta} B_\rho(g, r_\rho(g, A_\beta))$$

we have $co_\beta(A_\beta) \subset B_\rho(g, r_\rho(g, A_\beta))$ which implies $r_\rho(g, A_\beta) \leq r_\rho(g, A_\alpha)$. Since $A_\alpha \subset co_\beta(A_\beta)$, then

$$r_\rho(g, A_\beta) \leq r_\rho(g, A_\alpha) \leq r_\rho(g, co_\beta(A_\beta)) \leq r_\rho(g, A_\beta).$$

Therefore, we have $r_\rho(g, A_\alpha) = r_\rho(g, A_\beta)$ for every $g \in C_\beta$. Using the definition of the ρ-Chebyshev radius R_ρ, we get

$$R_\rho(A_\alpha) \leq R_\rho(A_\beta).$$

Let $f \in A_\alpha$ and set $s = r_\rho(f, A_\alpha)$. Then $f \in co_\beta(A_\beta)$ since $A_\alpha \subset co_\beta(A_\beta)$. Hence,

$$f \in \left(\bigcap_{g \in A_\beta} B_\rho(g, s) \right) \bigcap co_\beta(A_\beta).$$

Since C_β is a one-local retract of C, then

$$S_\beta = C_\beta \cap \left(\bigcap_{g \in A_\beta} B_\rho(g, s) \right) \cap co_\beta(A_\beta) \neq \emptyset.$$

Since $A_\beta = C_\beta \cap co_\beta(A_\beta)$, then we have

$$S_\beta = A_\beta \cap \left(\bigcap_{g \in A_\beta} B_\rho(g, s) \right).$$

Let $h \in S_\beta$, then $h \in \bigcap_{g \in A_\beta} B_\rho(g, s)$. Hence, $r_\rho(h, A_\beta) \leq s$ which implies $R_\rho(A_\beta) \leq s = r_\rho(f, A_\alpha)$, for every $f \in A_\alpha$. Hence $R_\rho(A_\beta) \leq R_\rho(A_\alpha)$. Therefore we have

$$R_\rho(A_\beta) = R_\rho(A_\alpha), \quad \text{for every } \alpha, \beta \in \Gamma.$$

Since $\delta_\rho(A_\beta) > 0$ for every $\beta \in \Gamma$, set A_β'' to be the ρ-Chebyshev center of A_β, i.e., $A_\beta'' = \mathscr{C}_\rho(A_\beta)$, for every $\beta \in \Gamma$. Since $R_\rho(A_\beta) = R_\rho(A_\alpha)$, for every $\alpha, \beta \in \Gamma$, then the family (A_β'') is decreasing. Indeed, let $\alpha \prec \beta$ and $f \in A_\beta''$. Then we have $r_\rho(f, A_\beta) = R_\rho(A_\beta)$. Since we proved that $r_\rho(g, A_\beta) = r_\rho(g, A_\alpha)$, for every $g \in C_\beta$, then

$$r_\rho(f, A_\alpha) = r_\rho(f, A_\beta) = R_\rho(A_\beta) = R_\rho(A_\alpha),$$

which implies that $f \in A_\alpha''$. Therefore, we have $A'' = \prod_{\beta \in \Gamma} A_\beta'' \in \mathscr{F}$. Since $A'' \subset A$ and A is minimal, we get $A = A''$. Therefore, we have $\mathscr{C}_\rho(A_\beta) = A_\beta$ for every $\beta \in \Gamma$. This contradicts the fact that $\mathscr{A}(C_\beta)$ is normal for every $\beta \in \Gamma$. Hence, there exists $\beta_0 \in \Gamma$ such that

$$\delta(A_\beta) = 0, \quad \text{for every } \beta \succ \beta_0.$$

The proof of our claim is therefore complete. Then we have $A_\beta = \{f\}$, for every $\beta \succ \beta_0$. This clearly implies that $f \in \bigcap_{\beta \in \Gamma} C_\beta \neq \emptyset$. In order to complete the proof, we need to show that $S = \bigcap_{\beta \in \Gamma} C_\beta$ is a one-local retract of C. Let $(B_i)_{i \in I}$ be a family of ρ-balls centered in S such that $\bigcap_{i \in I} B_i \neq \emptyset$. Set $D_\beta = (\bigcap_{i \in I} B_i) \cap C_\beta$, for any $\beta \in \Gamma$. Since C_β is a one-local retract of C, and the family (B_i) is centered in C_β, then D_β is not empty and $D_\beta \in \mathscr{A}(C_\beta)$. Therefore, $D = \prod_{\beta \in \Gamma} D_\beta \in \mathscr{F}$. Let $A = \prod_{\beta \in \Gamma} A_\beta \subset D$ be a minimal element of $\mathscr{F}$. The above proof shows that

$$\emptyset \neq \bigcap_{\beta \in \Gamma} A_\beta \subset \bigcap_{\beta \in \Gamma} D_\beta.$$

The proof of Theorem 5.17 is therefore complete.
□

The next theorem is useful to prove the main common fixed point result of this section.

Theorem 5.18. *Let $\rho \in \mathfrak{R}$ and $C \subset L_\rho$ be nonempty, ρ-closed, and ρ-bounded. Assume that $\mathscr{A}(C)$ is compact and ρ-normal. Let $(C_\beta)_{\beta \in \Gamma}$ be a family of one-local retracts of C such that for any finite subset I of Γ, $\bigcap_{\beta \in I} C_\beta$ is not empty and is a one-local retract of C.*

Proof. Consider the family $\mathscr{F}$ of subsets $I \subset \Gamma$ such that for any finite subset $J \subset \Gamma$ (empty or not), we know that $\bigcap_{\alpha \in I \cup J} C_\alpha$ is a nonempty one-local retract of C. Note that $\mathscr{F}$ is not empty since any finite subset of Γ is in $\mathscr{F}$. Using Theorem 5.17, we can show that $\mathscr{F}$ satisfies the hypothesis of Zorn's lemma. Hence $\mathscr{F}$ has a maximal element $I \subset \Gamma$. Assume $I \neq \Gamma$. Let $\alpha \in \Gamma \setminus I$. Obviously we have $I \cup \{\alpha\} \in \mathscr{F}$. This is a clear contradiction with the maximality of I. Therefore we have $I = \Gamma \in \mathscr{F}$, i.e., $\bigcap_{\beta \in \Gamma} C_\beta$ is not empty and is a one-local retract of C.
□

In order to show that any family of commutative ρ-nonexpansive mappings have a common fixed point, we need to establish that it is the case for any finite family.

Theorem 5.19. *Let $\rho \in \mathfrak{R}$ and $C \subset L_\rho$ be nonempty, ρ-closed, and ρ-bounded. Assume that $\mathscr{A}(C)$ is compact and ρ-normal. Then for any finite family $\mathscr{F} = \{T_1, T_2, \cdots, T_n\}$ of commutative ρ-nonexpansive mappings defined on C has a common fixed point, i.e., $F(T_1) \cap \cdots \cap F(T_n) \neq \emptyset$. Moreover, the set of common fixed point set, denoted $F(\mathscr{F}) = F(T_1) \cap \cdots \cap F(T_n)$, is a one-local retract of C.*

Proof. Let us first prove Theorem 5.19 for two mappings T_1 and T_2. Using Theorem 5.14, we know that $F(T_1)$ is a nonempty one-local retract of C. Since T_1 and T_2 are commutative, then $T_2(F(T_1)) \subset F(T_1)$. Theorems 5.8 and 5.16 show that the restriction of T_2 to $F(T_1)$ has a fixed point. Again Theorem 5.14 implies that the common fixed point set $F(T_1) \cap F(T_2)$ is a nonempty one-local retract of C. Using the same argument shows that the conclusion of Theorem 5.19 is valid for any finite number of mappings.

$\square$

We are now ready to prove the main result of this section.

Theorem 5.20. *Let $\rho \in \Re$ and $C \subset L_\rho$ be nonempty, ρ-closed, and ρ-bounded. Assume that $\mathscr{A}(C)$ is compact and ρ-normal. Then for any family $\mathscr{F} = \{T_i : i \in I\}$ of commutative ρ-nonexpansive mappings defined on C has a common fixed point, i.e., $\bigcap_{i \in I} F(T_i) \neq \emptyset$. Moreover the set of common fixed point set, denoted $F(\mathscr{F}) = \bigcap_{i \in I} F(T_i)$, is a one-local retract of C.*

Proof. Let $\Gamma = \{\beta : \beta$ be a nonempty finite subset of $I\}$. Theorem 5.19 implies that for every $\beta \in \Gamma$, the set F_β of common fixed point set of the mappings T_i, $i \in \beta$, is a nonempty one-local retract of C. Clearly the family $(F_\beta)_{\beta \in \Gamma}$ is decreasing and satisfies the assumptions of Theorem 5.18. Therefore, we have that $\bigcap_{\beta \in \Gamma} F_\beta$ is nonempty and is a one-local retract of C. The proof of Theorem 5.20 is complete.

$\square$

Using Corollary 4.1 we get the following result.

Corollary 5.2. *Let $\rho \in \Re$ and $C \subset L_\rho$ be nonempty, convex, ρ-closed, and ρ-bounded. Assume ρ is (UUC2). Then for any family $\mathscr{F} = \{T_i : i \in I\}$ of commutative ρ-nonexpansive mappings defined on C has a common fixed point, i.e.,*

$$F(\mathscr{F}) = \bigcap_{i \in I} F(T_i) \neq \emptyset.$$

Moreover the set of common fixed points is a one-local retract of C.

5.3.5 Asymptotic Nonexpansive Mappings in Modular Function Spaces Satisfying Δ_2-type Condition

Throughout this section, we assume that ρ has strong Opial property, and satisfies the Δ_2-type condition and σ-finite, i.e., there exists an increasing sequence of sets $K_n \in \mathscr{P}$ such that $0 < \rho(1_{K_n}) < \infty$ and $\Omega = \bigcup_{n \geq 1} K_n$.

Let C be a convex, strongly ρ-bounded and ρ-a.e. compact nonempty subset of the modular function space L_ρ. Recall that the mapping $T : C \to C$ is said to be a asymptotic ρ-nonexpansive mapping if there exists a sequence of positive integers $\{k_n\}_n$ which converge to 1 such that for every $n \in \mathbb{N}$ and $f, g \in C$ we have

$$\rho(T^n(f) - T^n(g)) \leq k_n \, \rho(f - g).$$

Denote $\mathscr{F}$ the family of all subsets K of C satisfying the following property: K is nonempty, convex and ρ-a.e. closed such that

$$f \in K \text{ implies } \Omega_{a.e.}(f) \subset K \tag{5.15}$$

where $\Omega_{a.e.}(f)$ is the set of all ρ-a.e. limits of subsequences of $\{T^n(f)\}$. Note that $\Omega_{a.e.}(f)$ is not empty since C is ρ-a.e. compact, and $\mathscr{F}$ is nonempty since $C \in \mathscr{F}$. Ordering $\mathscr{F}$ by inclusion, there exists a nonempty minimal element H in $\mathscr{F}$ which satisfies (5.15) since C is ρ-a.e. compact for the topology generated by d (see Proposition 3.15).

The following lemma is the counterpart in modular function spaces of Lemma 2.1 in [202] for Banach spaces.

Lemma 5.10. *Under the above assumptions, for each $f \in H$ define the functional*

$$r_f(g) = \limsup_{n \to \infty} \rho(T^n(f) - g),$$

for any $g \in L_\rho$. Then the functional $r_f(\cdot)$ is constant on H and this constant is independent of f in H.

Proof. Let $t > 0$ and $f \in H$. Set

$$H_t(f) = \{ g \in H : \ r_f(g) \leq t \} .$$

It is easily seen that $H_t(f)$ is convex because ρ is convex. We claim that $H_t(f)$ is ρ-a.e. closed. Indeed, assume that $\{g_m\} \in H_t(f)$ ρ-a.e. converges to $g \in H$. Using Theorem 4.8, we get

$$r_f(g) = \limsup_{n \to \infty} \rho(T^n(f) - g) \leq \liminf_{m \to \infty} \left(\limsup_{n \to \infty} \rho(T^n(f) - g_m) \right) = \liminf_{m \to \infty} r_f(g_m),$$

which implies $r_f(g) \leq t$, i.e., $g \in H_t(f)$. Hence $H_t(f)$ is ρ-a.e. closed. Since H is ρ-a.e. compact, we deduce that $H_t(f)$ is ρ-a.e. compact. Next, we claim that $H_t(f)$ satisfies property (5.15). Indeed, let $g \in H_t(f)$ and $h \in \Omega_{a.e.}(g)$. We need to check that $h \in H_t(f)$. By definition of $\Omega_{a.e.}(g)$, there exists an increasing sequence of integers $\{n_i\}_i$ such that $T^{n_i}(g) \to h$ ρ-a.e. Theorem 4.8 implies that

$$
\begin{aligned}
r_f(h) &= \limsup_{n\to\infty} \rho(T^n(f) - h) \leq \liminf_{i\to\infty} \left(\limsup_{n\to\infty} \rho(T^n(f) - T^{n_i}(g)) \right) \\
&\leq \liminf_{i\to\infty} r_f\left(T^{n_i}(g)\right) \leq \limsup_{i\to\infty} r_f\left(T^{n_i}(g)\right) \leq \limsup_{m\to\infty} r_f\left(T^m(g)\right) \\
&\leq \limsup_{m\to\infty} \left(\limsup_{n\to\infty} \rho(T^n(f) - T^m(g)) \right) \\
&\leq \limsup_{m\to\infty} \left(k_m \limsup_{n\to\infty} \rho(T^{n-m}(f) - g) \right) \\
&\leq \limsup_{m\to\infty} \left(\limsup_{n\to\infty} \rho(T^n(f) - g) \right) \leq t.
\end{aligned}
$$

Hence $h \in H_t(f)$ as claimed. The minimality of H implies that $H_t(f)$ is empty or equal to H. From this, it is clear that $r_t(.)$ is constant on H. In order to complete the proof of this lemma, we need to prove that r_f is independent of f. Let $f, g \in H$. Since C is ρ-a.e. compact, there exists a subsequence $\{T^{n_i}(g)\}_i$ of $\{T^n(g)\}_n$ which ρ-a.e. converges to $h \in C$. Since H satisfies property (5.15), we have $h \in H$. Theorem 4.8 implies

$$
\rho(T^n(f) - h) \leq \liminf_{i\to\infty} \rho(T^n(f) - T^{n_i}(g)).
$$

Hence,

$$
\begin{aligned}
r_f = r_f(h) &= \limsup_{n\to\infty} \rho(T^n(f) - h) \\
&\leq \limsup_{n\to\infty} \left(\liminf_{i\to\infty} \rho(T^n(f) - T^{n_i}(g)) \right) \\
&\leq \limsup_{n\to\infty} \left(\limsup_{m\to\infty} \rho(T^n(f) - T^m(g)) \right) \\
&\leq \limsup_{m\to\infty} \rho(f - T^m(g)) = r_g(f) = r_g,
\end{aligned}
$$

which implies $r_g = r_f$.
$\square$

Keeping in mind that if ρ satisfies the Δ_2-type condition, then the ρ-convergence and the Luxemburg norm convergence coincide, and we have the following result as Lemma 5.11.

Lemma 5.11. *Let ρ be a convex function modular satisfying the Δ_2-type condition. Let S be a nonempty, norm compact subset of L_ρ with $\operatorname{diam}_\rho(S) > 0$. Then there exists $f \in \overline{conv}(S)$ such that*

$$
\sup\{\rho(g - f) : g \in S\} < \operatorname{diam}_\rho(S).
$$

Proof. Since S is compact and ρ is norm continuous, there exist $f_0, f_1 \in S$ such that $\rho(f_0 - f_1) = \operatorname{diam}_\rho(S)$. Let S_0 be a maximal subset of S such that $f_0, f_1 \in S_0$ and for any $f, g \in S_0$, $f \neq g$, we have $\rho(f - g) = \text{diam}_\rho(S)$. Since S is compact, S_0 must be finite. Write $S_0 = \{f_0, f_1, f_2, \ldots, f_n\}$ and define

$$h = \frac{f_0 + f_1 + \cdots + f_n}{n + 1}.$$

Since S is compact, there exists $g_0 \in S$ such that

$$\rho(g_0 - h) = \sup\{\rho(g - h) : g \in S\}.$$

On the other hand, using the convexity of ρ, we get

$$\rho(g_0 - h) = \rho\left(\sum_{k=0}^{k=n} \left(\frac{1}{n+1}\right) g_0 - \sum_{k=0}^{k=n} \left(\frac{1}{n+1}\right) f_k\right)$$

$$\leq \sum_{k=0}^{k=n} \left(\frac{1}{n+1}\right) \rho(g_0 - f_k) \leq \text{diam}_\rho(S).$$

If $\rho(g_0 - h) = \text{diam}_\rho(S)$, then we must have $\rho(g_0 - f_k) = \text{diam}_\rho(S)$, for $k = 0, 1, \ldots, n$. This contradicts the maximality of S_0. Hence,

$$\sup\{\rho(g - h) : g \in S\} = \rho(g_0 - h) < \text{diam}_\rho(S).$$

$\square$

Theorem 5.21. *Let ρ be a convex σ-finite function modular with strong Opial property satisfying the Δ_2-type condition and C be a strongly ρ-bounded, ρ-a.e. compact subset of L_ρ. Let $T : C \to C$ be an asymptotic ρ-nonexpansive mapping. Let H be a convex subset of C such that:*

(i) If $f \in H$ then $\Omega_{\rho-a.e}(f) \subset H$.
(ii) For each $f \in H$, any subsequence $\{T^{n_i}(f)\}_i$ of $\{T^n(f)\}_n$, has a ρ-convergent subsequence.

Then T has a fixed point.

Proof. Consider the family $\mathscr{F}$ of nonempty ρ-a.e. compact subset of H, which satisfies property (5.15). $\mathscr{F}$ is not empty since $H \in \mathscr{F}$. By the previous results, $\mathscr{F}$ has a minimal element. Let K be a minimal element of $\mathscr{F}$. Assume that K has more than one point, i.e., $\delta_\rho(K) > 0$. Let $f \in K$. Set

$$S = \Omega_{\|\cdot\|_\rho}(f) = \{g \in H : T^{n_i}(f) \ \|\cdot\|_\rho\text{-converges to } g \text{ for some } n_i \uparrow \infty\}.$$

It is easy to see that $S \subset K$. We claim that $S = T(S)$. Indeed, let $g \in S$. Then there exists a sequence $\{T^{n_i}(f)\}_i$ which $\|\cdot\|_\rho$-converges to g. Since T is continuous, we have $T^{n_i+1}(f) \to T(g)$ in norm. By the definition of S, we get $T(g) \in S$, i.e., $T(S) \subset S$. Let us show the other inclusion, i.e., $S \subset T(S)$. Let $g \in S$. Again by the definition of S, there exists a sequence $\{T^{n_i}(f)\}_i$ which $\|\cdot\|_\rho$-converges to g. The sequence $\{T^{n_i-1}(f)\}_i$ has a norm convergent subsequence, say $\{T^{n_{\phi(i)}-1}(f)\}_i$. Let

h be its $\|\cdot\|_\rho$-limit. Since T is continuous, we get

$$T(h) = T\big(\lim_{i\to\infty} T^{n_{\phi(i)}-1}(f)\big) = \lim_{i\to\infty} T^{n_{\phi(i)}}(f) = g.$$

Hence $g \in T(S)$, i.e., $S \subset T(S)$. Therefore, $T(S) = S$ is true. Next, notice that the assumption (ii) implies that S is norm compact. Lemma 5.11 implies the existence of $f_0 \in \overline{conv}(S) \subset K$ such that

$$\sup\{\rho(g - f_0) : g \in S\} < \delta_\rho(S). \qquad (A)$$

Let $r = \sup\{\rho(g - f_0) : g \in S\}$. Set $D = \{h \in K; \ \sup_{g\in S}\rho(g - h) \leq r\}$. Since $f_0 \in D$ and ρ is convex, D is a nonempty convex subset of K. We claim that $D = K$. Indeed, let us first show that D is ρ-a.e. compact. By the assumption (ii), it is enough to show that D is ρ-a.e. closed. Let $\{h_n\}_n$ be a sequence in D such that $h_n \to h \in L_\rho$ ρ-a.e. Fix $g \in S$. Since $g - h_n \to g - h$ ρ-a.e. Theorem 4.8 implies

$$\rho(g - h) \leq \liminf_{n\to\infty} \rho(g - h_n),$$

which yields

$$\rho(g - h) \leq \liminf_{n\to\infty} \Big(\sup\{\rho(f - h_n) : f \in S\} \Big) \leq r.$$

Hence, $\sup\{\rho(h - g) : g \in S\} \leq r$, i.e., $h \in D$. Next, we check that D satisfies property (5.15)). Indeed, let $f \in D$ and $g \in \Omega_{a.e}(f)$. Then there exists a sequence of natural numbers n_i such that $T^{n_i}(f) \to g$ ρ-a.e.. Using Theorem 4.8 we obtain

$$\rho(g - h) \leq \liminf_{n\to\infty} \rho(T^{n_i}(f) - h) \leq \limsup_{n\to\infty} \rho(T^n(f) - h),$$

for any $h \in S$. Since $T(S) = S$, there exists a sequence $\{u_n\}_n$ in S such that $h = T^n(u_n)$, for any $n \geq 1$. Hence,

$$\begin{aligned}
\rho(g - h) &\leq \limsup_{n\to\infty} \rho(T^n(f) - T^n(u_n)) \\
&\leq \limsup_{n\to\infty} k_n\, \rho(f - u_n) \\
&\leq \limsup_{n\to\infty} \rho(f - u_n) \\
&\leq \sup\{\rho(f - u); \ u \in S\} \leq r.
\end{aligned}$$

So $\sup\{\rho(g - h) : h \in S\} \leq r$ which implies $g \in D$. Thus D satisfies property (5.15) and by minimality of K, we obtain $D = K$. On the other hand,

$$\delta_\rho(D) \leq r < \delta_\rho(S) \leq \delta_\rho(K),$$

which is a contradiction. Therefore, K is reduced to one point. Property (5.15) forces this point to be a fixed point of T.

□

Now we are ready to state and prove the main result of this section.

Theorem 5.22. *[65] Let ρ be a convex, σ-finite function modular with strong Opial property satisfying the Δ_2-type condition. Let C be a convex, strongly ρ-bounded and ρ-a.e. compact, nonempty subset of L_ρ. Let $T : C \to C$ be asymptotic ρ-nonexpansive. Then T has a fixed point.*

Proof. Let $\mathscr{F}$ be the family of nonempty convex subsets of C, which satisfy the property (5.15). $\mathscr{F}$ is not empty since $C \in \mathscr{F}$. By Zorn's lemma, $\mathscr{F}$ has a minimal element H. Let us show that H satisfies the hypothesis of Theorem 5.21. It suffices to check that H satisfies property (ii). Let r be the function defined on H as in Lemma 5.10. If $r = 0$ we have $\lim_{n \to \infty} T^n(f) = g$, for any $f, g \in H$, which implies (ii). Otherwise, assume that $r > 0$. Let $f \in H$ be such that there exists a sequence $\{T^{n_i}(f)\}_i$, which has no norm convergent subsequence. Thus, there exists $\varepsilon > 0$ and a subsequence $\{T^{n(k)}(f)\}_k$ such that

$$\mathrm{sep}(\{T^{n(k)}(f)\}_k) = \inf\{\rho(T^{n(k)}(f) - T^{n(k')}(f)) : k \neq k'\} \geq \varepsilon.$$

Since H is ρ-a.e. compact, there exists $f_\infty \in H$ such that $T^{n(k)}(f) \overset{\rho-a.e}{\longrightarrow} f_\infty \in H$ as $k \to \infty$. Without loss of generality, we may assume the existence of

$$\lim_{k \to \infty} \rho(T^{n(k)}(f) - f_\infty) = l.$$

Since $\limsup_{n \to \infty} \rho(T^n(f) - f) = r$, we choose $\eta > 0$ such that $\eta < \dfrac{\varepsilon}{2}$, and an integer $n_0 \geq 1$, such that for all $n \geq n_0$ we have $\rho(T^n(f) - f) < r + \eta$. Fix $n \geq n_0$. There exists $k_0 \geq 1$ such that for all $k \geq k_0$, we have $n(k) \geq n + n_0$ and

$$\rho(T^n f - T^{n(k)} f) = \rho\left(T^n(f) - T^{n+(n(k)-n)}(f)\right)$$
$$\leq k_n \rho\left(f - T^{n(k)-n}(f)\right) < k_n (r + \eta).$$

Note that if $f_n \overset{\rho-a.e}{\longrightarrow} f$ and $\mathrm{sep}\{f_n\}_n \geq \varepsilon$, then by Theorem 4.8, we have

$$\varepsilon \leq \liminf_{m \to \infty} \liminf_{n \to \infty} \rho(f_n - f_m) \leq 2 \liminf_{n \to \infty} \rho(f_n - f).$$

Using again Theorem 4.8, we get

$$\liminf_{n \to \infty} \rho(f_n) = \liminf_{n \to \infty} \rho(f_n - f) + \rho(f) \geq \frac{\varepsilon}{2} + \rho(f).$$

In particular, since $\{T^{n(k)}f - T^n f\}_k$ is ρ-a.e. convergent to $f_\infty - T^n f$ as $k \to \infty$ and satisfies $\mathrm{sep}(\{T^{n(k)}f - T^n f\}_k) \geq \varepsilon$, we get

$$\rho(T^n f - f_\infty) \leq \liminf_{k \to \infty} \rho(T^{n(k)}f - T^n f) - \frac{\varepsilon}{2}.$$

Hence,

$$\rho(f_\infty - T^n f) \leq r + \eta - \frac{\varepsilon}{2},$$

which implies

$$r = \limsup_{n \to \infty} \rho(f_\infty - T^n f) \leq r + \eta - \frac{\varepsilon}{2} < r.$$

This contradiction completes the proof of Theorem 5.22.
□

Consider $L_\rho = L_p(\Omega, \mu)$, $p \geq 1$, for a σ-finite measure μ. Let C be a convex, bounded and closed nonempty subset of L_p, for $1 < p < \infty$. Let $T : C \to C$ be an asymptotic nonexpansive mapping. Using the uniform convexity of L_p when $p > 1$, one can see that T has a fixed point. However the result does not hold for $p = 1$ (even for nonexpansive mappings, see [8]). Since L_1 is a modular function space, Theorem 5.22 implies the existence of fixed point if $p = 1$ when C is ρ-a.e. compact. Thus we can state.

Corollary 5.3. *Let C be a convex, bounded, and closed nonempty subset of $L_1(\Omega, \mu)$, where μ is σ-finite measure. Assume that C is compact for the topology of the convergence local in measure. Let $T : C \to C$ be an asymptotic nonexpansive mapping. Then, T has a fixed point.*

Proof. Under the above hypothesis, ρ-a.e. compact sets and compact sets in the topology of convergence local in measure are identical. □

5.3.6 KKM and Ky Fan Theorems in Modular Function Spaces

Among the results equivalent to the Brouwer's fixed point theorem, the theorem of *Knaster–Kuratowski–Mazurkiewicz* occupies a special place [67]. Let $\rho \in \mathfrak{R}$ and let $C \subset L_\rho$ be nonempty. The set of all subsets of C is denoted as 2^C. The notation $conv(A)$ describes the convex hull of A, while $\overline{conv}_\rho(A)$ describes the smallest ρ-closed convex subset of L_ρ which contains A. Recall that a family $\{A_\alpha : A_\alpha \in 2^{L_\rho}, \alpha \in \Gamma\}$ is said to have the finite intersection property if the intersection of each finite subfamily is not empty.

Definition 5.10. Let $\rho \in \mathfrak{R}$ and let $C \subset L_\rho$ be nonempty. A multivalued mapping $G : C \to 2^{L_\rho}$ is called a *Knaster–Kuratowski–Mazurkiewicz mapping* (in short *KKM-mapping*) if

$$conv(\{f_1, \ldots, f_n\}) \subset \bigcup_{1 \leq i \leq n} G(f_i)$$

for any $f_1, \ldots, f_n \in C$.

Theorem 5.23. *Let $\rho \in \Re$. Let $C \subset L_\rho$ be nonempty and $G : C \to 2^{L_\rho}$ be a KKM-mapping such that for any $f \in C$, $G(f)$ is nonempty and finitely ρ-closed. Then, the family $\{G(f) : f \in C\}$ has the finite intersection property.*

Proof. Assume this is not the case, i.e., there exist $f_1, \ldots, f_n \in C$ such that

$$\bigcap_{1 \leq i \leq n} G(f_i) = \emptyset.$$

Define $L = \overline{conv}_\rho(\{f_i\})$ in L_ρ. Our assumptions imply that $L \cap G(f_i)$ is ρ-closed for every $i = 1, 2, \ldots, n$. Since the norm convergence implies the modular convergence, we deduce that $L \cap G(f_i)$ is closed for the Luxemburg norm $\|.\|_\rho$ for any $i \in \{1, \ldots, n\}$. Thus for every $f \in L$, there exists i_0 such that f does not belong to $L \cap G(f_{i_0})$ since $L \cap \left(\bigcap_{1 \leq i \leq n} G(f_i) \right) = \emptyset$. Hence,

$$d\left(f, L \cap G(f_{i_0})\right) = \inf\{\|f - g\|_\rho : g \in L \cap G(f_{i_0})\} > 0,$$

because $L \cap G(f_{i_0})$ is closed. We use the function

$$\alpha(f) = \sum_{1 \leq i \leq n} d\left(f, L \cap G(f_i)\right) > 0,$$

where $f \in K = conv\{f_1, \ldots, f_n\}$ to define the map $T : K \to K$ by

$$T(f) = \frac{1}{\alpha(f)} \sum_{1 \leq i \leq n} d\left(f, L \cap G(f_i)\right) f_i.$$

Clearly, T is a continuous map. Since K is a compact convex subset of the Banach space $(L_\rho, \|f\|_\rho)$, Brouwer's theorem implies the existence of a fixed point $f_0 \in K$ of T, i.e., $T(f_0) = f_0$. Set

$$I = \left\{ i : d\left(f_0, L \cap G(f_i)\right) \neq 0 \right\}.$$

Clearly,

$$f_0 = \frac{1}{\alpha(f_0)} \sum_{i \in I} d\left(f_0, L \cap G(f_i)\right) f_i.$$

Hence, $f_0 \notin \bigcup_{i \in I} G(f_i)$ and $f_0 \in conv(\{f_i : i \in I\})$ as this contradicts the assumption

$$conv\left(\{f_i : i \in I\}\right) \subset \bigcup_{i \in I} G(f_i).$$

$\square$

As an immediate consequence, we obtain the following result.

Theorem 5.24. *Let $\rho \in \mathfrak{R}$. Let $C \subset L_\rho$ be nonempty and $G : C \to 2^{L_\rho}$ be a KKM-mapping such that for any $f \in C$, $G(f)$ is nonempty and ρ-closed. Assume that there exists $f_0 \in C$ such that $G(f_0)$ is ρ-compact. Then, we have*

$$\bigcap_{f \in C} G(f) \neq \emptyset.$$

Notice that the ρ-compactness of $G(f_0)$ may be weakened, i.e., we can still reach the conclusion if one involves an auxiliary multivalued map and a suitable topology on L_ρ, as shown in the next theorem whose proof is immediate.

Theorem 5.25. *Let $\rho \in \mathfrak{R}$. Let $C \subset L_\rho$ be nonempty and $G : C \to 2^{L_\rho}$ be a KKM-mapping such that for any $f \in C$, $G(f)$ is nonempty and finitely ρ-closed. Assume there is a multivalued map $K : C \to 2^{L_\rho}$ such that $G(f) \subset K(f)$ for every $f \in C$ and*

$$\bigcap_{f \in C} K(f) = \bigcap_{f \in C} G(f).$$

If there is some topology τ on L_ρ such that each $K(f)$ is τ-compact, then

$$\bigcap_{f \in C} G(f) \neq \emptyset.$$

Before we state an analog to Ky Fan fixed point result [67], we need the following definition.

Definition 5.11. Let $\rho \in \mathfrak{R}$. Let $C \subset L_\rho$ be a nonempty ρ-closed subset. Let $T : C \to L_\rho$ be any map. T is called ρ-*continuous* if $\{T(f_n)\}$ ρ-converges to $T(f)$ whenever $\{f_n\}$ ρ-converges to f. Also T is called *strongly ρ-continuous* if T is ρ-continuous and

$$\liminf_{n \to \infty} \rho(g - T(f_n)) = \rho(g - T(f)),$$

for any sequence $\{f_n\} \subset C$, which ρ-converges to f and for any $g \in C$.

In general, it is unknown for what type of function modular ρ, ρ-continuity implies strong ρ-continuity. However, the Δ_2-property is enough to provide this implication. Lemma 5.12 is needed to prove the modular version of the Ky Fan fixed point theorem.

Lemma 5.12. *Let $\rho \in \mathfrak{R}$. Let $K \subset L_\rho$ be nonempty convex and ρ-compact. Let $T : K \to L_\rho$ be strongly ρ-continuous. Then, there exists $f_0 \in K$ such that*

$$\rho(f_0 - T(f_0)) = \inf_{f \in K} \rho\left(f - T(f_0)\right).$$

Proof. Consider the map $G : K \to 2^{L_\rho}$ defined by

$$G(g) = \left\{ f \in K : \rho(f - T(f)) \le \rho(g - T(f)) \right\} .$$

Since T is strongly ρ-continuous, then for any sequence $\{f_n\} \subset G(g)$ which ρ-converges to f, we have

$$\rho(f - T(f)) \le \liminf_{n \to \infty} \rho(f_n - T(f_n)) \le \liminf_{n \to \infty} \rho(g - T(f_n)) = \rho(g - T(f)),$$

on the basis of the Fatou property and the continuity of T. Clearly, this implies that $G(g)$ is ρ-closed for any $g \in K$. Next, we show that G is a KKM-mapping. Assume not. Then, there exists $\{g_1, \ldots, g_n\} \subset K$ and $f \in conv(\{g_i\})$ such that $f \notin \bigcup_{1 \le i \le n} G(g_i)$. This clearly implies

$$\rho(g_i - T(f)) < \rho(f - T(f)) , \quad \text{for } i = 1, \ldots, n .$$

Let $\varepsilon > 0$ be such that $\rho(g_i - T(f)) \le \rho(f - T(f)) - \varepsilon$, for $i = 1, 2, \ldots, n$. Since ρ is convex, for any $g \in conv(\{g_i\})$, we have

$$\rho(g - T(f)) \le \rho(f - T(f)) - \varepsilon.$$

As $f \in conv(\{g_i\})$, so we get

$$\rho(f - T(f)) \le \rho(f - T(f)) - \varepsilon.$$

Contradiction. Therefore, G is a KKM-mapping. By the ρ-compactness of K, we deduce that $G(g)$ is compact for any $g \in K$. Theorem 5.24 implies the existence of $f_0 \in \bigcap_{g \in K} G(g)$. Hence,

$$\rho(f_0 - T(f_0)) \le \rho(g - T(f_0))$$

for any $g \in K$. In particular, we have

$$\rho(f_0 - T(f_0)) = \inf_{g \in K} \rho\left(g - T(f_0)\right).$$

$\square$

We are now ready to state *Ky Fan Fixed Point Theorem* in modular function spaces.

Theorem 5.26. *Let $\rho \in \mathfrak{R}$. Let $K \subset L_\rho$ be nonempty convex and ρ-compact. Let $T : K \to L_\rho$ be strongly ρ-continuous. Assume that for any $f \in K$, with $f \ne T(f)$, there exists $\alpha \in (0, 1)$ such that*

$$K \cap B_\rho\left(f, \alpha\rho(f - T(f))\right) \cap B_\rho\left(T(f), (1 - \alpha)\rho(f - T(f))\right) \ne \emptyset. \qquad (5.16)$$

Then, T has a fixed point, i.e., $T(g) = g$ for some $g \in K$.

Proof. From the previous lemma, there exists $f_0 \in K$ such that

$$\rho(f_0 - T(f_0)) = \inf_{g \in K} \rho\left(g - T(f_0)\right). \tag{5.17}$$

We claim that f_0 is a fixed point of T. Assume it is not, i.e., $f_0 \neq T(f_0)$. Then, our assumption on K implies the existence of $\alpha \in (0,1)$ such that

$$K_0 = K \cap B_\rho\left(f_0, \alpha\rho(f_0 - T(f_0))\right) \cap B_\rho\left(T(f_0), (1-\alpha)\rho(f_0 - T(f_0))\right) \neq \emptyset.$$

Let $g \in K_0$. Then, $\rho(g - T(f_0)) \leq (1-\alpha)\rho(f_0 - T(f_0))$. This implies a contradiction with (5.17). $\square$

Note that the condition (5.16) is satisfied if $T(K) \subset K$, which implies the following important result.

Theorem 5.27. *Let $\rho \in \Re$. Let $K \subset L_\rho$ be nonempty convex and ρ-compact. Let $T : K \to K$ be strongly ρ-continuous. Then, T has a fixed point, i.e., $T(g) = g$ for some $g \in K$.*

5.4 Applications to Differential Equations

Let $\rho \in \Re$. To the end of this section we consider the following initial value problem for an unknown function $u : [0,A] \to C$, where $C \subset E_\rho$:

$$\begin{cases} u(0) = f \\ u'(t) + (I - T)u(t) = 0, \end{cases} \tag{5.18}$$

where $f \in C$ and $A > 0$ are fixed and $T : C \to C$ is ρ-nonexpansive.

Let us introduce the following convenient notations that are used throughout this section. For any $t > 0$ we define

$$K(t) = 1 - e^{-t} = \int_0^t e^{s-t} ds. \tag{5.19}$$

For any function $v : [0,A] \to L_\rho$, where $A > 0$, and any $t \in [0,A]$, we define

$$S(v)(t) = \int_0^t e^{s-t} v(s) ds. \tag{5.20}$$

Similarly, we denote

$$S_\tau(v)(t) = \sum_{i=0}^{n-1}(t_{i+1}-t_i)e^{t_i-t}v(t_i), \qquad (5.21)$$

for any $\tau = \{t_0,\ldots,t_n\}$, a subdivision of the interval $[0,A]$.

Let us start with the following technical lemma.

Lemma 5.13. *Let* $\rho \in \mathfrak{R}$ *be separable. Let* $x,y : [0,A] \to L_\rho$ *be two Bochner-integrable* $\|\cdot\|_\rho$*-bounded functions, where* $A > 0$*. Then for every* $t \in [0,A]$ *we have*

$$\rho\left(e^{-t}y(t) + \int_0^t e^{s-t}x(s)ds\right) \le e^{-t}\rho(y(t)) + K(t)\sup_{s\in[0,t]}\rho(x(s)). \qquad (5.22)$$

Proof. Without any loss of generality we can assume that $\sup_{s\in[0,t]}\rho(x(s)) < \infty$. Let $\tau = \{t_0,\ldots,t_n\}$ be a subdivision of the interval $[0,A]$. Define

$$Z_\tau(t) = e^{-t}y(t) + S_\tau(x)(t) = e^{-t}y(t) + \sum_{i=0}^{n-1}(t_{i+1}-t_i)e^{t_i-t}x(t_i).$$

It follows from the Lebesgue dominated convergence theorem for Bochner integrals, that for every $t \in [0,A]$, $Z_\tau(t)$ converges in $\|.\|_\rho$ to $Z(t)$ defined by

$$Z(t) = e^{-t}y(t) + S(x)(t) = e^{-t}y(t) + \int_0^t e^{s-t}x(s)ds,$$

whenever $|\tau| = \sup_{0\le i\le n}|t_{i+1}-t_i| \to 0$. Using the Fatou property of ρ and Proposition 3.6, we can easily see that

$$\rho(Z(t)) \le \liminf_{|\tau|\to 0}\rho(Z_\tau(t)). \qquad (5.23)$$

Since,

$$e^{-t} + \sum_{i=0}^{n-1}(t_{i+1}-t_i)e^{t_i-t} \le 1,$$

and ρ is convex, it follows that

$$\rho(Z_\tau(t)) \le e^{-t}\rho(y(t)) + \left(\sum_{i=0}^{n-1}(t_{i+1}-t_i)e^{t_i-t}\right)\sup_{s\in[0,t]}\rho(x(s)). \qquad (5.24)$$

By combining (5.24) with (5.23) and letting $|\tau| \to 0$ we get the desired inequality (5.22).
$\square$

We are now ready to prove our main result.

Theorem 5.28. *Let $\rho \in \mathfrak{R}$ be separable. Let $C \subset E_\rho$ be a nonempty, convex, ρ-bounded, ρ-closed set with the Vitali property. Let $T : C \to C$ be a ρ-nonexpansive mapping. Let us fix $f \in C$ and $A > 0$ and define the sequence of functions $u_n : [0,A] \to C$ by the following inductive formula:*

$$\begin{cases} u_0(t) = f \\ u_{n+1}(t) = e^{-t}f + \int_0^t e^{s-t}T(u_n(s))ds. \end{cases} \tag{5.25}$$

Then for every $t \in [0,A]$ there exists $u(t) \in C$ such that

$$\rho(u_n(t) - u(t)) \to 0 \tag{5.26}$$

and the function $u : [0,A] \to C$ defined by (5.26) is a solution of the initial value problem (5.18). Moreover,

$$\rho(f - u_n(t)) \leq K^{n+1}(A)\delta_\rho(C). \tag{5.27}$$

Proof. We show that $u_n(t) \in C$ for every $n \in \mathbb{N}$ and every $t \in [0,A]$. Indeed, for a given subdivision τ of $[0,A]$, define

$$u_{n+1}^\tau(t) = e^{-t}f + S_\tau\Big(T(u_n(t_i))\Big)(t).$$

Since C is a convex set, it is easy to prove by induction that $u_n^\tau(t) \in C$. Note that, by the properties of the Bochner integral, $\|u_n^\tau(t) - u_n(t)\|_\rho \to 0$. Since C is $\|\cdot\|_\rho$-closed it is also ρ-closed. Hence, $u_n(t) \in C$ as claimed. Now we prove that

$$\rho(u_{n+p}(t) - u_n(t)) \leq K^{n+1}(A)\delta_\rho(C) \tag{5.28}$$

for all $t \in [0,A]$ and every $n, p \in \mathbb{N}$. We prove (5.28) by induction with respect to n. For $n = 0$ we have

$$u_p(t) - u_0(t) = \int_0^t e^{s-t}T(u_{p-1}(s))ds.$$

Applying Lemma 5.13 with $y(t) = 0$ and $x(t) = T(u_{p-1}(s)) - f$ we have

$$\rho(u_p(t) - u_0(t)) \leq K(t) \sup_{s \in [0,t]} \rho(T(u_{p-1}(s)) - f) \leq K(A)\delta_\rho(C),$$

because both f, and $T(u_{p-1}(s))$ belong to C. Now, assume that (5.28) holds for an $n \in \mathbb{N}$. Then,

$$u_{n+1+p}(t) - u_{n+1}(t) = \int_0^t e^{s-t}\Big(T(u_{n+p}(s)) - T(u_n(s))\Big)ds.$$

Applying Lemma 5.13 again, with $y(t) = 0$ and $x(t) = T(u_{n+p}(s)) - T(u_n(s))$ we have

$$\rho(u_{n+1+p}(t) - u_{n+1}(t)) \leq K(A) \sup_{s \in [0,\,t]} \rho\Big(T(u_{n+p}(s)) - T(u_n(s))\Big)$$

$$\leq K(A) \sup_{s \in [0,\,t]} \rho(u_{n+p}(s) - u_n(s))$$

$$\leq K^{n+1}(A)\, \delta_\rho(C),$$

where we used the ρ-nonexpansiveness of T and the inductive assumption. It follows directly from (5.28) that for every $t \in [0,A]$, the sequence $\{u_n(t)\}$ is a ρ-Cauchy sequence, the latter implies that

$$\|u_n(t) - u(t)\|_\rho \to 0. \tag{5.29}$$

Remember that T is ρ-nonexpansive and hence it is ρ-continuous in C, which, via the Vitali property, implies that T is $\|\cdot\|_\rho$-continuous in C. Using this fact, it is easy to prove by induction that u_n is a $\|\cdot\|_\rho$-continuous mapping from $[0,A]$ to C. Consequently, functions $\lambda_n : [0,A] \to C$ defined by $\lambda_n(s) = T(u_n(s))$ are $\|\cdot\|_\rho$-continuous and bounded. Hence, it follows from the Lebesgue dominated convergence theorem for Bochner integrals that

$$\lim_{|\tau| \to 0} \|S_\tau(T(u_k(t))) - \int_0^t e^{s-t} T(u_k(s)) ds\|_\rho = 0, \tag{5.30}$$

for every $k \in \mathbb{N}$. Using the ρ-nonexpansiveness of T and the fact that $\rho(u_n(t) - u(t)) \to 0$, we have the following:

$$\rho\Big(S_\tau(T(u(t))) - S_\tau(T(u_n(t)))\Big) \leq \sum_{i=0}^{n-1} (t_{i+1} - t_i) e^{t_i - t} \rho\Big(T(u(t)) - T(u_n(t))\Big)$$

$$\leq A\, e^{A-t} \rho(u(t) - u_n(t)) \to 0,$$

which, via the Vitali property, implies that

$$\lim_{n \to \infty} \|S_\tau(T(u(t))) - S_\tau(T(u_n(t)))\|_\rho = 0. \tag{5.31}$$

For every $t \in [0,A]$ we have now

$$\|S_\tau(T(u(t))) - u(t) + e^{-t} f\|_\rho \leq \|S_\tau(T(u(t))) - S_\tau(T(u_n(t)))\|_\rho$$

$$+ \|S_\tau(T(u_n(t))) - \int_0^t e^{s-t} T(u_n(s)) ds\|_\rho$$

$$+ \|\int_0^t e^{s-t} T(u_n(s)) ds - u(t) + e^{-t} f\|_\rho.$$

Observing that

$$\int_0^t e^{s-t}T(u_n(s))ds = u_{n+1} - e^{-t}f$$

and using (5.31), (5.30), and (5.29) we obtain that

$$\lim_{|\tau|\to 0} \|S_\tau(T(u(t)) - u(t) + e^{-t}f\|_\rho = 0.$$

Hence, the function $s \mapsto e^{s-t}T(u(s))$ is Bochner-integrable on $[0,A]$ and

$$\int_0^t e^{s-t}T(u(s))ds = u(t) - e^{-t}f. \tag{5.32}$$

Using (5.32) it can be easily proved, via the standard calculus methods, that u is differentiable and is a solution of the initial value problem (5.18). The proof of Theorem 5.28 is now complete.
□

Remark 5.9. It is clear that the solution of the the initial value problem (5.18) can be extended to a solution $u(t)$ defined on $[0,+\infty)$ such that its restriction to the interval $[0,A]$ is the ρ-limit of the sequence $\{u_n\}$ defined as in the proof of Theorem 5.28.

Remark 5.10. Theorem 5.28 extends results of [106] proven for norm continuous, nonexpansive mappings acting in Musielak–Orlicz spaces with the Δ_2 property.

Example 5.6. Let us recall an example of a modular generated by the Urysohn nonlinear integral operator:

$$T(f)(x) = \int_0^1 k(x, y, |f(y)|)dy + f_0(x),$$

where f_0 is an arbitrarily chosen function and the unknown $f : [0,1] \to \mathbb{R}$ is Lebesgue measurable. For the kernel k we assume that,

(a) $k : [0,1] \times [0,1] \times \mathbb{R}_+ \to \mathbb{R}_+$ is Lebesgue measurable,
(b) $k(x,y,0) = 0$,
(c) $k(x,y,.)$ is continuous, convex and increasing to $+\infty$,
(d) $\int_0^1 k(x,y,t)dx > 0$ for $t > 0$ and $y \in (0,1)$.

In addition, assume that for almost all $t \in [0,1]$ and measurable functions f, g there holds

$$\int_0^1 \left\{ \int_0^1 k(t,u,|k(u,v,|f(v)|) - k(u,v,|g(v)|)|)dv \right\}du \le \int_0^1 k(t,u,|f(u) - g(u)|)du.$$

Setting $\rho(f) = \int_0^1 \left\{ \int_0^1 k(x,y,|f(y)|)dy \right\}dx$ and using Jensen's inequality it is easy to show that ρ is a convex function modular, and that T is nonexpansive with respect to ρ. Let us fix an $r > 0$ and set $C = \{f \in E_\rho : \rho(f - f_0) \leq r\}$. It is easy to see that $T : C \to C$. If we assume additionally that there exists a constant $M > 0$ and a Bochner-integrable function $h : [0,1] \times [0,1] \to [0,\infty)$ such that for every $u \geq 0$ and $x, y \in [0,1]$

$$k(x,y,2u) \leq M \, k(x,y,u) + h(x,y), \tag{5.33}$$

then the modular ρ has the property Δ_2 in the sense of Definition 3.5. Using Theorem 5.28 it is then easy to see that the corresponding initial value problem

$$\begin{cases} u(0) = f_0 \\ u'(t) + (I - T)u(t) = 0, \end{cases} \tag{5.34}$$

has a solution in C that can be calculated as the ρ-limit of the sequence $\{u_n\}$ as defined in Theorem 5.28.

Still in the context of Theorem 5.28, let us introduce the following notation: for $f \in C$ we write $u_f(t) \equiv f$ for all $t \geq 0$. Let us ask a question for which $f \in C$, u_f are the solutions of (5.18). The answer is provided in our next result.

Theorem 5.29. *Under the assumptions of Theorem 5.28, u_f is a solution of (5.18) if and only if $f \in F(T)$.*

Proof. It is obvious that for every fixed point $f \in F(T)$, u_f is a solution of (5.18). Conversely, let us assume that u_f is a solution of (5.18). Arguing like in the proof of Theorem 5.28, we obtain that

$$\rho(f - u_n(t)) \leq K^{n+1}(A)\delta_\rho(C) \tag{5.35}$$

for any $n \in \mathbb{N}$, any $A > 0$ and any $t \in [0,A]$. Noting that

$$e^{-t}f + K(t)T(f) - u_{n+1}(t) = \int_0^t e^{s-t}\Big(T(f) - T(u_n(s))\Big)ds$$

and applying Lemma 5.13 we obtain

$$\rho\Big(e^{-t}f + K(t)T(f) - u_{n+1}(t)\Big) \leq K(t) \sup_{s \in [0,t]} \rho\Big(T(f) - T(u_n(s))\Big).$$

Using the fact that T is ρ-nonexpansive and inequality (5.35) we get

$$\rho\Big(e^{-t}f + K(t)T(f) - u_{n+1}(t)\Big) \leq K^{n+2}(A)\delta_\rho(C),$$

hence, $\{u_n(t)\}$ ρ-converges to $e^{-t}f + K(t)T(f)$ for any $t \geq 0$. The uniqueness of the ρ-limit implies that

$$f = u_f(t) = e^{-t}f + K(t)T(f),$$

which yields $T(f) = f$, as claimed.
$\square$

Remark 5.11. Theorem 5.29 establishes a connection between the theory of differential equations in modular function spaces and the fixed point theory for nonlinear mappings acting in modular function spaces. It is clear that, for example, that applying it in conjunction with our fixed point theorem 5.7 we obtain a strong characterization of stationary points of the system (5.18).

Remark 5.12. There is a solid body of work on the subject of ordinary and partial differential equations in Orlicz and Musielak–Orlicz spaces, especially in the context of Orlicz–Sobolev and Musielak–Orlicz–Sobolev spaces, see for instance [48, 86, 87, 93, 196], and more recent works often related to applications to modeling of "smart fluids," see e.g., [6, 22, 53, 88, 89, 92, 187] and the papers referenced there. Frequently in these applications, a nonlinear extension of modular function spaces are used leading to the concepts of the modular metric spaces, see e.g., [3, 46, 47]. Typically these results use the classical techniques of differential equations with values in Banach spaces. As outlined above, our approach is to use the modular notions, like the ρ-nonexpansiveness, whenever this is practical. Our results generalize the results of Khamsi from [106] for norm continuous, nonexpansive mappings acting in Musielak–Orlicz spaces L^φ with φ satisfying the Musielak–Orlicz version of the Δ_2 condition, in relation to the problem of existence of ρ-nonexpansive semigroups of mappings, which, in their turn, extended classical Banach space results of [51, 183].

Chapter 6
Fixed Point Construction Processes

Assume $\rho \in \Re$ is $(UUC1)$. Let C be a ρ-closed, ρ-bounded convex nonempty subset of L_ρ. Let $T : C \rightarrow C$ be a pointwise asymptotically nonexpansive mapping. According to Theorem 5.7 the mapping T has a fixed point. The proof of this important theorem is of the existential nature and does not describe any algorithm for constructing a fixed point of an asymptotic pointwise ρ-nonexpansive mapping. This chapter aims at filling this gap. Therefore, we will define iterative processes for the fixed point construction in modular function spaces and we will prove their convergence. These algorithms will be based on classical iterative methods introduced originally by Mann in [161] and Ishikawa in [97]. The results of this section draw mostly on the research exposed in [54].

6.1 Preliminaries

Denoting $a_n(x) = \max(\alpha_n(x), 1)$, we observe that without loss of generality we can assume that T is asymptotic ρ-pointwise nonexpansive if

$$\rho(T^n(f) - T^n(g)) \leq a_n(f)\rho(f - g), \quad \text{for all } f, g \in C, \ n \in \mathbb{N},$$

$$\lim_{n \to \infty} a_n(f) = 1, \ a_n(f) \geq 1, \quad \text{for all } f \in C, \ n \in \mathbb{N}.$$

Define $b_n(f) = a_n(f) - 1$. In view of the above, we have

$$\lim_{n \to \infty} b_n(f) = 0.$$

The above notation will be consistently used throughout this section.

By $\mathscr{T}(C)$ we will denote the class of all asymptotic pointwise nonexpansive mappings $T : C \rightarrow C$.

© Springer International Publishing Switzerland 2015

M. A. Khamsi, W. M. Kozlowski, *Fixed Point Theory in Modular Function Spaces*,
DOI 10.1007/978-3-319-14051-3_6

In this section, we will impose some restrictions on the behavior of a_n and b_n.

Definition 6.1. Define $\mathscr{T}_r(C)$ as a class of all $T \in \mathscr{T}(C)$ such that a_n is a bounded function for every $n \geq 1$ and $\sum_{n=1}^{\infty} b_n(x) < \infty$ for every $x \in C$.

Remark 6.1. The restrictions described in Definition 6.1 are typical for controlling the convergence of iterative processes for asymptotically nonexpansive mappings, see, for example, [133].

Let us start with the following technical result.

Lemma 6.1. *Let $\rho \in \mathfrak{R}$. Let $C \subset L_\rho$ be a convex set, and let $T \in \mathscr{T}_r(C)$. If $\{x_k\}$ is a ρ-approximate fixed point sequence for T, that is, $\rho(T(x_k) - x_k) \to 0$ as $k \to \infty$, then for every fixed $m \in \mathbb{N}$ there holds*

$$\rho\left(\frac{T^m(x_k) - x_k}{m}\right) \to 0, \quad as \ k \to \infty.$$

Proof. It follows from the fact that a_n is a bounded function for every $n \geq 1$ that there exists a finite constant $M > 0$ such that

$$\sum_{j=1}^{m-1} sup\{a_j(x) : x \in C\} \leq M.$$

Using the convexity of ρ and the ρ-nonexpansiveness of T we get

$$\rho\left(\frac{T^m(x_k) - x_k}{m}\right) = \rho\left(\frac{1}{m}\sum_{j=0}^{m-1}(T^{j+1}(x_k) - T^j(x_k))\right)$$

$$\leq \frac{1}{m}\sum_{j=0}^{m-1}\rho(T^{j+1}(x_k) - T^j(x_k)) \leq \rho(T(x_k) - x_k)\left(\sum_{j=1}^{m-1}a_j(x_n) + 1\right)$$

$$\leq \frac{1}{m}(M+1)\rho(T(x_k) - x_k) \to 0,$$

as $k \to \infty$.
□

Corollary 6.1. *If, under the hypothesis of Lemma 6.1, ρ satisfies additionally the Δ_2 condition, then $\rho(T^m(x_k) - x_k) \to 0$ as $k \to \infty$.*

6.2 Demiclosedness Principle

The following modular version of the Demiclosedness Principle will be used in the proof of our convergence Theorem 6.2. Our proof, the Demiclosedness Principle, uses the parallelogram inequality valid in the modular spaces with the $(UUC1)$

property (Lemma 4.5). For the discussion of the Demiclosedness Principle in the Banach space context, see Sect. 2.5.3.

Theorem 6.1 (Demiclosedness Principle). *[54] Let $\rho \in \Re$. Assume that*

(a) ρ is (UUC1),
(b) ρ has Strong Opial Property,
(c) ρ has Δ_2 property and is uniformly continuous.

Let $C \subset L_\rho$ be nonempty, convex, strongly ρ-bounded, and ρ-closed, and let $T \in \mathcal{T}_r(C)$. Let $\{x_n\} \subset C$, and $x \in C$. If $x_n \to x$ $\rho - a.e.$ and $\rho(T(x_n) - x_n) \to 0$ then $x \in F(T)$.

Proof. Let us recall that by definition of the uniform continuity of ρ to every $\varepsilon > 0$ and $L > 0$, there exists $\delta > 0$ such that

$$|\rho(g) - \rho(g + h)| \leq \varepsilon, \tag{6.1}$$

provided $\rho(h) < \delta$ and $\rho(g) \leq L$. Fix any $m \in \mathbb{N}$. Noting that $\rho(x_n - x) \leq M < \infty$ due to the strong ρ-boundedness of C and that $\rho(T^m(x_n) - x_n) \to 0$ by Corollary 6.1, it follows then from (6.1) with $g = x_n - x$ and $h = T^m(x_n) - x_n$ that

$$|\rho(x_n - x) - \rho(x_n - x + T^m(x_n) - x_n)| \to 0, \quad \text{as } n \to \infty.$$

Hence,

$$\limsup_{n \to \infty} \rho(x_n - x) = \limsup_{n \to \infty} \rho(T^m(x_n) - x). \tag{6.2}$$

Define the ρ-type φ by

$$\varphi(x) = \limsup_{n \to \infty} \rho(x_n - x).$$

By (6.2) we get

$$\varphi(x) = \limsup_{n \to \infty} \rho(T^m(x_n) - x).$$

Hence, for every $y \in C$ there holds

$$\varphi(T^m(y)) = \limsup_{n \to \infty} \rho(T^m(x_n) - T^m(y)) \tag{6.3}$$

$$\leq a_m(y) \limsup_{n \to \infty} \rho(x_n - y) = a_m(y)\varphi(y)$$

Using (6.3) with $y = x$ and by passing with m to infinity we conclude that

$$\limsup_{m \to \infty} \varphi(T^m(x)) \leq \varphi(x). \tag{6.4}$$

Since ρ satisfies the Strong Opial property, it also satisfies the Opial property. Since $x_n \to x$ $\rho - a.e.$, it follows via the Opial property that for any $y \neq x$

$$\varphi(x) = \limsup_{n \to \infty} \rho(x_n - x) < \limsup_{n \to \infty} \rho(x_n - y) = \varphi(y),$$

which implies that

$$\varphi(x) = \inf\{\varphi(y) : y \in C\}. \tag{6.5}$$

Combining (6.4) with (6.5), we have

$$\varphi(x) \leq \limsup_{m \to \infty} \varphi(T^m(x)) \leq \varphi(x),$$

that is,

$$\limsup_{m \to \infty} \varphi(T^m(x)) = \varphi(x).$$

We claim that

$$\lim_{m \to \infty} \rho\left(T^m(x) - x\right) = 0. \tag{6.6}$$

Assume to the contrary that (6.6) does not hold, that is

$$\rho\left(T^m(x) - x\right) \text{ does not tend to zero.} \tag{6.7}$$

By Δ_2, then it follows from (6.7) that $\rho\left(\dfrac{T^m(x) - x}{2}\right)$ does not tend to zero. By passing to a subsequence if necessary we can assume that there exists $0 < t < M$ such that

$$\rho\left(\frac{T^m(x) - x}{2}\right) > t > 0, \quad \text{for } m \in \mathbb{N},$$

which implies that

$$\rho(x_n - x) + \rho\left(x_n - T^m(x)\right) > \frac{t}{2}, \quad \text{for every } m, n \in \mathbb{N}.$$

Hence,

$$\max\left\{\rho(x_n - x), \rho\left(x_n - T^m(x)\right)\right\} \geq \frac{t}{4}, \quad \text{for every } m, n \in \mathbb{N}.$$

Applying the parallelogram inequality from Lemma 4.5,

$$\rho^2\left(\frac{z+y}{2}\right) \leq \frac{1}{2}\rho^2(z) + \frac{1}{2}\rho^2(y) - \Psi\left(r, s, \frac{1}{r}\rho(z - y)\right),$$

where $\rho(z) \leq r, \rho(y) \leq r$ and $\max\{\rho(z), \rho(y)\} \geq s$ for $0 < s < r$, with $r = M, s = \frac{t}{4}$, $z = x_n - x,\ y = T^m(x)$, we get

$$\rho^2\left(x_n - \frac{x + T^m(x)}{2}\right) \leq \frac{1}{2}\rho^2(x_n - x) + \frac{1}{2}\rho^2(x_n - T^m(x)) - \Psi\left(M, \frac{t}{4}, \frac{1}{M}\rho(x - T^m(x))\right).$$

Note that by (6.5)

$$\varphi^2(x) \leq \varphi^2\left(\frac{x + T^m(x)}{2}\right) = \limsup_{n \to \infty} \rho^2\left(x_n - \frac{x + T^m(x)}{2}\right).$$

Combining the last two formulas we obtain

$$\varphi^2(x) \le \frac{1}{2} \limsup_{n \to \infty} \rho^2(x_n - x) + \frac{1}{2} \limsup_{n \to \infty} \rho^2(x_n - T^m(x)) - \Psi\left(M, \frac{t}{4}, \frac{1}{M}\rho(x - T^m(x))\right),$$

which implies

$$0 \le \Psi\left(M, \frac{t}{4}, \frac{1}{M}\rho(x - T^m(x))\right) \le \frac{1}{2}\varphi^2(T^m(x)) - \frac{1}{2}\varphi^2(x).$$

Hence,

$$0 \le \limsup_{m \to \infty} \Psi\left(M, \frac{t}{4}, \frac{1}{M}\rho(x - T^m(x))\right)$$
$$\le \frac{1}{2}\limsup_{m \to \infty} \varphi^2(T^m(x)) - \frac{1}{2}\varphi^2(x) \le 0.$$

Using the properties of Ψ, we conclude that $\rho(x - T^m(x))$ tends to zero itself, which contradicts our assumption (6.7). Hence, $\rho(x - T^m(x)) \to 0$ as $m \to \infty$ as claimed in (6.6). Clearly, then $\rho(x - T^{m+1}(x)) \to 0$ as $m \to \infty$, that is, $T^{m+1}(x) \to x(\rho)$ while $T^{m+1}(x) \to T(x)(\rho)$ as well as by the ρ-continuity of T. By the uniqueness of the ρ-limit, we obtain $T(x) = x$, that is, $x \in F(T)$, which completes the proof.
□

6.3 Generalized Mann Iteration Process

Let us define our main iterative process for the construction of fixed points in modular function spaces.

Definition 6.2. Let $T \in \mathscr{T}_r(C)$ and let $\{n_k\}$ be an increasing sequence of natural numbers. Let $\{t_k\} \subset (0, 1)$ be bounded away from 0 and 1. The *generalized Mann iteration process* generated by the mapping T, the sequence $\{t_k\}$, and the sequence $\{n_k\}$, denoted by $gM(T, \{t_k\}, \{n_k\})$ is defined by the following iterative formula:

$$x_{k+1} = t_k T^{n_k}(x_k) + (1 - t_k)x_k, \text{ where } x_1 \in C \text{ is chosen arbitrarily.}$$

Definition 6.3. We say that a generalized Mann iteration process $gM(T, \{t_k\}, \{n_k\})$ is *well-defined* if

$$\limsup_{k \to \infty} a_{n_k}(x_k) = 1. \tag{6.8}$$

Remark 6.2. Observe that by the definition of asymptotic pointwise nonexpansiveness, $\lim_{k \to \infty} a_k(x) = 1$ for every $x \in C$. Hence, we can always select a subsequence $\{a_{n_k}\}$ such that (6.8) holds. In other words, by a suitable choice of $\{n_k\}$ we can always make $gM(T, \{t_k\}, \{n_k\})$ well-defined.

Let us quote an elementary lemma about real numbers, see for example [36].

Lemma 6.2. *Suppose* $\{r_k\}$ *is a bounded sequence of real numbers and* $\{d_{k,n}\}$ *is a doubly-index sequence of real numbers which satisfy*

$$\limsup_{k\to\infty}\limsup_{n\to\infty} d_{k,n} \leq 0, \text{ and } r_{k+n} \leq r_k + d_{k,n}$$

for each $k,n \geq 1$. *Then* $\{r_k\}$ *converges to an* $r \in \mathbb{R}$.

The following result provides an important technique which will be used in this chapter.

Lemma 6.3. *Let* $\rho \in \mathfrak{R}$ *be* $(UUC1)$. *Let* $C \subset L_\rho$ *be a* ρ-closed, ρ-bounded, and convex set. Let $T \in \mathcal{T}_r(C)$ and let $\{n_k\} \subset \mathbb{N}$. Assume that a sequence $\{t_k\} \subset (0,1)$ is bounded away from 0 and 1. Let w be a fixed point of T and $gM(T,\{t_k\},\{n_k\})$ be a generalized Mann process. Then, there exists $r \in \mathbb{R}$ such that $\lim_{k\to\infty} \rho(x_k - w) = r$.

Proof. Let $w \in F(T)$. Since

$$\begin{aligned}
\rho(x_{k+1} - w) &\leq t_k\rho(T^{n_k}(x_k) - w) + (1-t_k)\rho(x_k - w) \\
&= t_k\rho(T^{n_k}(x_k) - T^{n_k}(w)) + (1-t_k)\rho(x_k - w) \\
&\leq t_k(1 + b_{n_k}(w))\rho(x_k - w) + (1-t_k)\rho(x_k - w) \\
&= t_k b_{n_k}(w)\rho(x_k - w) + \rho(x_k - w) \\
&\leq b_{n_k}(w)diam_\rho(C) + \rho(x_k - w),
\end{aligned}$$

it follows that for every $n \in \mathbb{N}$,

$$\rho(x_{k+n} - w) \leq \rho(x_k - w) + diam_\rho(C) \sum_{i=k}^{k+n-1} b_{n_i}(w).$$

Denote $r_p = \rho(x_p - w)$ for every $p \in \mathbb{N}$ and $d_{k,n} = diam_\rho(C) \sum_{i=k}^{k+n-1} b_{n_i}(w)$. Observe that since $T \in \mathcal{T}_r(C)$, it follows that $\limsup_{k\to\infty}\limsup_{n\to\infty} d_{k,n} = 0$. By Lemma 6.2, there exists an $r \in \mathbb{R}$ such that $\lim_{k\to\infty} \rho(x_k - w) = r$ as claimed. $\quad\square$

The next result will be essential for proving the convergence theorems for iterative process.

Lemma 6.4. *Let* $\rho \in \mathfrak{R}$ *be* $(UUC1)$. *Let* $C \subset L_\rho$ *be a* ρ-closed, ρ-bounded, and convex set, and $T \in \mathcal{T}_r(C)$. Assume that a sequence $\{t_k\} \subset (0,1)$ is bounded away from 0 and 1. Let $\{n_k\} \subset \mathbb{N}$ and $gM(T,\{t_k\},\{n_k\})$ be a generalized Mann iteration process. Then

$$\lim_{k\to\infty} \rho(T^{n_k}(x_k) - x_k) = 0, \tag{6.9}$$

and

$$\lim_{k\to\infty} \rho(x_{k+1} - x_k) = 0. \tag{6.10}$$

Proof. By Theorem 5.7, T has at least one fixed point $w \in C$. In view of Lemma 6.3, there exists $r \in \mathbb{R}$ such that

$$\lim_{k \to \infty} \rho(x_k - w) = r. \tag{6.11}$$

Note that

$$\limsup_{k \to \infty} \rho(T^{n_k}(x_k) - w) = \limsup_{k \to \infty} \rho(T^{n_k}(x_k) - T^{n_k}(w)) \tag{6.12}$$

$$\leq \limsup_{k \to \infty} a_{n_k}(w)\rho(x_k - w) \leq r,$$

and that

$$\lim_{k \to \infty} \rho(t_k(T^{n_k}(x_k) - w) + (1 - t_k)(x_k - w)) = \lim_{k \to \infty} \rho(x_{k+1} - w) = r.$$

Set $f_k = T^{n_k}(x_k) - w$, $g_k = x_k - w$, and note that $\limsup_{k \to \infty} \rho(g_k) \leq r$ by (6.11), and $\limsup_{k \to \infty} \rho(f_k) \leq r$ by (6.12). Observe also that

$$\lim_{k \to \infty} \rho(t_k f_k + (1 - t_k)g_k) = \lim_{k \to \infty} \rho(t_k T^{n_k}(x_k) + (1 - t_k)x_k - w) = \lim_{k \to \infty} \rho(x_{k+1} - w) = r.$$

Hence, it follows from Lemma 4.2 that

$$\lim_{k \to \infty} \rho(T^{n_k}(x_k) - x_k) = \lim_{k \to \infty} \rho(f_k - g_k) = 0,$$

which by the construction of the sequence $\{x_k\}$ is equivalent to

$$\lim_{k \to \infty} \rho(x_{k+1} - x_k) = 0,$$

as claimed. $\quad\square$

In the next lemma, we prove that under suitable assumptions, the sequence $\{x_k\}$ becomes an approximate fixed point sequence, which will provide an important step in the proof of the generalized Mann iteration process convergence. First, we need to define the following notions.

Definition 6.4. A strictly increasing sequence $\{n_i\} \subset \mathbb{N}$ is called *quasiperiodic* if the sequence $\{n_{i+1} - n_i\}$ is bounded, or equivalently if there exists a number $p \in \mathbb{N}$ such that any block of p consecutive natural numbers must contain a term of the sequence $\{n_i\}$. The smallest of such numbers p will be called a *quasiperiod* of $\{n_i\}$.

Lemma 6.5. *Let* $\rho \in \mathfrak{R}$ *be (UUC1) satisfying* Δ_2. *Let* $C \subset L_\rho$ *be a* ρ-*closed,* ρ-*bounded, and convex set, and* $T \in \mathcal{T}_r(C)$. *Let* $\{t_k\} \subset (0, 1)$ *be bounded away from 0 and 1. Let* $\{n_k\} \subset \mathbb{N}$ *be such that the generalized Mann process* $gM(T, \{t_k\}, \{n_k\})$ *is well-defined. If, in addition, the set of indices* $\mathcal{J} = \{j : n_{j+1} = 1 + n_j\}$ *is quasiperiodic, then* $\{x_k\}$ *is an approximate fixed point sequence, that is,*

$$\lim_{k \to \infty} \rho(T(x_k) - x_k) = 0. \tag{6.13}$$

Proof. Let $p \in \mathbb{N}$ be a quasiperiod of $\mathcal{J}$. Observe that it is enough to prove that $\rho(T(x_k) - x_k) \to 0$ as $k \to \infty$ through $\mathcal{J}$. Indeed, from this fact it follows that if we fix $\varepsilon > 0$ then

$$\rho(T(x_k) - x_k) \le \varepsilon$$

for sufficiently large $k \in \mathcal{J}$. By the quasiperiodicity of $\mathcal{J}$, to every positive integer k there exists $j_k \in \mathcal{J}$ such that $|k - j_k| \le p$. Assume that $k - p \le j_k \le k$ (the proof for the other case is identical). Since T is ρ-Lipschitzian with the Lipschitz constant $M = sup\{a_1(x) : x \in C\}$, there exist a $0 < \delta < \dfrac{\varepsilon}{3}$ such that

$$\rho(T(x) - T(y)) < \varepsilon \ if \ \rho(x - y) < \delta.$$

Note that by (6.10) and by Δ_2, $\rho(p(x_{k+1} - x_k)) < \dfrac{\delta}{p}$ for k sufficiently large. This implies that

$$\rho(x_k - x_{j_k}) \le \frac{1}{p}\left(\rho(p(x_k - x_{k-1})) + \ldots + \rho(p(x_{j_k+1} - x_{j_k}))\right) \le p\frac{\delta}{p} = \delta,$$

and therefore

$$\rho\left(\frac{x_k - T(x_k)}{3}\right) \le \frac{1}{3}\rho(x_k - x_{j_k}) + \frac{1}{3}\rho(x_{j_k} - T(x_{j_k})) + \frac{1}{3}\rho(T(x_{j_k}) - T(x_k)) \le \delta + \frac{\varepsilon}{3} + \frac{\varepsilon}{3} < \varepsilon,$$

which demonstrates that

$$\rho\left(\frac{x_k - T(x_k)}{3}\right) \to 0$$

as $k \to \infty$. By Δ_2 we get $\rho(T(x_k) - x_k) \to \infty$.

To prove that $\rho(T(x_k) - x_k) \to 0$ as $k \to \infty$ through $\mathcal{J}$, observe that, since $n_{k+1} = n_k + 1$ for such k, there holds

$$\rho\left(\frac{x_k - T(x_k)}{4}\right) \le \frac{1}{4}\rho(x_k - x_{k+1}) + \frac{1}{4}\rho(x_{k+1} - T^{n_{k+1}}(x_{k+1}))$$

$$+ \frac{1}{4}\rho(T^{n_{k+1}}(x_{k+1}) - T^{n_{k+1}}(x_k)) + \frac{1}{4}\rho(TT^{n_k}(x_k) - T(x_k))$$

$$\le \frac{1}{4}\rho(x_k - x_{k+1}) + \frac{1}{4}\rho(x_{k+1} - T^{n_{k+1}}(x_{k+1}))$$

$$+ \frac{1}{4}a_{n_{k+1}}(x_{k+1})\rho(x_k - x_{k+1}) + \frac{1}{4}M\rho(T^{n_k}(x_k) - x_k)$$

which tends to zero in view of (6.9), (6.10), and (6.8). $\square$

We are now ready to prove our main convergence result, the proof follows [54].

Theorem 6.2. *Let $\rho \in \mathfrak{R}$. Assume that*

(i) ρ is (UUC1),
(ii) ρ has Strong Opial Property,
(iii) ρ has Δ_2 property and is uniformly continuous.

Let $C \subset L_\rho$ be nonempty, $\rho - a.e.$ compact, convex, strongly ρ-bounded, and ρ-closed, and let $T \in \mathscr{T}_r(C)$. Assume that a sequence $\{t_k\} \subset (0,1)$ is bounded away from 0 and 1. Let $\{n_k\} \subset \mathbb{N}$ and $gM(T, \{t_k\}, \{n_k\})$ be a well-defined generalized Mann iteration process. Assume, in addition, that the set of indices

$$\mathscr{J} = \{j : n_{j+1} = 1 + n_j\}$$

is quasiperiodic. Then, there exists $x \in F(T)$ such that $x_n \to x$ ρ-a.e.

Proof. Observe that by Theorem 5.7 the set of fixed points $F(T)$ is nonempty, and ρ-closed. Also Note that by Theorem 4.8, it follows from the Strong Opial property of ρ that any ρ-type attains its minimum in C. By Lemma 6.5 the sequence $\{x_k\}$ is an approximate fixed point sequence, that is

$$\rho(T(x_k) - x_k) \to 0$$

as $k \to \infty$. Consider $y, z \in C$, two ρ-a.e. cluster points of $\{x_k\}$. There exits then $\{y_k\}$, $\{z_k\}$ subsequences of $\{x_k\}$ such that $y_k \to y$ ρ-a.e., and $z_k \to z$ ρ-a.e. By Theorem 6.1, $y \in F(T)$ and $z \in F(T)$. By Lemma 6.3, there exist $r_y, r_z \in \mathbb{R}$ such that

$$r_y = \lim_{k \to} \rho(x_k - y), \quad r_z = \lim_{k \to} \rho(x_k - z).$$

We claim that $y = z$. Assume to the contrary that $y \neq z$. Then by the Strong Opial property we have

$$r_y = \liminf_{k \to \infty} \rho(y_k - y) < \liminf_{k \to \infty} \rho(y_k - z)$$
$$= \liminf_{k \to \infty} \rho(z_k - z) < \liminf_{k \to \infty} \rho(z_k - y) = r_y.$$

The contradiction implies that $y = z$. Therefore, $\{x_k\}$ has at most one ρ-a.e. cluster point. Since C is ρ-a.e. compact, it follows that the sequence $\{x_k\}$ has exactly one ρ-a.e. cluster point, which means that $\rho(x_k) \to x$ ρ-a.e. Using Theorem 6.1 again, we get $x \in F(T)$ as claimed.
□

Remark 6.3. It is easy to see that we can always construct a sequence $\{n_k\}$ with the quasiperiodic properties specified in the assumptions of Theorem 6.2. When constructing concrete implementations of this algorithm, the difficulty will be to ensure that the constructed sequence $\{n_k\}$ is not "too sparse" in the sense that the generalized Mann process $gM(T, \{t_k\}, \{n_k\})$ remains well-defined. The similar, quasiperiodic type assumptions are common in the asymptotic fixed point theory, see, for example, [36, 133, 144].

If T is ρ-nonexpansive then obviously it is also asymptotic pointwise ρ-nonexpansive with $a_n(x) = 1$ for any $x \in Ca$ and any natural n. Hence, one can take $n_k = k$ as the defining sequence for the generalized Mann process and consequently $\mathscr{J} = \{j : n_{j+1} = 1 + n_j\}$ is always quasiperiodic. Thus, we have the following result for ρ-nonexpansive mappings.

Theorem 6.3. *Let $\rho \in \mathfrak{R}$. Assume that*

(i) ρ is (UUC1),
(ii) ρ has Strong Opial Property,
(iii) ρ has Δ_2 property and is uniformly continuous.

Let $C \subset L_\rho$ be nonempty, $\rho - a.e.$ compact, convex, strongly ρ-bounded, and ρ-closed, and let $T : C \to C$ be ρ-nonexpansive. Assume that a sequence $\{t_k\} \subset (0,1)$ is bounded away from 0 and 1. Let $gM(T, \{t_k\}, k)$ be generalized Mann iteration process, which must be well-defined because of ρ-nonexpansiveness of T. Then, there exists $x \in F(T)$ such that $x_n \to x$ ρ-a.e.

6.4 Generalized Ishikawa Iteration Process

The two-step Ishikawa iteration process is a generalization of the one-step Mann process. The Ishikawa iteration process provides more flexibility in defining the algorithm parameters which is important from the numerical implementation perspective.

Definition 6.5. Let $T \in \mathscr{T}_r(C)$ and let $\{n_k\}$ be an increasing sequence of natural numbers. Let $\{t_k\} \subset (0,1)$ be bounded away from 0 and 1, and $\{s_k\} \subset (0,1)$ be bounded away from 1. The *generalized Ishikawa iteration process* generated by the mapping T, the sequences $\{t_k\}$, $\{s_k\}$, and the sequence $\{n_k\}$, denoted by $gI(T, \{t_k\}, \{s_k\}, \{n_k\})$ is defined by the following iterative formula:

$$x_{k+1} = t_k T^{n_k} \left(s_k T^{n_k}(x_k) + (1 - s_k)x_k \right) + (1 - t_k)x_k,$$

where $x_1 \in C$ is chosen arbitrarily.

Definition 6.6. We say that a generalized Ishikawa iteration process

$$gI(T, \{t_k\}, \{s_k\}, \{n_k\})$$

is *well-defined* if

$$\limsup_{k \to \infty} a_{n_k}(x_k) = 1. \tag{6.14}$$

Remark 6.4. Observe that, by the definition of asymptotic pointwise nonexpansiveness, $\lim_{k \to \infty} a_k(x) = 1$ for every $x \in C$. Hence, we can always select a subsequence $\{a_{n_k}\}$ such that (6.14) holds. In other words, by a suitable choice of $\{n_k\}$ we can always make $gI(T, \{t_k\}, \{s_k\}, \{n_k\})$ well-defined.

Lemma 6.6. *Let* $\rho \in \Re$ *be* (UUC1). *Let* $C \subset L_\rho$ *be a* ρ-closed, ρ-bounded, and *convex set. Let* $T \in \mathcal{T}_r(C)$ *and let* $\{n_k\} \subset \mathbb{N}$. *Let* $\{t_k\} \subset (0,1)$ *be bounded away from* 0 *and* 1, *and* $\{s_k\} \subset (0,1)$ *be bounded away from* 1. *Let* $w \in F(T)$ *and* $gI(T, \{t_k\}, \{s_k\}, \{n_k\})$ *be a generalized Ishikawa process. There exists then an* $r \in \mathbb{R}$ *such that* $\lim_{k \to \infty} \rho(x_k - w) = r$.

Proof. Define $T_k : C \to C$ by

$$T_k(x) = t_k T^{n_k}\left(s_k T^{n_k}(x) + (1 - s_k)x\right) + (1 - t_k)x, \ x \in C.$$

It is easy to see that $x_{k+1} = T_k(x_k)$ and that $F(T) \subset F(T_k)$. Moreover, a straight calculation shows that each T_k satisfies

$$\rho(T_k(x) - T_k(y)) \leq A_k(x)\rho(x - y),$$

where

$$A_k(x) = 1 + t_k a_{n_k}(M_k(x))(1 + s_k a_{n_k}(x) - s_k) - t_k, \tag{6.15}$$

and

$$M_k(x) = s_k T^{n_k}(x) + (1 - s_k)x.$$

Note that $A_k(x) \geq 1$ which follows directly from the fact that $a_{n_k}(x) \geq 1$ and from (6.15). Using (6.15) and the fact that $M_k(w) = w$ we have

$$B_k(w) = A_k(w) - 1 = t_k(1 + s_k a_{n_k}(w))(a_{n_k}(w) - 1) \leq (1 + a_{n_k}(w))b_{n_k}(w).$$

Fix any $M > 1$. Since $\lim_{k \to \infty} a_{n_k}(w) = 1$, it follows that there exists a $k_0 \geq 1$ such that for $k > k_0$, $a_{n_k}(w) \leq M$. Therefore, by the same argument as in the proof of Lemma 6.3, we deduce that for $k > k_0$ and $n > 1$

$$\rho(x_{k+n} - w) \leq \rho(x_k - w) + diam_\rho(C) \sum_{i=k}^{k+n-1} B_i(w)$$

$$\leq \rho(x_k - w) + diam_\rho(C)(1 + M) \sum_{i=k}^{k+n-1} b_{n_i}(w).$$

Arguing like in the proof of Lemma 6.3, we conclude that there exists an $r \in \mathbb{R}$ such that $\lim_{k \to \infty} \rho(x_k - w) = r$. $\square$

Lemma 6.7. *Let* $\rho \in \Re$ *be* (UUC1). *Let* $C \subset L_\rho$ *be a* ρ-closed, ρ-bounded, and *convex set. Let* $T \in \mathcal{T}_r(C)$ *and let* $\{n_k\} \subset \mathbb{N}$. *Let* $\{t_k\} \subset (0,1)$ *be bounded away from* 0 *and* 1, *and* $\{s_k\} \subset (0,1)$ *be bounded away from* 1. *Let* $gI(T, \{t_k\}, \{s_k\}, \{n_k\})$ *be a generalized Ishikawa process. Define* $y_k = s_k T^{n_k}(x_k) + (1 - s_k)x_k$. *Then*

$$\lim_{k \to \infty} \rho(T^{n_k}(y_k) - x_k) = 0, \tag{6.16}$$

or equivalently

$$\lim_{k\to\infty} \rho(x_{k+1} - x_k) = 0, \tag{6.17}$$

Proof. By Theorem 5.7, $F(T) \neq \emptyset$. Let us fix $w \in F(T)$. By Lemma 6.6, the limit $\lim_{k\to\infty} \rho(x_k - w)$ exists. Let us denote it by r. Since $w \in F(T)$, $T \in \mathscr{T}_r(C)$, and $\lim_{k\to\infty} \rho(x_k - w) = r$, by Lemma 6.6 we have the following

$$\limsup_{k\to\infty} \rho\left(T^{n_k}(y_k) - w\right) = \limsup_{k\to\infty} \rho\left(T^{n_k}(y_k) - T^{n_k}(w)\right)$$

$$\leq \limsup_{k\to\infty} a_{n_k}(w)\rho(y_k - w) = \limsup_{k\to\infty} a_{n_k}(w)\rho\left(s_k T^{n_k}(x_k) + (1-s_k)x_k - w\right)$$

$$\leq \limsup_{k\to\infty}[s_k a_{n_k}(w)\rho(T^{n_k}(x_k) - w) + (1-s_k)a_{n_k}(w)\rho(x_k - w)]$$

$$\leq \limsup_{k\to\infty}[s_k a_{n_k}^2(w)\rho(x_k - w) + (1-s_k)a_{n_k}(w)\rho(x_k - w)] \leq r.$$

Note that

$$\lim_{k\to\infty} \rho\left(t_k(T^{n_k}(y_k) - w) + (1-t_k)(x_k - w)\right)$$

$$= \lim_{k\to\infty} \rho\left(t_k T^{n_k}(y_k) + (1-t_k)x_k - w\right) = \lim_{k\to\infty} \rho(x_{k+1} - w) = r.$$

Applying Lemma 4.2 with $u_k = T^{n_k}(y_k) - w$ and $v_k = x_k - w$, we obtain the desired equality $\lim_{k\to\infty} \rho(T^{n_k}(y_k) - x_k) = 0$, while (6.17) follows from (6.16) via the construction formulas for x_{k+1} and y_k.
□

Lemma 6.8. *Let $\rho \in \mathfrak{R}$ be (UUC1) satisfying Δ_2. Let $C \subset L_\rho$ be a ρ-closed, ρ-bounded, and convex set. Let $T \in \mathscr{T}_r(C)$ and let $\{n_k\} \subset \mathbb{N}$. Let $\{t_k\} \subset (0,1)$ be bounded away from 0 and 1, and $\{s_k\} \subset (0,1)$ be bounded away from 1. Let $gI(T, \{t_k\}, \{s_k\}, \{n_k\})$ be a well-defined generalized Ishikawa process. Then*

$$\lim_{k\to\infty} \rho\left(T^{n_k}(x_k) - x_k\right) = 0. \tag{6.18}$$

Proof. Let $y_k = s_k T^{n_k}(x_k) + (1-s_k)x_k$. Hence

$$T^{n_k}(x_k) - x_k = \frac{1}{1-s_k}\left(T^{n_k}(x_k) - y_k\right).$$

Since $\{s_k\} \subset (0,1)$ is bounded away from 1, there exists $0 < s < 1$ such that $s_k \leq s$ for every $k \geq 1$. Hence,

$$\rho(T^{n_k}(x_k) - x_k) = \rho\left(\frac{1}{1-s_k}(T^{n_k}(x_k) - y_k)\right) \leq \rho\left(\frac{1}{1-s}(T^{n_k}(x_k) - y_k)\right).$$

The right-hand side of this inequality tends to zero because $\rho\left(T^{n_k}(x_k) - y_k\right) \to 0$ by Lemma 6.7 and the fact that ρ satisfies Δ_2.

□

Lemma 6.9. *Let $\rho \in \Re$ be (UUC1) satisfying Δ_2. Let $C \subset L_\rho$ be a ρ-closed, ρ-bounded, and convex set, and $T \in \mathscr{T}_r(C)$. Let $\{t_k\} \subset (0,1)$ be bounded away from 0 and 1 and $\{s_k\} \subset (0,1)$ be bounded away from 1. Let $\{n_k\} \subset \mathbb{N}$ be such that the generalized Ishikawa process $gI(T, \{t_k\}, \{s_k\}, \{n_k\})$ is well-defined. If, in addition, the set $\mathscr{J} = \{j : n_{j+1} = 1 + n_j\}$ is quasiperiodic, then $\{x_k\}$ is an approximate fixed point sequence, i.e.,*

$$\lim_{k \to \infty} \rho(T(x_k) - x_k) = 0.$$

Proof. The proof is analogous to that of Lemma 6.5 with (6.17) used instead of (6.10) and (6.18) replacing (6.9).

□

Theorem 6.4. *Let $\rho \in \Re$. Assume that*

(i) ρ is (UUC1),
(ii) ρ has Strong Opial Property,
(iii) ρ has Δ_2 property and is uniformly continuous.

Let $C \subset L_\rho$ be a nonempty, $\rho - a.e.$ compact, convex, strongly ρ-bounded, and ρ-closed set, and let $T \in \mathscr{T}_r(C)$. Let $T \in \mathscr{T}_r(C)$. Let $\{t_k\} \subset (0,1)$ be bounded away from 0 and 1, and $\{s_k\} \subset (0,1)$ be bounded away from 1. Let $\{n_k\}$ be such that the generalized Ishikawa process $gI(T, \{t_k\}, \{s_k\}, \{n_k\})$ is well-defined. If, in addition, the set $\mathscr{J} = \{j : n_{j+1} = 1 + n_j\}$ is quasiperiodic, then $\{x_k\}$ generated by $gI(T, \{t_k\}, \{s_k\}, \{n_k\})$ converges $\rho - a.e.$ to a fixed point $x \in F(T)$.

Proof. The proof is analogous to that of Theorem 6.2 with Lemma 6.5 replaced by Lemma 6.9, and Lemma 6.3 replaced by Lemma 6.6.   □

Similarly as with the Generalized Mann process, we can argue that if T is ρ-nonexpansive then obviously it is also asymptotic pointwise ρ-nonexpansive with $a_n(x) = 1$ for any $x \in C$ and any natural n. Hence, one can take $n_k = k$ as the defining sequence for the generalized Ishikawa process and consequently $\mathscr{J} = \{j : n_{j+1} = 1 + n_j\}$ is always quasiperiodic. Thus, we have the following result for ρ-nonexpansive mappings.

Theorem 6.5. *Let $\rho \in \Re$. Assume that*

(i) ρ is (UUC1),
(ii) ρ has Strong Opial Property,
(iii) ρ has Δ_2 property and is uniformly continuous.

Let $C \subset L_\rho$ be nonempty, $\rho - a.e.$ compact, convex, strongly ρ-bounded, and ρ-closed, and let $T : C \to C$ be ρ-nonexpansive. Let $\{t_k\} \subset (0,1)$ be bounded away from 0 and 1, and $\{s_k\} \subset (0,1)$ be bounded away from 1. Then $\{x_k\}$ generated by $gI(T, \{t_k\}, \{s_k\}, \{k\})$ converges $\rho - a.e.$ to a fixed point $x \in F(T)$.

6.5 Strong Convergence

It is interesting that, assuming C is ρ-compact (which is a strong form of compactness in modular function spaces), instead of ρ-a.e. compactness (which is a weak form of compactness), both generalized Mann and Ishikawa processes converge strongly to a fixed point of T even without assuming the Opial property or any quasiperiodicity assumptions. The next result proves exactly that.

Theorem 6.6. *Let $\rho \in \mathfrak{R}$ satisfy conditions $(UUC1)$ and Δ_2. Let $C \subset L_\rho$ be a ρ-compact, ρ-bounded, and convex set, and let $T \in \mathcal{T}_r(C)$. Let $\{t_k\} \subset (0,1)$ be bounded away from 0 and 1, and $\{s_k\} \subset (0,1)$ be bounded away from 1. Let $\{n_k\}$ be such that the generalized Mann process $gM(T, \{t_k\}, \{n_k\})$ (respectively, Ishikawa process $gI(T, \{t_k\}, \{s_k\}, \{n_k\})$) is well-defined. Then, there exists a fixed point $x \in F(T)$ such that the sequence $\{x_k\}$ generated by $gM(T, \{t_k\}, \{n_k\})$ (respectively, $gI(T, \{t_k\}, \{s_k\}, \{n_k\})$) converges strongly to a fixed point of T, that is, $\lim_{k \to \infty} \rho(x_k - x) = 0$.*

Proof. By the ρ-compactness of C we can select a subsequence $\{x_{p_k}\}$ of $\{x_k\}$ such that there exists $x \in C$ with

$$\lim_{k \to \infty} \rho(T(x_{p_k}) - x) = 0. \tag{6.19}$$

Note that

$$\rho\left(\frac{x_{p_k} - x}{2}\right) \le \frac{1}{2}\rho\left(x_{p_k} - T(x_{p_k})\right) + \frac{1}{2}\rho\left(T(x_{p_k}) - x\right), \tag{6.20}$$

which tends to zero by Lemma 6.4 (respectively Lemma 6.9) and by (6.19). By Δ_2 if follows from (6.20) that

$$\rho(x_{p_k} - x) \to 0 \text{ as } k \to \infty. \tag{6.21}$$

Observe that by the convexity of ρ and by ρ-nonexpansiveness of T, we have

$$\rho\left(\frac{T(x) - x}{3}\right) \le \frac{1}{3}\rho\left(T(x) - T(x_{p_k})\right) + \frac{1}{3}\rho\left(T(x_{p_k}) - x_{p_k}\right) + \frac{1}{3}\rho(x_{p_k} - x)$$

$$\le \frac{1}{3}\rho(x - x_{p_k}) + \frac{1}{3}\rho\left(T(x_{p_k}) - x_{p_k}\right) + \frac{1}{3}\rho(x_{p_k} - x),$$

which tends to zero by (6.21) and by Lemma 6.4 (respectively Lemma 6.9). From there we get $\rho(T(x) - x) = 0$ which implies that $x \in F(T)$. Applying Lemma 6.3 (respectively Lemma 6.6) we conclude that $\lim_{k \to \infty} \rho(x_k - x)$ exists. By (6.21) this limit must be equal to zero which implies that $\lim_{k \to \infty} \rho(x_k - x) = 0$, as claimed.

$\square$

Remark 6.5. Observe that in view of the Δ_2 assumption, the ρ-compactness of the set C assumed in Theorem 6.6 is equivalent to the compactness in the sense of the Luxemburg norm defined by ρ.

Chapter 7
Semigroups of Nonlinear Mappings in Modular Function Spaces

Let us recall that a family $\{T_t\}_{t\geq 0}$ of mappings forms a semigroup if $T_0(x) = x$, and $T_{s+t} = T_s(T_t(x))$. Such a situation is quite typical in mathematics and applications. For instance, in the theory of dynamical systems, the modular function space L_ρ would define the state space and the mapping $(t,x) \to T_t(x)$ would represent the evolution function of a dynamical system. The question about the existence of common fixed points, and about the structure of the set of common fixed points, can be interpreted as a question whether there exist points that are fixed during the state space transformation T_t at any given point of time t, and if yes—what the structure of a set of such points may look like. In the setting of this chapter, the state space may be an infinite dimensional. Therefore, it is natural to apply these result to not only to deterministic dynamical systems but also to stochastic dynamical systems. Because of wide body of potential applications, the theory of semigroups of nonlinear mappings in modular function spaces, initiated in the 1992 paper by Khamsi [106], has become recently a subject of an intensive development, see [9, 55, 56, 135, 139, 140, 141]. There is, however, a lot of space for future research in this area.

7.1 Definitions

Let us start with the modular definitions of Lipschitzian—in the modular sense—mappings, and of associated definitions of semigroups of nonlinear mappings.

Definition 7.1. Let $\rho \in \Re$ and let $C \subset L_\rho$ be nonempty and ρ-closed. A mapping $T : C \to C$ is called a ρ-*Lipschitzian* if there exists a constant $0 < L$ such that

$$\rho(T(f) - T(g)) \leq L\rho(f - g), \quad \text{for any } f, g \in L_\rho.$$

Definition 7.2. A one-parameter family $\mathscr{F} = \{T_t : t \geq 0\}$ of mappings from C into itself is said to be a ρ-*Lipschitzian (respectively ρ-nonexpansive) semigroup* on C if $\mathscr{F}$ satisfies the following conditions:

© Springer International Publishing Switzerland 2015   185
M. A. Khamsi, W. M. Kozlowski, *Fixed Point Theory in Modular Function Spaces*,
DOI 10.1007/978-3-319-14051-3_7

(i) $T_0(x) = x$ for $x \in C$;

(ii) $T_{t+s}(x) = T_t(T_s(x))$ for $x \in C$ and $t, s \geq 0$;

(iii) for each $t \geq 0$, T_t is ρ-Lipschitzian (respectively ρ-nonexpansive).

Definition 7.3. A one-parameter family $\mathscr{F} = \{T_t : t \geq 0\}$ of mappings from C into itself is said to be a ρ-*contractive semigroup* on C if $\mathscr{F}$ satisfies the following conditions:

(i) $T_0(x) = x$ for $x \in C$;

(ii) $T_{t+s}(x) = T_t(T_s(x))$ for $x \in C$ and $t, s \geq 0$;

(iii) for each $t \geq 0$, T_t is a ρ-contraction with a constant $0 < L_t < 1$ such that $\limsup_{t \to \infty} L_t < 1$.

Definition 7.4. A semigroup $\mathscr{F} = \{T_t : t \geq 0\}$ is called *strongly continuous* if for every $z \in C$, the following function

$$\Lambda_z(t) = \rho\left(T_t(z) - z\right) \tag{7.1}$$

is continuous at every $t \in [0, \infty)$.

Definition 7.5. A semigroup $\mathscr{F} = \{T_t : t \geq 0\}$ is called *continuous* if for every $z \in C$, the mapping $t \longmapsto T_t(z)$ is ρ-continuous at every $t \in [0, \infty)$, i.e., $\rho\left(T_{t_n}(z) - T_t(z)\right) \to 0$ as $t_n \to t$.

7.2 Fixed Point Existence for Semigroups of Nonexpansive Mappings

Following [135], we will prove the existence of common fixed points for contractive and nonexpansive semigroups.

Theorem 7.1. *Let* $\rho \in \mathfrak{R}$. *Assume that* L_ρ *has the* ρ-*a.e. strong Opial property. Let* $C \subset E_\rho$ *be a nonempty, strongly* ρ-*bounded,* ρ-*a.e. compact convex set. Let* $\mathscr{F}$ *be a* ρ-*contractive semigroup on* C. *Then,* $\mathscr{F}$ *has a unique common fixed point* $z \in C$ *and for each* $u \in C$, $\rho\left(T_t(u) - z\right) \to 0$ *as* $t \to \infty$.

Proof. First, let us prove the uniqueness. Assume that $z, w \in \mathscr{F}$. Then we have

$$\rho(z - w) = \rho(T_t(z) - T_t(w)) \leq L_t \rho(z - w),$$

for any $t \geq 0$. If we let $t \to \infty$, we will get

$$\rho(z - w) \leq L\rho(z - w),$$

where

$$L = \limsup_{t \to \infty} L_t.$$

Since $L < 1$, we conclude that $\rho(z - w) = 0$ or equivalently that $z = w$, hence there must exist at most one common fixed point for $\mathscr{F}$.

To prove the existence of the common fixed point, let us fix any $x \in C$ and define the ρ-type τ by

$$\tau(u) = \limsup_{t \to \infty} \rho(T_t(x) - u), \tag{7.2}$$

for $u \in C$. By Theorem 4.8, there exists $z \in C$ such that

$$\tau(z) = \inf\{\tau(y) : y \in C\}.$$

Let us prove that $\tau(z) = 0$. To this end, take $s, t \geq 0$ and observe that by the nonexpansiveness of T_t we have

$$\rho(T_{s+t}(x) - T_t(z)) \leq L_t \rho(T_s(x) - z),$$

and then letting $s \to \infty$

$$\tau(T_t(z)) \leq L_t \tau(z)$$

which implies the following

$$\tau(z) = \inf\{\tau(y) : y \in C\} \leq \tau(T_t(z)) \leq L_t \, \tau(z).$$

Passing with $t \to \infty$ we get

$$\tau(z) \leq L\tau(z).$$

Since $0 < L < 1$ it follows that $\tau(z) = 0$. Hence, by (7.2),

$$\lim_{t \to \infty} \rho(T_t(x) - z) = 0. \tag{7.3}$$

Since T_s is a ρ-contraction for any given $s \geq 0$, it follows that

$$\lim_{t \to \infty} \rho(T_t(x) - T_s(z)) = \lim_{t \to \infty} \rho(T_{s+t}(x) - T_s(z)) \leq \lim_{t \to \infty} \rho(T_t(x) - z) = 0. \tag{7.4}$$

By the uniqueness of the ρ-limit, we conclude from (7.3) and (7.4) that $T_s(z) = z$ for any $s \geq 0$, i.e $z \in \mathscr{F}$.

To prove the convergence of orbits, let us fix any $u \in C$. We shall prove that $\{T_t(u)\}$ converges to z. Indeed, we have

$$\rho(T_{t+s}(u) - z) = \rho(T_{t+s}(u) - T_t(z)) \leq L_t \rho(T_s(u) - z),$$

for any $t, s \geq 0$. Hence

$$\limsup_{s \to \infty} \rho(T_{t+s}(u) - z) \leq \limsup_{s \to \infty} L_t \, \rho(T_s(u) - z).$$

Since $\limsup_{s\to\infty}\rho(T_{t+s}(u)-z) = \limsup_{t\to\infty}\rho(T_s(u)-z)$, we get

$$\limsup_{s\to\infty}\rho(T_s(u)-z) \le L_t \limsup_{s\to\infty}\rho(T_s(u)-z),$$

for any $t \ge 0$. If we let $t \to \infty$, we obtain

$$\limsup_{s\to\infty}\rho(T_s(u)-z) \le L\limsup_{s\to\infty}\rho(T_s(u)-z).$$

Since $L < 1$, we get

$$\limsup_{s\to\infty}\rho(T_s(u)-z) = 0.$$

Clearly, we can derive the same equality where $\limsup$ is replaced by $\liminf$ which implies the desired conclusion: $\lim_{s\to\infty}\rho(T_s(u)-z) = 0$.

□

Let us prove now the existence of common fixed points for ρ-nonexpansive semigroups.

Theorem 7.2. *Assume $\rho \in \Re$ is (UUC1). Let C be a ρ-closed ρ-bounded convex nonempty subset. Let $\mathscr{F}$ be a ρ-nonexpansive semigroup on C. Then, the set $F(\mathscr{F})$ of common fixed points is nonempty, ρ-closed, and convex.*

Proof. Let us fix an $x \in C$ and define the ρ-type

$$\tau(y) = \limsup_{t\to\infty}\rho(T_t(x)-y),$$

where $y \in C$. Let $\tau_0 = \inf\{\tau(y) : y \in C\}$ and let $\{z_n\}$ be a minimizing sequence for τ, i.e.,

$$\lim_{n\to\infty}\tau(z_n) = \tau_0.$$

By Lemma 4.6, there exists a $z \in C$ such that

$$\lim_{n\to\infty}\rho(z_n-z) = 0.$$

We are going to prove that $z \in F(\mathscr{F})$. Noting that

$$\rho(T_{s+t}(x) - T_t(y)) \le \rho(T_s(x) - t)$$

and passing with $s \to \infty$ we get

$$\tau(T_t(y)) \le \tau(y).$$

In particular, for any $n \ge 1$ we have

$$\tau(T_t(z_n)) \le \tau(z_n). \tag{7.5}$$

Let us fix any sequence $t_k \to \infty$ and note that for every $s > 0$ the sequence $\{T_{t_k+s}(z_k)\}$ is a minimizing sequence for τ. Indeed, using (7.5) we obtain

$$\tau_0 \le \tau(T_{t_k+s}(z_k)) \le \tau(z_k) \to \tau_0.$$

Hence, Lemma 4.6 implies that

$$\lim_{k \to \infty} \rho(T_{t_k+s}(z_k) - z) = 0, \tag{7.6}$$

and in particular for $s = 0$,

$$\lim_{k \to \infty} \rho(T_{t_k}(z_k) - z) = 0. \tag{7.7}$$

Using (7.7) we get

$$\rho(T_{t_k+s}(z_k) - T_s(z)) \le \rho(T_{t_k}(z_k) - z) \to 0.$$

Since the ρ-limit is unique, (7.6) and (7.2) give $T_s(z) = z$, i.e., $z \in F(\mathscr{F})$ as claimed. Let us prove that $F(\mathscr{F})$ is ρ-closed. Let $x_n \in F(\mathscr{F})$ and $\rho(x_n - x) \to 0$. Observe that for every $t \ge 0$,

$$\rho\left(\frac{1}{3}(T_t(x) - x)\right) \le \rho(T_t(x) - T_t(x_n)) + \rho(T_t(x_n) - x_n) + \rho(x_n - x)$$

$$\le \rho(x_n - x) + \rho(x_n - x) \to 0.$$

Hence, $x \in F(T_t)$ proving that $x \in F(\mathscr{F})$, and consequently, that $F(\mathscr{F})$ is ρ-closed.

To prove convexity of $F(\mathscr{F})$, we need to demonstrate that

$$w = \frac{u+v}{2} \in F(\mathscr{F})$$

provided $u, v \in F(\mathscr{F})$. Indeed, let $t > 0$. Define $x = T_t(w) - u$, $y = T_t(w) - v$. Note that

$$\frac{x+y}{2} = T_t(w) - w.$$

Define

$$r = \rho\left(\frac{v-u}{2}\right)$$

and observe that

$$\rho(x) = \rho(T_t(w) - u) = \rho(T_t(w) - T_t(u)) \le \rho(w - u) = \rho\left(\frac{v-u}{2}\right) = r.$$

Similarly, $\rho(y) \leq r$. Hence, $x, y \in D_2(r, 1)$ and therefore

$$\delta_2(r, 1) \leq 1 - \frac{1}{r}\rho\left(\frac{x+y}{2}\right) = 1 - \frac{1}{r}\rho(T_t(w) - w).$$

Using the assumed $(UUC1)$ and Proposition 4.1, we conclude that ρ satisfies $(UC2)$. Hence by Lemma 4.4, $\delta_2(r, 1) = 1$, which yields

$$\frac{1}{r}\rho(T_t(w) - w) \leq 1 - \delta_2(r, 1) = 0.$$

Therefore, $T_t(w) = w$, i.e., $w \in F(\mathscr{F})$, completing the proof.
□

Remark 7.1. Recently, Al-Mezel, Al-Roqi, and Khamsi, provided in [7] a partial generalization of the second theorem showing, under suitable assumptions, the existence of common fixed points of a commutative family of ρ-nonexpansive mappings defined on a ρ-closed ρ-bounded convex nonempty subset of L_ρ.

7.3 Characterization of the Set of Common Fixed Points

In this section, we will show how, under suitable assumptions, the set of common fixed points of a ρ-nonexpansive semigroup can be represented as the set of fixed points of just one ρ-nonexpansive mapping or—in a different context—by an intersection of the fixed point sets of two suitably chosen ρ-nonexpansive mappings. We will use these representation theorems to show how to construct common fixed points of semigroups of nonlinear mappings. The idea of such representations of semigroups in Banach spaces was developed by Suzuki, [192, 193] and then extended to the setting of modular function spaces by Kozlowski [140] and by Alsulami and Kozlowski [9].

Let $0 < \alpha < \beta$, and $\alpha \leq s \leq \beta$. Define inductively a sequence $\{A_n(s)\}$ of subsets of $[\alpha, \beta]$ by

$$A_1(s) = \{s\}, \quad A_{n+1}(s) = \bigcup_{t \in A_n(s)} \{|\alpha - t|, |\beta - t|\} \ for \ n \in \mathbb{N}.$$

Set

$$A(s) = \bigcup_{n=1}^{\infty} A_n(s).$$

We will use the following three results concerning real numbers.

Lemma 7.1. *[193] If α/β is an irrational number, then for every $s \in [\alpha, \beta]$, the closure of $A(s)$ is equal to $[0, \beta]$.*

Lemma 7.2. *[193] If α/β is a rational number, then for every $s \in [\alpha,\beta]$, the set $A(s)$ is finite.*

Lemma 7.3. *[192] Let t be a nonnegative real number and let $\{\beta_n\}$ be a sequence in $(0,\infty)$ converging to 0. Define sequences $\{\delta_n\}$ in $(0,\infty)$ and $\{k_n\}$ in $\mathbb{N} \cup \{0\}$ as follows:*

1. $\delta_1 = t$;
2. $k_n = [\delta_n/\beta_n]$, for $n \in \mathbb{N}$;
3. $\delta_{n+1} = \delta_n - k_n\beta_n$, for $n \in \mathbb{N}$.

Then for all $n \in \mathbb{N}$ the following hold:

1. $0 \leq \delta_{n+1} < \beta_n$;
2. $\delta_n \xrightarrow{n} 0$;
3. $\displaystyle\sum_{j=1}^{n} k_j\beta_j + \delta_{n+1} = 1$;
4. $\displaystyle\sum_{j=1}^{\infty} k_j\beta_j = t$.

We are now ready to prove our main theorem about representation of a set of common fixed points by a set of fixed points of one nonexpansive mapping.

Theorem 7.3. *[140] Let $\mathscr{F} = \{T_t : t \geq 0\}$ be a strongly continuous nonexpansive semigroup on a ρ-bounded subset C of a modular function space L_ρ, where $\rho \in \mathfrak{R}$. Let α and β be positive real numbers such that α/β is an irrational number. Let $\lambda \in (0,1)$ be arbitrary. Then*

$$F(\mathscr{F}) = F\left(\lambda T_\alpha + (1-\lambda)T_\beta\right).$$

Proof. Note first that the mapping $\lambda T_\alpha + (1-\lambda)T_\beta$ is nonexpansive with respect to the convex modular ρ. Without loosing generality we can assume that $\alpha < \beta$. It suffices to prove

$$F\left(\lambda T_\alpha + (1-\lambda)T_\beta\right) \subset F(\mathscr{F}),$$

as the opposite direction inclusion is trivial. Also, this inclusion is trivial when $F(\lambda T_\alpha + (1-\lambda)T_\beta) = \emptyset$. We can assume, therefore, that there exists $w \in C$ such that

$$\lambda T_\alpha(w) + (1-\lambda)T_\beta(w) = w. \qquad (7.8)$$

Since $\mathscr{F}$ is a strongly continuous semigroup it follows that the function Λ_w is a continuous real-valued function on the interval $[0,\beta]$ and hence it attains its maximum at a number $\tau \in [0,\beta]$ which means that

$$\rho\left(T_\tau(w) - w\right) = \max\left\{\rho\left(T_t(w) - w\right) : t \in [0,\beta]\right\}.$$

Since $\tau \in A(\tau) \subset [0, \beta]$, we have

$$\rho\left(T_\tau(w) - w\right) = \max\left\{\rho\left(T_s(w) - w\right) : s \in A(\tau)\right\}. \tag{7.9}$$

Let us prove by induction that for every $n \in \mathbb{N}$ and any $s \in A_n(\tau)$,

$$\rho\left(T_\tau(w) - w\right) = \rho\left(T_s(w) - w\right). \tag{7.10}$$

Since $A_1(\tau) = \{\tau\}$, (7.10) is true for $n = 1$. Let us make an inductive assumption that (7.10) holds for $n = k$. Fix arbitrary $t \in A_k(\tau)$. By the inductive assumption we have

$$\rho\left(T_\tau(w) - w\right) = \rho\left(T_t(w) - w\right). \tag{7.11}$$

Substituting (7.8) into the right hand side of (7.11) and using the convexity of ρ, and then using nonexpansiveness, we obtain the following

$$\rho\left(T_\tau(w) - w\right) \leq \lambda\rho\left(T_t(w) - T_\alpha w\right) + (1 - \lambda)\rho\left(T_t(w) - T_\beta w\right)$$
$$\leq \lambda\rho\left(T_{|\alpha-t|}(w) - w\right) + (1 - \lambda)\rho\left(T_{|\beta-t|}(w) - w\right)$$
$$\leq \rho\left(T_\tau(w) - w\right),$$

where the last inequality comes from the fact that $|\alpha - t|$ and $|\beta - t|$ belong to $A_{k+1}(\tau) \subset A(\tau)$ and from (7.9). From (7.3) it follows easily that

$$\rho\left(T_\tau(w) - w\right) = \rho\left(T_{|\alpha-t|}(w)\right) + \rho\left(T_{|\beta-t|}(w)\right).$$

By arbitrariness of $t \in A_k(\tau)$, we conclude that (7.10) holds for $k + 1$ and hence by induction it also holds for all natural n. From (7.10) it follows that for any $s \in A(\tau)$,

$$\rho\left(T_\tau(w) - w\right) = \rho\left(T_s(w) - w\right). \tag{7.12}$$

By Lemma 7.1, $A(\tau)$ is dense in $[0, \beta]$. Hence, by the continuity of Λ_w we deduce from (7.12) that

$$\rho\left(T_\tau(w) - w\right) = \rho\left(T_s(w) - w\right),$$

for every $s \in [0, \beta]$. Substituting $s = 0$ and remembering that $T_0(w) = w$ we have

$$\rho\left(T_\tau(w) - w\right) = 0,$$

and consequently $\rho\left(T_s(w) - w\right) = 0$ for every $s \in [0, \beta]$ implying $T_s(w) = w$ for any $s \in [0, \beta]$. Let t be any positive real number, hence $t = n\beta + s$ for a $n \in \mathbb{N} \cup \{0\}$ and

$s \in [0, \beta]$. Therefore

$$T_t(w) = T_\beta^n \circ T_s(w) = T_\beta^n(w) = w,$$

which means that $w \in F(\mathscr{F})$, as claimed.
$\square$

Please note that we did not assume in Theorem 7.3 that the common fixed points actually exist. This theorem, however, reduces the question of existence of common fixed points to the question of existence of fixed points for each ρ-nonexpansive mapping. We have, therefore, the following result which corresponds to the Banach space results from [34, 193].

Theorem 7.4. *Let* $\mathscr{F} = \{T_t : t \geq 0\}$ *be a strongly continuous nonexpansive semigroup on a ρ-closed, ρ-bounded, and convex subset C of a modular function space L_ρ, where $\rho \in \mathfrak{R}$. Assume that every ρ-nonexpansive mapping on C has a fixed point. Then the set of common fixed points of $\mathscr{F}$ is nonempty.*

Remark 7.2. Please note that our Theorem 7.4 combined with the existence theorem for ρ-nonexpansive mappings (Theorem 5.7) gives us an alternative proof of the existence part of Theorem 7.2.

As we saw, the strong continuity assumption of the semigroup was of critical importance. Let us give an important example when this condition is satisfied. It is easy to see that if a semigroup $\mathscr{F}$ is continuous and the modular ρ is uniformly continuous then $\mathscr{F}$ is strongly continuous. Consequently, we have the following result.

Theorem 7.5. *Let* $\mathscr{F} = \{T_t : t \geq 0\}$ *be a continuous nonexpansive semigroup on a subset C of a modular function space L_ρ, where $\rho \in \mathfrak{R}$ is uniformly continuous. Let α and β be positive real numbers such that α/β is an irrational number. Let $\lambda \in (0, 1)$ be arbitrary. Then*

$$F(\mathscr{F}) = F\left(\lambda T_\alpha + (1 - \lambda)T_\beta\right).$$

A natural question arises: what representation of common fixed point sets can be achieved without assuming that the semigroup $\mathscr{F}$ is strongly continuous or without the uniform continuity of ρ? It turns out that we can indeed characterize finite intersections of sets of fixed points. Let us limit ourselves to the case of two mappings. The proof of the general case follows the same pattern.

Theorem 7.6. *Let* $\mathscr{F} = \{T_t : t \geq 0\}$ *be a nonexpansive semigroup on a subset C of a modular function space L_ρ, where $\rho \in \mathfrak{R}$. Let α and β be positive real numbers such that α/β is a rational number. Let $\lambda \in (0, 1)$ be arbitrary. Then*

$$F(T_\alpha) \cap F(T_\beta) = F\left(\lambda T_\alpha + (1 - \lambda)T_\beta\right).$$

Proof. Assume that $\alpha < \beta$. It suffices to prove that

$$F\left(\lambda T_\alpha + (1 - \lambda)T_\beta\right) \subset F(T_\alpha) \cap F(T_\beta).$$

Fix an arbitrary $w \in F\left(\lambda T_\alpha + (1 - \lambda)T_\beta\right)$. By Lemma 7.2, the set $A(0)$ is finite. Therefore, there exists $\tau \in A(0)$ such that

$$\rho\left(T_\tau(w) - w\right) = \max\left\{\rho\left(T_s(w) - w\right) : s \in A(0)\right\}.$$

Arguing like in the proof of Theorem 7.3, we conclude that

$$\rho\left(T_\tau(w) - w\right) = \rho\left(T_s(w) - w\right),$$

for every $s \in A(0)$. Since $0 \in A(0)$ and $T_0(w) = w$, it follows that

$$\rho\left(T_\tau(w) - w\right) = 0.$$

Using this and the fact that both $\alpha \in A(0)$ and $\beta \in A(0)$, we have

$$\rho\left(T_\alpha(w) - w\right) = \rho\left(T_\beta(w) - w\right) = \rho\left(T_\tau(w) - w\right) = 0,$$

which implies that

$$T_\alpha(w) = T_\beta(w) = w.$$

The proof is complete.
□

Let us prove the following result which relates to Bruck's theorem in Banach spaces, see [34].

Theorem 7.7. *[9] Let $\rho \in \Re$ be a strictly convex function modular. Let $C \subset L_\rho$ and let T and S be two ρ-nonexpansive mappings from C into L_ρ with a common fixed point. Then for each $\lambda \in (0,1)$, a mapping $U : C \to L_\rho$ defined by*

$$U(x) = \lambda S(x) + (1 - \lambda)T(x),$$

for $x \in L_\rho$ is ρ-nonexpansive and $F(U) = F(S) \cap F(T)$.

Proof. A straightforward calculation shows that the mapping U is ρ-nonexpansive. It is also clear that $F(S) \cap F(T) \subset F(U)$. Therefore, to complete the proof we need only to prove the converse inclusion.

To this end, let us fix $x \in F(U)$ and $w \in F(S) \cap F(T)$. Let us calculate:

$$\rho(x-w) = \rho\left(\lambda S(x) + (1-\lambda)T(x) - w\right)$$
$$\leq \lambda\rho\left(S(x) - w\right) + (1-\lambda)\rho\left(T(x) - w\right)$$
$$= \lambda\rho\left(S(x) - S(w)\right) + (1-\lambda)\rho\left(T(x) - T(w)\right) \qquad (7.13)$$
$$\leq \lambda\rho(x-w) + (1-\lambda)\rho(x-w).$$

In particular, (7.13) yields the following

$$\rho(x-w) = \rho\left(S(x) - w\right) = \rho\left(T(x) - w\right). \qquad (7.14)$$

Indeed, let us assume to the contrary that

$$\rho\left(S(x) - w\right) < \rho(x-w). \qquad (7.15)$$

Combining (7.13) with (7.15) we have

$$\rho(x-w) \leq \lambda\rho\left(S(x) - w\right) + (1-\lambda)\rho\left(T(x) - w\right)$$
$$< \lambda\rho(x-w) + (1-\lambda)\rho(x-w) = \rho(x-w), \qquad (7.16)$$

which is impossible. Since the same reasoning can be applied assuming that

$$\rho\left(T(x) - w\right) < \rho(x-w),$$

we conclude that the claim (7.14) holds. Set $f = S(x) - w$ and $g = T(x) - w$ and observe that (7.14) implies that $\rho(f) = \rho(g)$. Straight calculation shows that

$$\rho\left(\lambda f + (1-\lambda)g\right) = \rho\left(U(x) - w\right). \qquad (7.17)$$

On the other hand, it follows from (7.14) and from the assumption $x \in F(U)$ that

$$\lambda\rho(f) + (1-\lambda)\rho(g) = \lambda\rho\left(S(x) - w\right) + (1-\lambda)\rho\left(T(x) - w\right)$$
$$= \lambda\rho(x-w) + (1-\lambda)\rho(x-w) = \rho(x-w) = \rho(U(x) - w). \qquad (7.18)$$

Comparing (7.17) to (7.18), we obtain immediately

$$\rho\left(\lambda f + (1-\lambda)g\right) = \lambda\rho(f) + (1-\lambda)\rho(g), \qquad (7.19)$$

which by the strict convexity of ρ implies that $f = g$, and consequently that $S(x) = T(x)$. Compute

$$x = U(x) = \lambda S(x) + (1-\lambda)T(x) = \lambda S(x) + (1-\lambda)S(x) = S(x), \qquad (7.20)$$

and hence $x \in F(S)$. Similarly, we can prove that $x \in F(T)$. Hence, $x \in F(S) \cap F(T)$ as claimed.

□

Our next result, interesting by itself, will be used in the proof of our representation result, Theorem 7.9.

Theorem 7.8. *[9] Let $\rho \in \Re$ and let $\mathscr{F} = \{T_t : t \geq 0\}$ be a continuous semigroup of mappings on a subset C of L_ρ. Let $\{\alpha_n\}$ be a sequence of nonnegative numbers converging to $\alpha \in [0, \infty)$ such that $\alpha_n \neq \alpha$ for all $n \in \mathbb{N}$. Then we have the following representation of the set of all common fixed points of $\mathscr{F}$,*

$$F(\mathscr{F}) = \bigcap_{n=1}^{\infty} F(T_{\alpha_n}). \qquad (7.21)$$

Proof. We only need to prove that $\bigcap_{n=1}^{\infty} F(T_{\alpha_n}) \subset F(\mathscr{F})$ as the other direction is trivial. Let $z \in C$ be such that $T_{\alpha_n}(z) = z$ for every $n \in \mathbb{N}$. Observe first that if $\{t_n\}$ is a sequence of nonnegative real numbers such that $T_{t_n}(z) = z$ and $t_n \to t$, where $t \in [0, \infty)$, then $T_t(z) = z$. Indeed,

$$\rho\left(\frac{T_t(z) - z}{2}\right) \leq \frac{1}{2}\rho\left(T_t(z) - T_{t_n}(z)\right) + \frac{1}{2}\rho\left(T_{t_n}(z) - z\right)$$

$$= \frac{1}{2}\rho\left(T_t(z) - T_{t_n}(z)\right) \to 0,$$

by the continuity of $\mathscr{F}$. Hence $T_t(z) = z$ as claimed.

The above observation implies, in particular, that $T_\alpha(z) = z$. Let us define $\beta_n = |\alpha_n - \alpha| > 0$, where $n \in \mathbb{N}$. It follows from the assumptions that each β_n is a positive real number and that $\beta_n \to 0$. Note that

$$\max\{\alpha_n, \alpha\} = \min\{\alpha_n, \alpha\} + \beta_n.$$

Hence, denoting $m_n = \min\{\alpha_n, \alpha\}$ and $M_n = \max\{\alpha_n, \alpha\}$, we have

$$T_{\beta_n}(z) = T_{\beta_n} \circ T_{m_n}(z) = T_{\beta_n + m_n}(z) = T_{M_n}(z) = z.$$

Fix any $t > 0$. By Lemma 7.3, there exists a sequence $\{k_n\}$ in $\mathbb{N} \cup \{0\}$ such that

$$t = \sum_{i=1}^{\infty} k_i \beta_i.$$

Denoting

$$s_n = \sum_{i=1}^{n} k_i \beta_i,$$

we obtain for each $n \in \mathbb{N}$ with $s_n > 0$

$$T_{s_n}(z) = T_{\beta_n}^{k_n} \circ T_{\beta_{n-1}}^{k_{n-1}} \circ \cdots \circ T_{\beta_2}^{k_2} \circ T_{\beta_1}^{k_1}(z) = z.$$

Since $T_0(z) = z$, it follows that $T_{s_n}(z) = z$ for every $n \in \mathbb{N}$. Because $s_n \to t$ and $\mathscr{F}$ is a continuous semigroup, we conclude, as previously observed, that $T_t(z) = z$ which concludes the proof of the theorem.
$\square$

The following technical result about real numbers (Lemma 3 in [192]) will be used in the proof of our next representation theorem.

Lemma 7.4. *[192] Let α and β be positive real numbers satisfying $\alpha/\beta \notin \mathbb{Q}$. Define sequences $\{\alpha_n\}$ in $(0, \infty)$ and $\{k_n\}$ in $\mathbb{N}$ as follows:*

1. $\alpha_1 = \max\{\alpha, \beta\}$;
2. $\alpha_2 = \min\{\alpha, \beta\}$;
3. $k_n = [\alpha_n/\alpha_{n+1}]$, for all $n \in \mathbb{N}$;
4. $\alpha_{n+2} = \alpha_n - k_n \alpha_{n+1}$, for all $n \in \mathbb{N}$.

   Then the following hold:

1. $0 < \alpha_{n+1} < \alpha_n$, for all $n \in \mathbb{N}$;
2. $k_n \in \mathbb{N}$, for all $n \in \mathbb{N}$;
3. $\alpha_n/\alpha_{n+1} \notin \mathbb{Q}$, for all $n \in \mathbb{N}$;
4. $\{\alpha_n\}$ converges to 0.

In our next result, we will represent the set of all common fixed points of a semigroup by an intersection of just two mappings from this semigroup. In contrast to Theorem 7.6 we will assume only that the semigroup is continuous, i.e., we will not require the strong continuity.

Theorem 7.9. *[9] Let $\rho \in \mathfrak{R}$ and let $\mathscr{F} = \{T_t : t \geq 0\}$ be a continuous semigroup of mappings on a subset C of L_ρ. Let $\alpha > 0$ and $\beta > 0$ be two real numbers such that $\alpha/\beta \notin \mathbb{Q}$. Then*

$$F(\mathscr{F}) = F(T_\alpha) \cap F(T_\beta).$$

Proof. We only need to prove that

$$F(T_\alpha) \cap F(T_\beta) \subset F(\mathscr{F}) \tag{7.22}$$

as the converse inclusion is obvious. To this end, let us fix $z \in C$ such that $z \in F(T_\alpha) \cap F(T_\beta)$. Let $\{\alpha_n\}$ in $(0, \infty)$ and $\{k_n\}$ in $\mathbb{N}$ be two sequences defined as in Lemma 7.4. We will show that

$$T_{\alpha_n}(z) = T_{\alpha_{n+1}}(z) = z, \tag{7.23}$$

for every $n \in \mathbb{N}$. Observe that

$$T_{\alpha_1}(z) = z \quad \text{and} \quad T_{\alpha_2}(z) = z.$$

Thus, Eq. (7.23) holds for $n = 1$. Now, suppose that $T_{\alpha_n}(z) = T_{\alpha_{n+1}}(z) = z$. Then, we have

$$T_{\alpha_{n+2}}(z) = T_{\alpha_{n+2}} \circ T^{k_n}_{\alpha_{n+1}}(z) = T_{\alpha_{n+2}+k_n\alpha_{n+1}}(z) = T_{\alpha_n}(z) = T_{\alpha_{n+1}}(z) = z.$$

Hence, by induction, $T_{\alpha_n}(z) = z$ for every $n \in \mathbb{N}$ and consequently

$$F(T_\alpha) \cap F(T_\beta) \subset \bigcap_{n=1}^{\infty} F\left(T_{\alpha_n}\right). \tag{7.24}$$

Since, by construction, $\{\alpha_n\}$ is a sequence of positive numbers converging to $0 \in [0, \infty)$, it follows from Theorem 7.8 that

$$F(\mathscr{F}) = \bigcap_{n=1}^{\infty} F\left(T_{\alpha_n}\right). \tag{7.25}$$

Combining (7.24) with (7.25) we obtain the desired inclusion (7.22) which completes the proof.
□

The next result is an immediate consequence of the fact that the uniform convexity $(UUC1)$ implies the strict convexity (SC) of ρ, and of Theorems 7.2, 7.7, and 7.9.

Theorem 7.10. *[9] Let $\rho \in \mathfrak{R}$ be $(UUC1)$, and let $\mathscr{F} = \{T_t : t \geq 0\}$ be a continuous semigroup of ρ-nonexpansive mappings on a ρ-closed, ρ-bounded, convex, nonempty subset of L_ρ. Let $\alpha > 0$ and $\beta > 0$ be two real numbers such that $\alpha/\beta \notin \mathbb{Q}$. Fix an arbitrary $\lambda \in (0,1)$. Then $F(\mathscr{F}) \neq \emptyset$ and*

$$F(\mathscr{F}) = F\left(\lambda T_\alpha + (1 - \lambda)T_\beta\right). \tag{7.26}$$

Remark 7.3. Note that we already obtained the conclusion of Theorem 7.10 in the case when the assumption of the uniform convexity was replaced by that the strong continuity of the semigroup $\mathscr{F}$, see Theorem 7.5.

7.4 Convergence of Mann Iteration Processes

We concluded the previous section with Theorem 7.10 which says that, under suitable assumptions, the set of all common fixed points of a continuous semigroup of ρ-nonexpansive mappings is nonempty and can be represented as the set of all fixed points of just one ρ-nonexpansive mapping. In this section, we will demonstrate how this result can be applied to the construction of such a common fixed point. This idea can be summarized as follows: using the results of the previous sections, we can reduce a problem of constructing a common fixed point for a semigroup of mappings to a problem of constructing a fixed point for just one ρ-nonexpansive mapping.

In the current section, we will prove the convergence of the Mann iterative process to a common fixed point of a continuous semigroup. Let us start with the definition of the Mann process, see [161] for the original definition. Compare also with Sect. 2.6 of this book.

Definition 7.6. Let $\rho \in \mathfrak{R}$, $C \subset L_\rho$, and let T be a ρ-nonexpansive self-mapping on C. Let $\sigma \in (0,1)$. The *Mann iteration process* generated by the mapping T and the constant σ, denoted by $M(T,\sigma)$, is defined by the following iterative formula:

$$x_{k+1} = \sigma T(x_k) + (1-\sigma)x_k,$$

where $x_1 \in C$ is chosen arbitrarily.

Let us start with an important technical result which corresponds to Lemma 6.3. The latter one was proved in a more complex settings of asymptotic ρ-nonexpansive mappings.

Lemma 7.5. *Let $\rho \in \mathfrak{R}$ be (UUC1), $C \subset L_\rho$ be a ρ-closed, ρ-bounded, and convex set. Let $T : C \to C$ be ρ-nonexpansive and let $\sigma \in (0,1)$. Denote by $\{x_k\}$ a sequence of elements of C generated by a Mann process $M(T,\sigma)$. Assume that w is a fixed point of T. Then there exists $r \in \mathbb{R}$ such that*

$$\lim_{k \to \infty} \rho(x_k - w) = r.$$

Proof. Since

$$\begin{aligned}
\rho(x_{k+1} - w) &\le \sigma\rho(T(x_k) - w) + (1-\sigma)\rho(x_k - w) \\
&= \sigma\rho(T(x_k) - T(w)) + (1-\sigma)\rho(x_k - w) \\
&\le \sigma\rho(x_k - w) + (1-\sigma)\rho(x_k - w) \\
&= \rho(x_k - w),
\end{aligned}$$

it follows that $\{\rho(x_k - w)\}$ is a nonincreasing sequence of nonnegative numbers, hence it is convergent to a number $r \in \mathbb{R}$.
□

Lemma 7.6. *Let $\rho \in \mathfrak{R}$ be (UUC1), $C \subset L_\rho$ be a ρ-closed, ρ-bounded, and convex set. Let $T : C \to C$ be ρ-nonexpansive and let $\sigma \in (0,1)$. Denote by $\{x_k\}$ a sequence of elements of C generated by a Mann process $M(T,\sigma)$. Then*

$$\lim_{k \to \infty} \rho(T(x_k) - x_k) = 0,$$

and

$$\lim_{k \to \infty} \rho(x_{k+1} - x_k) = 0.$$

Proof. By Theorem 5.7, T has at least one fixed point $w \in C$. In view of Lemma 7.5, there exists $r \in \mathbb{R}$ such that

$$\lim_{k \to \infty} \rho(x_k - w) = r. \qquad (7.27)$$

Note that

$$\limsup_{k \to \infty} \rho(T(x_k) - w) = \limsup_{k \to \infty} \rho(T(x_k) - T(w)) \leq \limsup_{k \to \infty} \rho(x_k - w) \leq r, \quad (7.28)$$

and that

$$\lim_{k \to \infty} \rho(\sigma(T(x_k) - w) + (1 - \sigma)(x_k - w)) = \lim_{k \to \infty} \rho(x_{k+1} - w) = r.$$

Set $f_k = T(x_k) - w$, $g_k = x_k - w$, and note that $\limsup_{k \to \infty} \rho(g_k) \leq r$ by (7.27), and $\limsup_{k \to \infty} \rho(f_k) \leq r$ by (7.28). Observe also that

$$\lim_{k \to \infty} \rho(\sigma f_k + (1 - \sigma)g_k) = \lim_{k \to \infty} \rho(\sigma T(x_k) + (1 - \sigma)x_k - w)$$
$$= \lim_{k \to \infty} \rho(x_{k+1} - w) = r.$$

Hence, it follows from Lemma 4.2 that

$$\lim_{k \to \infty} \rho(T(x_k) - x_k) = \lim_{k \to \infty} \rho(f_k - g_k) = 0,$$

which by the construction of the sequence $\{x_k\}$ is equivalent to

$$\lim_{k \to \infty} \rho(x_{k+1} - x_k) = 0,$$

as claimed.
□

We are now ready to prove the following version of the Mann process convergence theorem for a single ρ-nonexpansive mapping.

Theorem 7.11. *[9] Let $\rho \in \mathfrak{R}$. Assume that*

1. *ρ is (UUC1),*
2. *ρ has strong Opial property,*
3. *ρ has Δ_2 property and is uniformly continuous.*

Let $C \subset L_\rho$ be nonempty, $\rho - a.e.$ compact, convex, strongly ρ-bounded and ρ-closed. Let $T : C \to C$ be ρ-nonexpansive and $\sigma \in (0, 1)$. Denote by $\{x_k\}$ a sequence of elements of C generated by a Mann process $M(T, \sigma)$. Then there exists $x \in F(T)$ such that $x_n \to x$ ρ-a.e.

Proof. Observe that by Theorem 5.7 the set of fixed points $F(T)$ is nonempty, convex, and ρ-closed. By Lemma 7.6, the sequence $\{x_k\}$ is an approximate fixed point sequence, that is,

$$\rho(T(x_k) - x_k) \to 0$$

as $k \to \infty$. Consider $y, z \in C$, two ρ-a.e. cluster points of $\{x_k\}$. There exits then $\{y_k\}$, $\{z_k\}$ subsequences of $\{x_k\}$ such that $y_k \to y$ ρ-a.e., and $z_k \to z$ ρ-a.e. By Theorem 6.1, $y \in F(T)$ and $z \in F(T)$. By Lemma 7.5, there exist $r_y, r_z \in \mathbb{R}$ such that

$$r_y = \lim_{k \to} \rho(x_k - y), \ r_z = \lim_{k \to} \rho(x_k - z).$$

We claim that $y = z$. Assume to the contrary that $y \neq z$. Then by the strong Opial property we have

$$r_y = \liminf_{k \to \infty} \rho(y_k - y) < \liminf_{k \to \infty} \rho(y_k - z)$$
$$= \liminf_{k \to \infty} \rho(z_k - z) < \liminf_{k \to \infty} \rho(z_k - y) = r_y.$$

The contradiction implies that $y = z$. Therefore, $\{x_k\}$ has at most one ρ-a.e. cluster point. Since C is ρ-a.e. compact it follows that the sequence $\{x_k\}$ has exactly one ρ-a.e. cluster point, which means that $\rho(x_k) \to x$ ρ-a.e. Using Theorem 6.1 again, we get $x \in F(T)$ as claimed. $\square$

Let us combine now Theorem 7.11 with Theorem 7.10 to demonstrate the convergence of an iterative algorithm to a common fixed point of a semigroup of nonlinear mappings in modular function spaces.

Theorem 7.12. *[9] Let $\rho \in \mathfrak{R}$. Assume that*

1. ρ is (UUC1),
2. ρ has strong Opial property,
3. ρ has Δ_2 property and is uniformly continuous.

Let $C \subset L_\rho$ be nonempty, $\rho - a.e.$ compact, convex, strongly ρ-bounded and ρ-closed. Let $\mathscr{F} = \{T_t : t \geq 0\}$ be a continuous semigroup of ρ-nonexpansive mappings on C. Assume that $\alpha > 0$ and $\beta > 0$ are two real numbers such that $\alpha/\beta \notin \mathbb{Q}$. Fix $\lambda, \kappa \in (0,1)$ such that $\kappa + \lambda < 1$. Define a sequence $\{x_n\}$ in C by $x_1 \in C$ and

$$x_{n+1} = \kappa T_\alpha(x_n) + \lambda T_\beta(x_n) + (1 - \kappa - \lambda)x_n, \quad (7.29)$$

for natural $n \geq 2$. Then $\{x_n\}$ ρ-a.e. converges to a common fixed point of the semigroup $\mathscr{F}$.

Proof. Define a mapping S by

$$S = \frac{\kappa}{\kappa + \lambda} T_\alpha + \frac{\lambda}{\kappa + \lambda} T_\beta$$

and observe that $S : C \to C$ is ρ-nonexpansive. Fix any $x_1 \in C$ and let $\{x_n\}$ be generated by the Mann process $M(S, \sigma)$, where $\sigma = \kappa + \lambda$:

$$x_{n+1} = \sigma S(x_n) + (1 - \sigma)x_n,$$

which is exactly the sequence defined by (7.29). By Theorem 7.11, there exists $x \in F(S)$ such that $x_n \to x$ ρ-a.e. By Theorem 7.10, $F(\mathscr{F}) = F(S)$, hence $x \in F(\mathscr{F})$. The proof is complete.

$\square$

7.5 Convergence of Ishikawa Iteration Processes

Definition 7.7. Let $\rho \in \mathfrak{R}$, $C \subset L_\rho$, and let T be a ρ-nonexpansive self-mapping on C. Let $\sigma, \tau \in (0, 1)$. The *Ishikawa iteration process* generated by the mapping T and the constants σ and τ, denoted by $I(T, \sigma, \tau)$, is defined by the following iterative formula:

$$x_{k+1} = \sigma T(\tau T(x_k) + (1 - \tau)x_k) + (1 - \sigma)x_k,$$

where $x_1 \in C$ is chosen arbitrarily.

Lemma 7.7. *Let $\rho \in \mathfrak{R}$ be (UUC1). Let $C \subset L_\rho$ be a ρ-closed, ρ-bounded, and convex set. Let $T : C \to C$ be ρ-nonexpansive and let $\sigma, \tau \in (0, 1)$. Denote by $\{x_k\}$ a sequence of elements of C generated by the Ishikawa process $I(T, \sigma, \tau)$. Assume that w is a fixed point of T. Then, there exists then an $r \in \mathbb{R}$ such that $\lim_{k \to \infty} \rho(x_k - w) = r$.*

Proof. Using a similar calculation to the one used in the proof of Lemma 7.5, it is not difficult to prove that $\{x_k\}$ is a bounded nonincreasing sequence of nonnegative numbers, hence it is convergent to a number $r \in \mathbb{R}$.

$\square$

Lemma 7.8. *Let $\rho \in \mathfrak{R}$ be (UUC1). Let $C \subset L_\rho$ be a ρ-closed, ρ-bounded, and convex set. Let $T : C \to C$ be ρ-nonexpansive and Let $\sigma, \tau \in (0, 1)$. Denote by $\{x_k\}$ a sequence of elements of C generated by the Ishikawa process $I(T, \sigma, \tau)$. Define*

$$y_k = \tau T(x_k) + (1 - \tau)x_k.$$

Then

$$\lim_{k \to \infty} \rho(T(y_k) - x_k) = 0, \tag{7.30}$$

or equivalently

$$\lim_{k \to \infty} \rho(x_{k+1} - x_k) = 0, \tag{7.31}$$

Proof. By Theorem 5.7, $F(T) \neq \emptyset$. Let us fix $w \in F(T)$. By Lemma 7.7, $\lim_{k \to \infty} \rho(x_k - w)$ exists. Let us denote it by r. Since $w \in F(T)$, $T \in \mathcal{T}_r(C)$, and $\lim_{k \to \infty} \rho(x_k - w) = r$ by Lemma 7.7, we have the following:

$$\limsup_{k \to \infty} \rho(T(y_k) - w) = \limsup_{k \to \infty} \rho(T(y_k) - T(w))$$
$$\leq \limsup_{k \to \infty} \rho(y_k - w)$$
$$= \limsup_{k \to \infty} \rho(\tau T(x_k) + (1 - \tau)x_k - w)$$
$$\leq \limsup_{k \to \infty} \left(\tau \rho(T(x_k) - w) + (1 - \tau)\rho(x_k - w) \right)$$
$$\leq \limsup_{k \to \infty} \left(\tau \rho(x_k - w) + (1 - \tau)\rho(x_k - w) \right) \leq r.$$

Note that

$$\lim_{k \to \infty} \rho(\sigma(T(y_k) - w) + (1 - \sigma)(x_k - w))$$
$$= \lim_{k \to \infty} \rho(\sigma T(y_k) + (1 - \sigma)x_k - w) = \lim_{k \to \infty} \rho(x_{k+1} - w) = r.$$

Applying Lemma 4.2 with $u_k = T(y_k) - w$ and $v_k = x_k - w$, we obtain the desired equality $\lim_{k \to \infty} \rho(T(y_k) - x_k) = 0$, while (7.31) follows from (7.30) via the construction formulas for x_{k+1} and y_k. $\qquad \square$

Lemma 7.9. *Let* $\rho \in \Re$ *be* (UUC1) *satisfying* Δ_2. *Let* $C \subset L_\rho$ *be a* ρ-*closed,* ρ-*bounded, and convex set. Let* $T : C \to C$ *be* ρ-*nonexpansive and let* $\sigma, \tau \in (0, 1)$. *Denote by* $\{x_k\}$ *a sequence of elements of* C *generated by the Ishikawa process* $I(T, \sigma, \tau)$. *Then*

$$\lim_{k \to \infty} \rho(T(x_k) - x_k) = 0.$$

Proof. Let $y_k = \tau T(x_k) + (1 - \tau)x_k$. Hence,

$$T(x_k) - x_k = \frac{1}{1 - \tau}(T(x_k) - y_k).$$

Since $\tau \in (0, 1)$, there exists $0 < s < 1$ such that $\tau \leq s$. Hence,

$$\rho(T(x_k) - x_k) = \rho\left(\frac{1}{1 - \tau}(T(x_k) - y_k)\right) \leq \rho\left(\frac{1}{1 - s}(T(x_k) - y_k)\right).$$

The right-hand side of this inequality tends to zero because $\rho(T(x_k) - y_k) \to 0$ by Lemma 7.8 and because ρ satisfies Δ_2. $\qquad \square$

Using Lemma 7.9 instead of Lemma 7.6, and Lemma 7.8 instead of Lemma 7.5, and arguing in a similar way as in the proof of Theorem 7.11, we can obtain the following convergence result for the Ishikawa process.

Theorem 7.13. *Let* $\rho \in \Re$. *Assume that*

1. ρ is (UUC1),
2. ρ has strong Opial property,
3. ρ has Δ_2 property and is uniformly continuous.

Let $C \subset L_\rho$ be nonempty, $\rho - a.e.$ compact, convex, strongly ρ-bounded, and ρ-closed. Let $T : C \to C$ be ρ-nonexpansive and $\sigma \in (0,1)$, $\tau \in (0,1)$. Denote by $\{x_k\}$ a sequence of elements of C generated by an Ishikawa process $I(T, \sigma, \tau)$. Then there exists $x \in F(T)$ such that $x_n \to x$ ρ-a.e.

Again, let us combine Theorem 7.13 with Theorem 7.10 to demonstrate the convergence of an Ishikawa-type, two-step iterative algorithm to a common fixed point of a semigroup of nonlinear mappings in modular function spaces. Please note that, as said before, typical implementations of the Ishikawa algorithm are convergent at a faster pace than the corresponding Mann schema.

Theorem 7.14. *Let* $\rho \in \Re$. *Assume that*

1. ρ is (UUC1),
2. ρ has strong Opial property,
3. ρ has Δ_2 property and is uniformly continuous.

Let $C \subset L_\rho$ be nonempty, $\rho - a.e.$ compact, convex, strongly ρ-bounded, and ρ-closed. Let $\mathscr{F} = \{T_t : t \geq 0\}$ be a continuous semigroup of ρ-nonexpansive mappings on C. Assume that $\alpha > 0$ and $\beta > 0$ are two real numbers such that $\alpha/\beta \notin \mathbb{Q}$. Fix $\lambda, \kappa \in (0,1)$ such that $\kappa + \lambda < 1$. Define a sequence $\{x_n\}$ in C by $x_1 \in C$ and

$$x_{n+1} = \kappa T_\alpha \Big(\kappa T_\alpha(x_n) + \lambda T_\beta(x_n) + (1 - \kappa - \lambda)x_n \Big)$$
$$+ \lambda T_\beta \Big(\kappa T_\alpha(x_n) + \lambda T_\beta(x_n) + (1 - \kappa - \lambda)x_n \Big) + (1 - \kappa - \lambda)x_n, \qquad (7.32)$$

for natural $n \geq 2$. Then $\{x_n\}$ ρ-a.e. converges to a common fixed point of the semigroup $\mathscr{F}$.

Proof. Similarly as in the Mann process case, let us define a mapping S by

$$S = \frac{\kappa}{\kappa + \lambda} T_\alpha + \frac{\lambda}{\kappa + \lambda} T_\beta$$

and note that $S : C \to C$ is ρ-nonexpansive. Fix any $x_1 \in C$ and let $\{x_n\}$ be generated by the two-step Ishikawa process $I(S, \sigma, \tau)$, where $\sigma = \kappa + \lambda$ and $\tau = \kappa + \lambda$:

$$x_{n+1} = \sigma S(\tau S(x_n) + (1 - \tau)x_n) + (1 - \sigma)x_n,$$

which is exactly the sequence defined by (7.32). By Theorem 7.13, there exists $x \in F(S)$ such that $x_n \to x$ ρ-a.e. By Theorem 7.10, $F(\mathscr{F}) = F(S)$, hence $x \in F(\mathscr{F})$, which completes the proof.

$\square$

7.6 Applications to Differential Equations

One can ask a legitimate question about existence of natural examples of semi-groups of nonlinear mappings in modular function spaces and their applications. Let us address this issue by providing a suitable generic example.

In the context of Sect. 5.4 and in particular of Theorem 5.28, we introduce the following notation: let $f \in C$, by u_f we will denote a solution of the initial value problem

$$\begin{cases} u(0) = f \\ u'(t) + (I - T)u(t) = 0, \end{cases} \tag{7.33}$$

where $f \in C$ and $A > 0$ are fixed and $T : C \to C$ is ρ-nonexpansive.

The solution u_f is assumed to be obtained by the ρ-limit (5.26). For any $t \geq 0$, let us define a mapping $S_t : C \to C$ by

$$S_t(g) = u_g(t).$$

Denote $\mathscr{S} = \{S_t\}_{t \geq 0}$ and $F(\mathscr{S}) = \{f \in C : S_t(f) = f, \text{ for all } t \geq 0\}$. In this section, we will prove that $\mathscr{S}$ is a continuous ρ-nonexpansive semigroup of nonlinear mappings. This result will be used in the following section for proving existence and constructing stationary points of the process defined by the system (7.33).

First, let us introduce the following convenient notation:

$$U : C \times [0, +\infty) \ni (g,t) \mapsto U(g,t) = u_g(t) = S_t(g) \in C.$$

Also, for any $n \in \mathbb{N}$ define $U_n : C \times [0, +\infty) \to C$ by following the recurrent system

$$\begin{cases} U_0(g,t) = g \\ U_{n+1}(g,t) = e^{-t}g + \int_0^t e^{s-t}T(U_n(g,s))ds. \end{cases}$$

First we need to prove the following important technical result.

Lemma 7.10. *Let $\rho \in \mathfrak{R}$ be separable. Let $C \subset E_\rho$ be a nonempty, convex, ρ-bounded, ρ-closed set with the Vitali property. Let $T : C \to C$ be a ρ-nonexpansive mapping. Then for any $g \in C$, $t \geq 0$, $\mu \geq 0$, $n \in \mathbb{N}$, $m \in \mathbb{N}$*

$$\rho\left(U_n(U(g,\mu),t) - U_{n+m}(g,t+\mu)\right) \leq \sum_{i=n+1}^{n+m+1} K^i(\mu)\delta_\rho(C) + K^{n+1}(t)\delta_\rho(C), \tag{7.34}$$

where

$$K(t) = 1 - e^{-t} = \int_0^t e^{s-t}ds.$$

Proof. The proof is by induction on $n \in \mathbb{N}$. Assume first that $n = 0$ and calculate

$$U_0(U(g,\mu),t) - U_m(g,t+\mu) \tag{7.35}$$
$$= U(g,\mu) - U_m(g,t+\mu)$$
$$= u_g(\mu) - U_m(g,t+\mu)$$
$$= u_g(\mu) - e^{-t-\mu}g - \int_0^{t+\mu} e^{s-t-\mu}T(U_{m-1}(g,s))ds.$$

Using straightforward calculus we get

$$\int_0^{t+\mu} e^{s-t-\mu}T(U_{m-1}(g,s))ds$$
$$= \int_0^{\mu} e^{s-\mu}T(U_{m-1}(g,s))ds + \int_0^{t} e^s T(U_{m-1}(g,s+\mu))ds.$$

Using the definition of U_m and the formula (7.36) we obtain the following

$$U_m(g,t+\mu) = e^{-t}\left(e^{-\mu} + \int_0^{\mu} e^{s-\mu}T(U_{m-1}(g,s))ds\right) \tag{7.36}$$
$$+ e^{-t}\int_0^{t} e^s T(U_{m-1}(g,s+\mu))ds$$
$$= e^{-t}U_m(g,\mu) + e^{-t}\int_0^{t} e^s T(U_{m-1}(g,s+\mu))ds.$$

Substituting (7.36) into (7.35) we get

$$U_0(U(g,\mu),t) - U_m(g,t+\mu) \tag{7.37}$$
$$= u_g(\mu) - e^{-t}U_m(g,\mu) - e^{-t}\int_0^{t} e^s T(U_{m-1}(g,s+\mu))ds$$
$$= e^{-t}\left(u_g(\mu) - U_m(g,\mu)\right) + \int_0^{t} e^{s-t}\left(u_g(\mu) - T(U_{m-1}(g,s+\mu))\right)ds,$$

where we used the fact that

$$e^{-t}u_g(\mu) + \int_0^{t} e^{s-t}u_g(\mu)ds = u_g(\mu).$$

Let us apply Lemma 5.13 with

$$x(t) = u_g(\mu) - T(U_{m-1}(g,t+\mu)),$$
$$y(t) = u_g(\mu) - U_m(g,\mu).$$

Hence, using (5.22) from Lemma 5.13 and (7.37) we get

$$\rho\Big(U_0(U(g,\mu),t)-U_m(g,t+\mu)\Big)=\rho\Big(e^{-t}y(t)+\int_0^t e^{s-t}x(s)ds\Big) \qquad (7.38)$$

$$\leq e^{-t}\rho(y(t))+K(t)\sup_{0\leq s\leq t}\rho(x(s))$$

$$=e^{-t}\rho(u_g(\mu)-U_m(g,\mu))+K(t)\sup_{0\leq s\leq t}\rho\Big(u_g(\mu)-T(U_{m-1}(g,s+\mu))\Big)$$

$$\leq e^{-t}\rho(u_g(\mu)-U_m(g,\mu))+K(t)\delta_\rho(C).$$

Applying to (7.38) inequality (5.27) from Theorem 5.28 we finally arrive at

$$\rho\Big(U_0(U(g,\mu),t)-U_m(g,t+\mu)\Big)\leq K^{m+1}(t)\delta_\rho(C)+K(t)\delta_\rho(C),$$

which gives us desired inequality (7.34) with $n=0$.

Assume now that (7.34) is true for $n\in\mathbb{N}$ and let us prove it for $n+1$. Using the definition of the recurrent sequence combined with (7.36) we have

$$U_{n+1}(U(g,\mu),t)-U_{n+m+1}(g,t+\mu)$$

$$=e^{-t}U(g,\mu)+\int_0^t e^{s-t}T(U_n(U(g,\mu),s))ds-U_{n+m+1}(g,t+\mu)$$

$$=e^{-t}U(g,\mu)+\int_0^t e^{s-t}T(U_n(U(g,\mu),s))ds$$

$$-\Big(e^{-t}U_{n+m+1}(g,\mu)+e^{-t}\int_0^t e^s T(U_{n+m}(g,s+\mu))ds\Big)$$

$$=e^{-t}\Big(U(g,\mu)-U_{n+m+1}(g,\mu)\Big)+\int_0^t e^{s-t}\Big(T(U_n(U(g,\mu),s))-T(U_{n+m}(g,s+\mu))\Big)ds.$$

Using Lemma 5.13 with the above formula and with

$$x(t)=T(U_n(U(g,\mu),t))-T(U_{n+m}(g,t+\mu)),$$
$$y(t)=U(g,\mu)-U_{n+m+1}(g,\mu),$$

we conclude that

$$\rho\Big(U_{n+1}(U(g,\mu),t)-U_{n+m+1}(g,t+\mu)\Big) \qquad (7.39)$$

$$\leq e^{-t}\rho(U(g,\mu)-U_{n+m+1}(g,\mu))$$

$$+K(t)\sup_{0\leq s\leq t}\rho\Big(T(U_n(U(g,\mu),s))-T(U_{n+m}(g,s+\mu))\Big).$$

Using the ρ-nonexpansiveness of T, the inductive assumption and (5.27) from Theorem 5.28 we conclude from (7.39) that

$$\rho\Big(U_{n+1}(U(g,\mu),t) - U_{n+m+1}(g,t+\mu)\Big)$$

$$\le e^{-t}\rho(U(g,\mu) - U_{n+m+1}(g,\mu)) + K(t)\Big(\sum_{i=m+1}^{n+m+1} K^i(\mu) + K^{n+1}(t)\Big)\delta_\rho(C)$$

$$\le e^{-t}K^{n+m+2}(\mu)\delta_\rho(C) + K(t)\Big(\sum_{i=m+1}^{n+m+1} K^i(\mu) + K^{n+1}(t)\Big)\delta_\rho(C)$$

$$\le \sum_{i=m+1}^{n+m+2} K^i(\mu)\delta_\rho(C) + K^{n+2}(t)\delta_\rho(C),$$

which completes the proof of Lemma 7.10.

$\square$

We are now ready to prove the main result of this section:

Theorem 7.15. *Let $\rho \in \mathfrak{R}$ be separable, $C \subset E_\rho$ be a nonempty, convex, ρ-bounded, ρ-closed set with the Vitali property. Assume $T : C \to C$ to be a ρ-nonexpansive mapping. Denote $S_t(f) = u_f(t)$, where $t \ge 0$, $f \in C$, and $u_f(t)$ is a solution of the initial value problem (5.25). Then $\{S_t\}_{t\ge0}$ is a continuous ρ-nonexpansive semigroup of nonlinear mappings on C.*

Proof. First let us fix $t \ge 0$ and demonstrate that S_t is ρ-nonexpansive. Fix $f,g \in C$ and recall that by Theorem 5.28,

$$\rho(U_n(f,t) - U(f,t)) \to 0 \qquad\qquad (7.40)$$
$$\rho(U_n(g,t) - U(g,t)) \to 0$$

as $n \to \infty$. By Proposition 3.14, it follows from (7.40) that

$$\rho(U(f,t) - U(g,t)) \le \liminf_{n\to\infty} \rho(U_n(f,t) - U_n(g,t)).$$

Applying Lemma 5.13 with

$$x(t) = U_n(f,t) - U_n(g,t),$$
$$y(t) = f - g,$$

we have

$$\rho(U_{n+1}(f,t) - U_{n+1}(g,t)) \qquad\qquad (7.41)$$
$$= \rho\Big(e^{-t}f + \int_0^t e^{s-t}U_n(f,s)ds - e^{-t}g - \int_0^t e^{s-t}U_n(g,s)ds\Big)$$
$$\le e^{-t}\rho(f-g) + K(t)\sup_{0\le s\le t}\rho(U_n(f,s) - U_n(g,s)).$$

.

By the obvious induction, keeping in mind that $K(t) = 1 - e^{-t}$ we can easily deduce from (7.41) that

$$\rho(U_n(f,t) - U_n(g,t)) \le \rho(f-g),$$

which together with (7.6) gives us finally

$$\rho(S_t(f) - S_t(g)) = \rho(U(f,t) - U(g,t)) \le \rho(f-g),$$

hence S_t is ρ-nonexpansive.

Now let us prove that $\{S_t\}_t \ge 0$ forms a semigroup of nonlinear mappings. Since it is obvious that $S_0(f) = u_f(0) = f$, it remains to prove that

$$S_{\mu+t} = S_\mu \circ S_t$$

for any $\mu, t \ge 0$. Let us fix temporarily $n \in \mathbb{N}$ and note that by Theorem 5.28,

$$\rho(U_{n+m}(f, t+\mu) - U(f, t+\mu)) \to 0$$

as $m \to \infty$. Hence, by Proposition 3.14 and by inequality (7.34) from Lemma 7.10, using also the fact that the series $\sum_i K^i(\mu)$ is convergent, we obtain the following

$$\rho\Big(U_n(U(f,\mu),t) - U(f,t+\mu)\Big) \tag{7.42}$$

$$\le \liminf_{m\to\infty} \rho\Big(U_n(U(f,\mu),t) - U_{n+m}(f,t+\mu)\Big)$$

$$\le \liminf_{m\to\infty} \sum_{i=n+1}^{n+m+1} K^i(\mu)\delta_\rho(C) + K^{n+1}(t)\delta_\rho(C)$$

$$\le K^{n+1}(t)\delta_\rho(C).$$

From (7.42) we see that

$$\rho\Big(U_n(U(f,\mu),t) - S_{t+\mu}(f)\Big) = \rho\Big(U_n(U(f,\mu),t) - U(f,t+\mu)\Big) \to 0$$

as $n \to \infty$. On the other hand, it follows from Theorem 5.28 that

$$\rho\Big(U_n(U(f,\mu),t) - S_t(S_\mu(f))\Big) = \rho\Big(U_n(S_\mu(f),t) - S_t(S_\mu(f))\Big) \to 0 \tag{7.43}$$

as well. The uniqueness of the ρ-limit yields to $S_{t+\mu}(f) = S_t(S_\mu(f))$ for every $t, \mu \ge 0$ and every $f \in C$. To prove ρ-continuity of the semigroup $\{S_t\}_{t \ge 0}$, let us observe that if $t_n \to t$ then

$$\rho(S_{t_n}(f) - S_t(f)) = \rho(u_f(t_n) - u_f(t)) \to 0$$

because u_f is continuous as a solution of the differential equation (7.33). This completes the proof of Theorem 7.15.

$\square$

We proved above that $\mathscr{S} = \{S_t\}_{t\geq 0}$, where $S_t(g) = u_g(t)$ for any $g \in C$, is a continuous ρ-nonexpansive semigroup of nonlinear mappings. Such a situation is quite typical in mathematics and its applications. For instance, in the theory of dynamical systems, the modular function space L_ρ would define the state space and the mapping $(t,x) \rightarrow S_t(x)$ would represent the evolution function of a dynamical system. The question about the existence of common fixed points, and about the structure of the set of common fixed points, can be interpreted as a question whether there exist stationary points for this process, that is, elements of C that are fixed during the state space transformation S_t at any given point of time t, and if yes—what the structure of a set of such points may look like and how such points can be constructed algorithmically. In the setting of this chapter, the state space may be an infinite dimensional. Therefore, it is natural to apply these result to not only to deterministic dynamical systems but also to stochastic dynamical systems.

It is immediate that if $f \in C$ is a common fixed point of $\mathscr{S}$ then $S_t(f) = f$ for all $t \geq 0$ which means that $u_f(t) = f$ for every such t. Therefore, the process defined by the Eq. (7.33) has a stationary point at such f. Let us now interpret our previous results in this context.

Combining Theorem 7.2, Theorem 5.28, and Theorem 7.15, we arrive at the following result.

Theorem 7.16. *Let $\rho \in \mathfrak{R}$ be separable and (UUC1), $C \subset E_\rho$ be a nonempty, convex, ρ-bounded, ρ-closed set with the Vitali property. Assume $T : C \rightarrow C$ to be a ρ-nonexpansive mapping. Denote $S_t(f) = u_f(t)$, where $t \geq 0$, $f \in C$, and $u_f(t)$ is a solution of the initial value problem (7.33). Then $\mathscr{S} = \{S_t\}_{t\geq 0}$ forms a continuous ρ-nonexpansive semigroup of nonlinear mappings on C. Moreover, the set of the stationary points for the process defined by the differential equation (7.33) with the evolution function $(t,x) \rightarrow S_t(x)$, is equal to $F(\mathscr{S})$, the set of all common fixed points of $\mathscr{S}$, which is nonempty, ρ-closed, and convex.*

Let us finish this section by providing an example how the results of the preceding sections can be utilized for constructing a stationary point of a process defined by the Urysohn operator:

$$T(f)(x) = \int_0^1 k(x,y,|f(y)|)dy + f_0(x),$$

with the assumptions as per Example 3.12. Using Theorem 5.28 we see that, given $f \in C$, the initial value problem

$$\begin{cases} u(0) = f \\ u'(t) + (I - T)u(t) = 0, \end{cases}$$

has a solution $u_f : [0, +\infty] \rightarrow C$. As proved in our Theorem 7.15 the formula

$$S_t(f) = u_f(t)$$

defines the semigroup of ρ-nonexpansive mappings. Note that ρ in this example is orthogonally additive and hence it has the strong Opial property. Therefore, assuming ρ is $(UUC1)$ and uniformly continuous, we can use the algorithmic methods previously established for the Mann process in Theorem 7.11, and for the Ishikawa process in Theorem 7.13 to construct a common fixed point of the semigroup $\{S_t\}$ which will be a stationary point of the Urysohn process defined by the evolution function $(t, f) \to u_f(t) \in C$.

7.7 Asymptotic Pointwise Nonexpansive Semigroups

In this section, we will extend some of the results obtained above for the ρ-nonexpansive semigroups to the case of asymptotic pointwise ρ-nonexpansive semigroups. This theory, initiated in [55, 56], is still in its infancy and the authors believe that there are a lot of interesting aspects of the theory that are waiting to be discovered.

Definition 7.8. A one-parameter family $\mathscr{F} = \{T_t : t \geq 0\}$ of mappings from C into itself is said to be an *asymptotic pointwise ρ-nonexpansive semigroup on C* if $\mathscr{F}$ satisfies the following conditions:

1. $T_0(f) = f$ for $f \in C$;
2. $T_{t+s}(f) = T_t(T_s(f))$ for $f \in C$ and $t, s \in [0, \infty)$;
3. for each $t \geq 0$, T_t is an asymptotic pointwise nonexpansive mapping, i.e., there exists a function $\alpha_t : C \to [0, \infty)$ such that

$$\rho(T_t(f) - T_t(g)) \leq \alpha_t(f)\rho(f - g) \text{ for all } f, g \in C.$$

   such that $\limsup_{t \to \infty} \alpha_t(f) \leq 1$ for every $f \in C$.
4. for each $f \in C$, the mapping $t \to T_t(f)$ is ρ-continuous.

Note that without loss of generality we may assume $\alpha_t(f) \geq 1$ for any $t \geq 0$ and $f \in C$ and $\limsup_{t \to \infty} \alpha_t(f) = \lim_{t \to \infty} \alpha_t(f) = 1$.

The definition of a ρ-type used in this book was based on a given sequence. In this section, we generalize this definition to be adapted to one-parameter family of mappings.

Definition 7.9. Let $K \subset L_\rho$ be convex and ρ-bounded.

1. A function $\tau : K \to [0, \infty]$ is called a ρ-type (or shortly a type) if there exists a one-parameter family $\{h_t\}_{t \geq 0}$ of elements of K such that for any $f \in K$ there holds

$$\tau(f) = \inf_{M > 0} \left(\sup_{t \geq M} \rho(h_t - f) \right).$$

2. Let τ be a type. A sequence $\{g_n\}$ is called a minimizing sequence of τ if

$$\lim_{n \to \infty} \tau(g_n) = \inf\{\tau(f) : f \in K\}.$$

Note that τ is convex provided ρ is convex.

The next result extends the minimizing sequence property for types defined by sequences to the case of one-parameter semigroup of nonlinear mappings, compare with Lemma 4.6.

Lemma 7.11. *[55] Assume $\rho \in \Re$ is (UUC1). Let C be a nonempty, ρ-bounded, ρ-closed, and convex subset of L_ρ. Let τ be a type defined by a one-parameter family $\{h_t\}_{t \ge 0}$ in C.*

(i) If $\tau(f_1) = \tau(f_2) = \inf_{f \in C} \tau(f)$, then $f_1 = f_2$.

(ii) Any minimizing sequence $\{f_n\}$ of τ is ρ-convergent. Moreover the ρ-limit of $\{f_n\}$ is independent of the minimizing sequence.

Proof. First let us prove (i). Let $f_1, f_2 \in C$ be such that $\tau(f_1) = \tau(f_2) = \inf_{f \in C} \tau(f)$.

Let us consider two cases.

Case 1: $\inf_{f \in C} \tau(f) = 0$. Since

$$\rho\left(\frac{f_1 - f_2}{2}\right) = \rho\left(\frac{f_1 - h_t + h_t - f_2}{2}\right) \le \rho(f_1 - h_t) + \rho(h_t - f_2)$$

for any $t \ge 0$, we get

$$\rho\left(\frac{f_1 - f_2}{2}\right) \le \sup_{t \ge M} \rho(f_1 - h_t) + \sup_{t \ge M} \rho(h_t - f_2)$$

for any $M > 0$. Since

$$\tau(f) = \inf_{M > 0}\left(\sup_{t \ge M} \rho(f - h_t)\right) = \lim_{M \to \infty} \sup_{t \ge M} \rho(f - h_t),$$

for any $f \in C$, we get

$$\rho\left(\frac{f_1 - f_2}{2}\right) \le \tau(f_1) + \tau(f_2) = 0,$$

which implies $f_1 = f_2$ as claimed.

Case 2: $\inf_{f \in C} \tau(f) > 0$. Assume to the contrary that $f_1 \ne f_2$. Set

$$R = \inf_{f \in C} \tau(f) \quad \text{and} \quad \varepsilon = \frac{\rho(f_1 - f_2)}{2R}.$$

Let $v \in (0,R)$. Then $\rho(f_1 - f_2) = 2R\varepsilon \geq (R+v)\varepsilon$. Using the definition of τ, we deduce that there exists $M_v > 0$ such that

$$\sup_{t \geq M_v} \rho(f_1 - h_t) \leq \tau(f_1) + v = R + v \quad \text{and} \quad \sup_{t \geq M_v} \rho(f_2 - h_t) \leq \tau(f_2) + v = R + v.$$

Since ρ is $(UUC1)$, there exists $\eta(R,\varepsilon) > 0$ such that

$$\delta(R+v,\varepsilon) \geq \eta(R,\varepsilon)$$

for any $v \in (0,R)$. So for any $t \geq M_v$, we have

$$\rho\left(\frac{f_1 + f_2}{2} - h_t\right) \leq (R+v)\left(1 - \delta(R+v,\varepsilon)\right) \leq (R+v)\left(1 - \eta(R,\varepsilon)\right).$$

Hence,

$$\tau\left(\frac{f_1 + f_2}{2}\right) \leq \sup_{t \geq M_v} \rho\left(\frac{f_1 + f_2}{2} - h_t\right) \leq (R+v)\left(1 - \eta(R,\varepsilon)\right).$$

Since C is convex, we get

$$R \leq \tau\left(\frac{f_1 + f_2}{2}\right) \leq (R+v)\left(1 - \eta(R,\varepsilon)\right).$$

If we let $v \to 0$, we will get

$$R \leq R\left(1 - \eta(R,\varepsilon)\right),$$

which is impossible since $R > 0$ and $\eta(R,\varepsilon) > 0$. Therefore, we must have $f_1 = f_2$.

Next we prove (ii). Denote $R = \inf_{g \in C} \tau(g)$. For any $n \geq 1$, let us set

$$K_n = \overline{conv}_\rho\left\{h_t : t \geq n\right\},$$

where $\overline{conv}_\rho(A)$ is the intersection of all ρ-closed convex subset of C which contains $A \subset C$. Since C is itself ρ-closed and convex, we get $K_n \subset C$ for any $n \geq 1$. Property (R) will then imply $\bigcap K_n \neq \emptyset$. Let us fix then arbitrary $f \in \bigcap K_n$, $g \in C$, and $\varepsilon > 0$. By definition of $\tau(g)$, there exists $M_\varepsilon > 0$ such that $\sup_{t \geq M_\varepsilon} \rho(g - h_t) \leq \tau(g) + \varepsilon$. Let $n \geq M_\varepsilon$. Then for any $t \geq n$, we have $\rho(g - h_t) \leq \tau(g) + \varepsilon$, i.e., $h_t \in B_\rho(g, \tau(g) + \varepsilon)$. Since $B_\rho(g, \tau(g) + \varepsilon)$ is ρ-closed and convex, we get $K_n \subset B_\rho(g, \tau(g) + \varepsilon)$. Hence $f \in B_\rho(g, \tau(g) + \varepsilon)$, i.e.,

$$\rho(g - f) \leq \tau(g) + \varepsilon. \tag{7.44}$$

Since ε was taken arbitrarily greater than 0, we get $\rho(g - f) \leq \tau(g)$, for any $g \in C$. Let $\{f_n\}$ be a minimizing sequence for τ. If $R = 0$ then, since $\{f_n\}$ is a minimizing sequence, we get $\lim_{n\to\infty} \tau(f_n) = R = 0$. Using (7.44) we can see that $\rho(f_n - f) \le \tau(f_n)$, for any $n \ge 1$. Hence $\{f_n\}$ is ρ-convergent to f. Since selection of f was independent of $\{f_n\}$, it follows that any minimizing sequence is ρ-convergent to f if $R = 0$. We can assume therefore that $R > 0$. For any $n \ge 1$, let us set

$$d_n = \sup_{i,j \ge n} \rho(f_i - f_j). \tag{7.45}$$

We claim that $\{f_n\}$ is ρ-Cauchy. Assume to the contrary that this is not the case. Since the sequence $\{d_n\}$ is decreasing and $\{f_n\}$ is not ρ-Cauchy, we get $d := \inf_{n \ge 1} d_n > 0$. Set $\varepsilon = \dfrac{d}{4R} > 0$. Let us fix arbitrary $v \in (0, R)$. Since $\lim_{n\to\infty} \tau(f_n) = R$, there exists $n_0 \ge 1$ such that for any $n \ge n_0$, we have

$$\tau(f_n) \le R + \frac{v}{2}. \tag{7.46}$$

Let $n \ge n_0$. By (7.45) there exists $i_n, j_n \ge 1$ such that

$$\rho(f_{i_n} - f_{j_n}) > d_n - \frac{d}{2} \ge \frac{d}{2} = 2R\varepsilon > (R+v)\varepsilon.$$

Using the definition of τ and (7.46), we deduce the existence of $M > 0$ such that

$$\sup_{t \ge M} \rho(f_{i_n} - h_t) \le \tau(f_{i_n}) + \frac{v}{2} \le R + v,$$

and

$$\sup_{t \ge M} \rho(f_{j_n} - h_t) \le \tau(f_{j_n}) + \frac{v}{2} \le R + v.$$

Hence

$$\rho\left(\frac{f_{i_n} + f_{j_n}}{2} - h_t\right) \le (R+v)\Big(1 - \delta(R+v, \varepsilon)\Big),$$

for any $t \ge M$. Since ρ is $(UUC1)$, there exists $\eta_1(R, \varepsilon) > 0$ such that $\delta_1(R+v, \varepsilon) \ge \eta_1(R, \varepsilon)$. Hence

$$\rho\left(\frac{f_{i_n} + f_{j_n}}{2} - h_t\right) \le (R+v)\Big(1 - \eta(R, \varepsilon)\Big),$$

for any $t \ge M$. Hence

$$R \le \tau\left(\frac{f_{i_n} + f_{j_n}}{2}\right) \le \sup_{t \ge M} \rho\left(\frac{f_{j_n} + f_{j_n}}{2} - h_t\right) \le (R+v)\Big(1 - \eta(R, \varepsilon)\Big),$$

for any $v \in (0, R)$. If we let $v \to 0$, we get $R \le R(1 - \eta_1(R, \varepsilon))$. This contradiction implies that $\{f_n\}$ is ρ-Cauchy. Since L_ρ is ρ-complete, we deduce that $\{f_n\}$ is ρ-convergent as claimed.

In order to finish the proof of (ii), let us show that the ρ-limit of $\{f_n\}$ is independent of the minimizing sequence. Indeed, let $\{g_n\}$ be another minimizing sequence of τ. The previous proof will show that $\{g_n\}$ is also ρ-convergent. In order to prove that the ρ-limits of $\{f_n\}$ and $\{g_n\}$ are equal, let us show that $\lim_{n\to\infty}\rho(f_n - g_n) = 0$. Assume not, i.e., $\lim_{n\to\infty}\rho(f_n - g_n) \neq 0$. Without loss of generality we may assume that there exists $d > 0$ such that $\rho(f_n - g_n) \geq d$, for any $n \geq 1$. Set $\varepsilon = \dfrac{d}{2R} > 0$. Let $v \in (0,R)$. Since $\lim_{n\to\infty}\tau(f_n) = \lim_{n\to\infty}\tau(g_n) = R$, there exists $n_0 \geq 1$ such that for any $n \geq 1$, we have $\tau(f_n) \leq R + \dfrac{v}{2}$, and $\tau(g_n) \leq R + \dfrac{v}{2}$. Fix $n \geq n_0$. Then

$$\rho(f_n - g_n) \geq d = 2R\varepsilon > (R+v)\varepsilon.$$

Using the definition of τ, we deduce the existence of $M > 0$ such that

$$\sup_{t\geq M}\rho(f_n - h_t) \leq \tau(f_n) + \frac{v}{2} \leq R+v,$$

and

$$\sup_{t\geq M}\rho(g_n - h_t) \leq \tau(g_n) + \frac{v}{2} \leq R+v.$$

Hence

$$\rho\left(\frac{f_n + g_n}{2} - h_t\right) \leq (R+v)\left(1 - \delta(R+v,\varepsilon)\right),$$

for any $t \geq M$. Since ρ is $(UUC1)$, there exists $\eta(R,\varepsilon) > 0$ such that $\delta(R+v,\varepsilon) \geq \eta(R,\varepsilon)$, for any $v > 0$. Hence

$$\rho\left(\frac{f_n + g_n}{2} - h_t\right) \leq (R+v)\left(1 - \eta(R,\varepsilon)\right),$$

for any $t \geq M$. Hence

$$\tau\left(\frac{f_n + g_n}{2}\right) \leq \sup_{t\geq M}\rho\left(\frac{f_n + g_n}{2} - h_t\right) \leq (R+v)\left(1 - \eta(R,\varepsilon)\right).$$

Using the definition of R, we get

$$R \leq (R+v)\left(1 - \eta(R,\varepsilon)\right),$$

for any $v \in (0,R)$. If we let $v \to 0$, we get $R \leq R(1 - \eta(R,\varepsilon))$. This contradiction implies $\lim_{n\to\infty}\rho(f_n - g_n) = 0$. The Fatou property will finally imply

$$\rho(f - g) \leq \liminf_{n\to\infty}\rho(f_n - g_n),$$

where f is the ρ-limit of $\{f_n\}$ and g is the ρ-limit of $\{g_n\}$. Hence $\rho(f-g) = 0$, i.e., $f = g$.

$\square$

Using Lemma 7.11, we can prove our common fixed point result for asymptotic pointwise nonexpansive semigroups.

Theorem 7.17. *[55] Assume $\rho \in \mathfrak{R}$ is (UUC1). Let C be a ρ-closed ρ-bounded convex nonempty subset of L_ρ. Let $\mathscr{F} = \{T_t : t \geq 0\}$ be an asymptotic pointwise nonexpansive semigroup on C. Then $\mathscr{F}$ has a common fixed point and the set $F(\mathscr{F})$ of common fixed points is ρ-closed and convex.*

Proof. Let us fix $f \in C$ and define the type function τ by

$$\tau(g) = \inf_{M>0} \left(\sup_{t \geq M} \rho(T_t(f) - g) \right).$$

Since C is ρ-bounded, we have $\tau(g) \leq diam_\rho(C) < +\infty$, for any $g \in C$. Hence $\tau_0 = \inf\{\tau(g) : g \in C\}$ exists and is finite. For any $n \geq 1$, there exists $g_n \in C$, such that

$$\tau_0 \leq \tau(g_n) < \tau_0 + \frac{1}{n}.$$

Therefore, $\lim_{n \to \infty} \tau(z_n) = \tau_0$, i.e., $\{g_n\}$ is a minimizing sequence for τ. By Lemma 7.11, there exists $g \in C$ such that $\{g_n\}$ ρ-converges to g. Let us now prove that $g \in F(\mathscr{F})$. Note that

$$\rho(T_{s+t}(f) - T_s(h)) \leq \alpha_s(h)\rho(T_t(f) - h),$$

for $s,t > 0$, and $h \in C$. Using the definition of τ, we get

$$\tau(T_s(h)) \leq \sup_{t+s \geq M} \rho(T_{s+t}(f) - T_s(h)) \leq \alpha_s(h) \sup_{t \geq M-s} \rho(T_t(f) - h),$$

for any $M > s$, which implies

$$\tau(T_s(h)) \leq \alpha_s(h)\tau(h). \tag{7.47}$$

Since $\lim_{s \to \infty} \alpha_s(g_1) = 1$, there exists $s_1 > 0$ such that for any $s \geq s_1$, we have $\alpha_s(g_1) < 1 + 1$. Repeating this argument, one will find $s_2 > s_1 + 1$ such that for any $s \geq s_2$, we have $\alpha_s(g_2) < 1 + \frac{1}{2}$. By induction, we will construct a sequence $\{s_n\}$ of positive numbers such that $s_{n+1} < s_n + \frac{1}{n}$ and for any $s \geq s_n$ we have $\alpha_s(g_n) < 1 + \frac{1}{n}$. Let us fix $t \geq 0$. Then the inequality (7.47) will imply

$$\tau(T_{s_n+t}(g_n)) \leq \alpha_{s_n+t}(g_n)\tau(g_n) \leq \left(1 + \frac{1}{n}\right)\tau(g_n),$$

for any $n \geq 1$. In particular we get $\{T_{s_n+t}(g_n)\}$ is a minimizing sequence of τ. Therefore, Lemma 7.11 implies that $\{T_{s_n+t}(g_n)\}$ ρ-converges to g, for any $t \geq 0$. In particular, we have $\{T_{s_n}(g_n)\}$ ρ-converges to g. Since

$$\rho\left(T_{s_n+t}(g_n) - T_t(g)\right) \leq \alpha_t(g)\rho(T_{s_n}(g_n) - g),$$

we get $\{T_{s_n+t}(g_n)\}$ ρ-converges to $T_t(g)$. Finally using

$$\rho\left(\frac{T_t(g)-g}{2}\right) \leq \rho\left(T_t(g) - T_{s_n+t}(g_n)\right) + \rho\left(T_{s_n+t}(g_n) - g\right),$$

we get $T_t(g) = g$. Since $t > 0$ was arbitrarily chosen, we get $g \in F(\mathscr{F})$, i.e., $F(\mathscr{F})$ is not empty.

Next let us prove that $F(\mathscr{F})$ is ρ-closed. Let $\{f_n\}$ in $F(\mathscr{F})$ ρ-convergent to f. Since

$$\rho(T_s(f_n) - T_s(f)) \leq \alpha_s(f)\rho(f_n - f),$$

for any $n \geq 1$ and $s > 0$, we get $\{T_s(f_n)\}$ ρ-converges to $T_s(f)$. Since $f_n \in F(\mathscr{F})$, we get $\{T_s(f_n)\} = \{f_n\}$. In other words, $\{f_n\}$ ρ-converges to $T_s(f)$ and f. The uniqueness of the ρ-limit, implies then that $T_s(f) = f$, for any $s \geq 0$, i.e., $f \in F(\mathscr{F})$. Therefore $F(\mathscr{F})$ is ρ-closed.

Let us finish the proof of Theorem 7.17 by showing that $F(\mathscr{F})$ is convex. It is sufficient to show that

$$h = \frac{f+g}{2} \in F(\mathscr{F}),$$

for any $f, g \in F(\mathscr{F})$. Without loss of generality, we will assume that $f \neq g$. Let $s > 0$. We have

$$\rho(f - T_s(h)) = \rho(T_s(f) - T_s(h)) \leq \alpha_s(f)\rho(f - h)$$

and

$$\rho(g - T_s(h)) = \rho(T_s(g) - T_s(h)) \leq \alpha_s(g)\rho(g - h).$$

Since $\rho(f - h) = \rho(g - h) = \rho\left(\frac{f-g}{2}\right)$, and

$$\rho\left(\frac{f-g}{2}\right) \leq \frac{1}{2}\rho(f - T_s(h)) + \frac{1}{2}\rho(g - T_s(h)),$$

we conclude that

$$\lim_{s \to \infty} \rho(f - T_s(h)) = \lim_{s \to \infty} \rho(g - T_s(h)) = \rho\left(\frac{f-g}{2}\right).$$

Similarly we have

$$\rho\left(f - \frac{h + T_s(h)}{2}\right) \leq \frac{1}{2}\rho\left(f - h\right) + \frac{1}{2}\rho(f - T_s(h)),$$

and

$$\rho\left(g - \frac{h + T_s(h)}{2}\right) \leq \frac{1}{2}\rho\left(g - h\right) + \frac{1}{2}\rho(g - T_s(h)).$$

Since

$$\rho\left(\frac{f - g}{2}\right) \leq \frac{1}{2}\rho\left(f - \frac{h + T_s(h)}{2}\right) + \frac{1}{2}\rho\left(g - \frac{h + T_s(h)}{2}\right),$$

we conclude that

$$\lim_{s \to \infty} \rho\left(f - \frac{h + T_s(h)}{2}\right) = \lim_{s \to \infty} \rho\left(g - \frac{h + T_s(h)}{2}\right) = \rho\left(\frac{f - g}{2}\right).$$

Therefore, we have

$$\lim_{s \to \infty} \rho(f - T_s(h)) = \lim_{s \to \infty} \rho\left(f - \frac{h + T_s(h)}{2}\right) = \rho(f - h).$$

Lemma 4.2 applied to $A_t = f - T_s(h)$ and $B_t = T_s(h) - g$ implies that $\rho(A_t - B_t) \to 0$. Hence

$$\lim_{s \to \infty} \rho(h - T_s(h)) = \lim_{s \to \infty} \rho\left(\frac{A_t - B_t}{2}\right) \leq \lim_{s \to \infty} \rho(A_t - B_t) = 0.$$

Clearly we will get $\lim_{s \to \infty} \rho(h - T_{s+t}(h)) = 0$, for any $t \geq 0$. Since

$$\rho(T_t(h) - T_{s+t}(h)) \leq \alpha_t(h)\rho(h - T_s(h)),$$

we have $\lim_{s \to \infty} \rho(T_t(h) - T_{s+t}(h)) = 0$. Finally using the inequality

$$\rho\left(\frac{h - T_t(h)}{2}\right) \leq \frac{1}{2}\rho(h - T_{s+t}(h)) + \frac{1}{2}\rho(T_t(h) - T_{s+t}(h)),$$

by letting $s \to \infty$, we get $T_t(h) = h$, for any $t \geq 0$, i.e., $h \in F(\mathscr{F})$.
$\square$

Chapter 8
Modular Metric Spaces

The concept of a metric space is closely related to our intuitive understanding of space in the 3-dimensional Euclidean space. In fact, the notion of metric is a generalization of the Euclidean metric arising from the basic long known properties of the Euclidean distance. Maurice Fréchet[1] is credited as the mathematician who introduced the abstract definition of a metric space [74]. Metric spaces are seen as a nonlinear version of vector spaces endowed with a norm. Following the same direction, one may think of a nonlinear version of modular function spaces. Indeed, throughout this book, we have seen that a modular function space is a vector space endowed with a modular function. Therefore, it is natural to consider a nonlinear version of function modular spaces. The first to consider such generalization was V. Chistyakov [46, 47]. Note that Chistyakov's generalization of modular function spaces had a physical interpretation. Informally speaking, whereas a metric on a set represents nonnegative finite distances between any two points of the set, a modular on a set attributes a nonnegative (possibly, infinite valued) "field of (generalized) velocities" to each "time" $\lambda > 0$ (the absolute value of) an average velocity $w_\lambda(x, y)$ which is associated in such a way that in order to cover the "distance" between points $x, y \in X$ it takes time λ to move from x to y with velocity $w_\lambda(x, y)$. The nonlinear approach to modular function spaces was initiated in [1, 2, 3].

8.1 Definitions

Let X be a nonempty set. Throughout this chapter, for a given function

$$\omega : (0, +\infty) \times X \times X \to [0, +\infty],$$

---

[1] The term *metric space* was in fact introduced by F. Hausdorff in *Grundzüge der Mengenlehre*, Leipzig, 1914.

© Springer International Publishing Switzerland 2015
M. A. Khamsi, W. M. Kozlowski, *Fixed Point Theory in Modular Function Spaces*,
DOI 10.1007/978-3-319-14051-3_8

we will write

$$\omega_\lambda(x,y) = \omega(\lambda,x,y),$$

for all $\lambda > 0$ and $x,y \in X$.

Definition 8.1. A function $\omega : (0,+\infty) \times X \times X \to [0,+\infty]$ is said to be a *modular metric* on X if it satisfies the following axioms:

(i) $x = y$ if and only if $\omega_\lambda(x,y) = 0$, for all $\lambda > 0$;
(ii) $\omega_\lambda(x,y) = \omega_\lambda(y,x)$, for all $\lambda > 0$, and $x,y \in X$;
(iii) $\omega_{\lambda+\mu}(x,y) \le \omega_\lambda(x,z) + \omega_\mu(z,y)$, for all $\lambda,\mu > 0$ and $x,y,z \in X$.

If instead of (i), we only have the condition

$$\text{(i')} \quad \omega_\lambda(x,x) = 0, \text{ for all } \lambda > 0, \ x \in X,$$

then ω is said to be a pseudomodular (metric) on X. A modular metric ω on X is said to be regular if the following weaker version of (i) is satisfied:

$$x = y \text{ if and only if } \omega_\lambda(x,y) = 0, \text{ for some } \lambda > 0.$$

Finally, ω is said to be convex if for $\lambda,\mu > 0$ and $x,y,z \in X$, it satisfies the inequality

$$\omega_{\lambda+\mu}(x,y) \le \frac{\lambda}{\lambda+\mu}\,\omega_\lambda(x,z) + \frac{\mu}{\lambda+\mu}\,\omega_\mu(z,y).$$

Remark 8.1. Note that for a metric pseudomodular ω on a set X, and any $x,y \in X$, the function $\lambda \to \omega_\lambda(x,y)$ is nonincreasing on $(0,+\infty)$. Indeed if $0 < \mu < \lambda$, then

$$\omega_\lambda(x,y) \le \omega_{\lambda-\mu}(x,x) + \omega_\mu(x,y) = w_\mu(x,y).$$

Definition 8.2. Let ω be a pseudomodular on X. Fix $x_0 \in X$. The two sets

$$X_\omega = X_\omega(x_0) = \{x \in X : \omega_\lambda(x,x_0) \to 0 \text{ as } \lambda \to +\infty\}$$

and

$$X_\omega^* = X_\omega^*(x_0) = \{x \in X : \exists \lambda = \lambda(x) > 0 \text{ such that } \omega_\lambda(x,x_0) < +\infty\}$$

are said to be *modular spaces (around x_0)* .

It is clear that $X_\omega \subset X_\omega^*$, but this inclusion may be proper in general. If ω is a modular on X, then the modular space X_ω can be equipped with a (nontrivial) metric, generated by ω and given by

$$d_\omega(x,y) = \inf\{\lambda > 0 : \omega_\lambda(x,y) \le \lambda\},$$

for any $x, y \in X_\omega$. If ω is a convex modular on X, then the two modular spaces coincide, i.e., $X_\omega^* = X_\omega$, and this common set can be endowed with the metric d_ω^* given by

$$d_\omega^*(x, y) = \inf\{\lambda > 0 : \omega_\lambda(x, y) \leq 1\},$$

for any $x, y \in X_\omega$. These distances will be called Luxemburg distances similarly as we did in the linear case. In the next example, we show how the linear modular function spaces are connected to modular metric spaces as defined above.

Example 8.1. A modular function space furnishes a wonderful example of a modular metric space. Indeed, define the function ω by

$$\omega_\lambda(f, g) = \rho\left(\frac{f - g}{\lambda}\right),$$

for all $\lambda > 0$, and $f, g \in L_\rho$. Then ω is a modular metric on L_ρ. Note that ω is convex if and only if ρ is convex. Moreover, we have

$$\|f - g\|_\rho = d_\omega^*(f, g),$$

for any $f, g \in L_\rho$.

Remark 8.2. Note that regular metric spaces are also modular metric spaces. Indeed, let (M, d) be a metric space. Then the functional $\omega : (0, +\infty) \times M \times M \to [0, +\infty)$ defined by

$$\omega_\lambda(x, y) = \frac{d(x, y)}{\lambda}$$

is a regular modular metric on M. Note that ω is convex. Note that in this case we have $d_\omega^* = d$.

For the regular metric spaces, a natural topology is introduced as follows.

Definition 8.3. Let (X, ω) be a modular metric space.

(1) The sequence $\{x_n\}_{n \in N}$ in X_ω is said to be ω-convergent to $x \in X_\omega$ if and only if $\omega_\lambda(x_n, x) \to 0$ as $n \to +\infty$, for some $\lambda > 0$.
(2) The sequence $\{x_n\}_{n \in N}$ in X_ω is said to be ω-Cauchy if $\omega_\lambda(x_m, x_n) \to 0$ as $m, n \to +\infty$, for some $\lambda > 0$.
(3) A subset C of X_ω is said to be ω-closed if the limit of a ω-convergent sequence of C always belong to C.
(4) A subset C of X_ω is said to be ω-complete if any ω-Cauchy sequence in C is a convergent sequence and its limit is in C.
(5) A subset C of X_ω is said to be ω-bounded if for some $\lambda > 0$, we have

$$\delta_\omega(C) = \sup\{\omega_\lambda(x, y) : x, y \in C\} < +\infty.$$

(6) ω is said to satisfy the Fatou property, if for any $\{x_n\}$ ω-convergent to x and $\{y_n\}$ ω-convergent to y, then

$$\omega_1(x,y) \leq \liminf_{n\to\infty} \omega_1(x_n,y_n).$$

In general if $\lim_{n\to\infty} \omega_\lambda(x_n,x) = 0$, for some $\lambda > 0$, then we may not have $\lim_{n\to\infty} w_\lambda(x_n,x) = 0$, for all $\lambda > 0$. This remark suggests the following definition.

Definition 8.4. Let (X,ω) be a modular metric space. ω is said to satisfy Δ_2-*condition* if and only if $\lim_{n\to\infty} \omega_\lambda(x_n,x) = 0$, for some $\lambda > 0$ implies $\lim_{n\to\infty} \omega_\lambda(x_n,x) = 0$, for all $\lambda > 0$.

As we did in the linear case, it is easy to show that

$$\lim_{n\to\infty} d_\omega(x_n,x) = 0 \text{ if and only if } \lim_{n\to\infty} \omega_\lambda(x_n,x) = 0, \text{ for all } \lambda > 0,$$

for any $\{x_n\} \in X_w$ and $x \in X_w$. In particular, we see that ω-convergence and d_ω convergence are equivalent if and only if the modular ω satisfies the Δ_2-condition. Moreover, if the modular ω is convex, then we know that d_ω^* and d_ω are equivalent which implies

$$\lim_{n\to\infty} d_\omega^*(x_n,x) = 0 \text{ if and only if } \lim_{n\to\infty} \omega_\lambda(x_n,x) = 0, \text{ for all } \lambda > 0,$$

for any $\{x_n\} \in X_\omega$ and $x \in X_\omega$.

Remark 8.3. Let (X,ω) be a modular metric space. Assume that ω is regular. Then we have the uniqueness of the ω-limit. Indeed, let $\{x_n\} \in X_\omega$ be a sequence such that $\{x_n\}$ ω-converges to $a \in X_\omega$ and $b \in X_\omega$. Then there exist $\lambda,\mu > 0$ such that

$$\lim_{n\to+\infty} \omega_\lambda(x_n,a) = 0 \text{ and } \lim_{n\to+\infty} \omega_\mu(x_n,b) = 0.$$

Since

$$\omega_{\lambda+\mu}(a,b) \leq \omega_\lambda(x_n,a) + \omega_\mu(x_n,b),$$

for any $n \geq 1$, we get $\omega_{\lambda+\mu}(a,b) = 0$. Since ω is regular, we must have $a = b$ as claimed.

Let us finish this section with the modular metric definitions of Lipschitzian mappings. The definitions are straightforward generalizations of their norm and metric equivalents.

Definition 8.5. Let (X,ω) be a modular metric space. Let C be a nonempty subset of X_ω. A mapping $T : C \to C$ is called an ω-Lipschitzian mapping if there exists a constant $k \geq 0$ such that

$$\omega_1(T(x),T(y)) \leq k\,\omega_1(x,y), \text{ for any } x,\ y \in C.$$

The smallest such constant k will be known as $Lip_\omega(T)$. If $Lip_\omega(T) < 1$, then T is called ω-contraction mapping. And if $Lip_\omega(T) \leq 1$, then T is called ω-nonexpansive mapping.

In the next definition, we introduce the concept of pointwise Lipschitzian mappings in modular metric spaces.

Definition 8.6. Let (X, ω) be a modular metric space. Let C be a nonempty subset of X_ω. A mapping $T : C \to C$ is called a pointwise ω-Lipschitzian mapping if there exists a function $\alpha : C \to [0, +\infty)$ such that

$$\omega_1(T(x), T(y)) \leq \alpha(x)\, \omega_1(x, y), \text{ for any } x, y \in C.$$

If $\alpha(C) \subset [0, 1)$, then T is called a pointwise ω-contraction mapping. And if $\alpha(C) \subset [0, 1]$, then T is called a pointwise ω-nonexpansive mapping.

In [46, 47], the author defined Lipschitzian mappings in modular metric spaces and proved some fixed point theorems. The above definition is more general. Indeed if we assume ω is convex, then we have

$$\omega_\lambda(T(x), T(y)) \leq \omega_\lambda(x, y), \text{ for any } \lambda > 0$$

if and only if

$$d_\omega^*(T(x), T(y)) \leq d_\omega^*(x, y).$$

Moreover, we have seen that in the modular function spaces, we may have

$$\omega_1(T(x), T(y)) \leq \omega_1(x, y),$$

but T is not Lipschitzian with respect to the Luxemburg distance d_ω^* with constant 1.

8.2 Banach Contraction Principle in Modular Metric Spaces

The statement of the Banach Contraction Principle in modular metric spaces goes as follow:

Theorem 8.1. *Let (X, ω) be a modular metric space. Assume ω is regular. Let C be a nonempty subset of X_ω. Assume that C is ω-complete and ω-bounded, i.e., $\delta_\omega(C) = \sup\{\omega_1(x, y) : x, y \in C\} < \infty$. Let $T : C \to C$ be an ω-contraction. Then T has a unique fixed point x_0. Moreover, the orbit $\{T^n(x)\}$ ω-converges to x_0, for each $x \in C$.*

Proof. Since T is a contraction, then there exists $k \in [0, 1)$ such that

$$\omega_1(T(x), T(y)) \leq k\, \omega_1(x, y), \text{ for any } x, y \in C.$$

First note that T has at most one fixed point. Indeed, let $a, b \in C$ be two fixed points of T. Then we have

$$\omega_1(a,b) = \omega_1(T(a), T(b)) \leq k\, \omega_1(a,b).$$

Our assumption on C implies $\omega_1(a,b) < \infty$. This will force $\omega_1(a,b) = 0$, which implies $a = b$ because ω is regular. Next we fix $x \in C$. Then we have

$$\omega_1(T^{n+h}(x), T^n(x)) \leq k^n\, \omega_1(T^h(x), x) \leq k^n\, \delta_\omega(C),$$

for any $n \geq 1$ and $h \geq 1$. This clearly implies that $\{T^n(x)\}$ is ω-Cauchy. Since C is ω-complete, therefore $\{T^n(x)\}$ ω-converges to some point $x_0 \in C$. Next, let us show that x_0 is a fixed point of T. Indeed, we have

$$\omega_2(x_0, T(x_0)) \leq \omega_1(x_0, T^n(x)) + \omega_1(T(x_0), T^n(x)),$$

which implies

$$\omega_2(x_0, T(x_0)) \leq \omega_1(x_0, T^n(x)) + k\, \omega_1(x_0, T^{n-1}(x)),$$

for any $n \geq 1$. Since $\{T^n(x)\}$ ω-converges to x_0, we get $\omega_2(x_0, T(x_0)) = 0$ which implies $T(x_0) = x_0$ since ω is regular. Since T has at most one fixed point, we conclude that any orbit of T, ω-converges to the only fixed point x_0 of T. $\square$

Remark 8.4. In the classical Banach contraction principle, the metric space is not supposed to be bounded. In fact, the contractive condition of the mapping implies that any orbit is bounded. In the case of a modular metric space, due to the failure of of the triangle inequality, it is not true that the contractive condition of the mapping will imply the boundedness of the orbit. Note that if T has a fixed point, then it is obvious that an orbit of T is bounded. Conversely, if T has a bounded orbit $\{T^n(x)\}$, for some $x \in C$, then we have

$$\omega_1(T^{n+h}(x), T^n(x)) \leq k^n\, \omega_1(T^h(x), x) \leq k^n \sup_{n,m \geq 1} \omega_1(T^n(x), T^m(x)),$$

for any $n \geq 1$ and $h \geq 1$. This clearly implies that $\{T^n(x)\}$ is ω-Cauchy. Since C is ω-complete, then $\{T^n(x)\}$ is ω-convergent to some $x_0 \in C$. Since

$$\omega_1(T^n(x), T(x_0)) \leq k\, \omega_1(T^{n-1}(x), x_0), \quad n = 1, 2, \cdots,$$

we conclude that $\{T^n(x)\}$ ω-converges to $T(x_0)$. Since ω is regular, we must have $T(x_0) = x_0$, i.e., x_0 is a fixed point of T. Next for any $z \in C$ such that $\omega_1(z, x_0) < \infty$, we have

$$\omega_1(T^n(z), x_0) = \omega_1(T^n(z), T^n(x_0)) \leq k^n\, \omega_1(z, x_0),$$

for any $n \geq 1$. Hence the orbit $\{T^n(z)\}$ ω-converges to x_0 as well. In this remark, we showed how one has to be careful when dealing with modulars. Indeed, a modular may take infinite value. This is the problem that the authors of [165] did not pay attention to.

Next we discuss the extension of Theorem 8.1 to the case of pointwise contraction mappings in modular metric spaces. Let (X, ω) be a modular metric space. Let C be a nonempty subset of X_ω. Assume that C is ω-complete. Let $T : C \to C$ be a pointwise ω-contraction. Let a be a fixed point of T. Then the orbit $\{T^n(x)\}$ ω-converges to a, for each $x \in C$ such that $\omega_1(x, a) < +\infty$. Indeed, we have

$$\omega_1(a, T^n(x)) = \omega_1(T^n(a), T^n(x)) \leq \alpha(a)^n \, \omega_1(a, x),$$

for any $n \geq 1$. Since $\alpha(a) < 1$, we conclude that $\{T^n(x)\}$ ω-converges to a. Moreover, if b is another fixed point of T such that $\omega_1(a, b) < +\infty$, then we must have $a = b$. But it is not clear how to prove the existence of any fixed point. In this case, we have to use a different technique than the one used in the proof of Theorem 8.1. Let (X, ω) be a modular metric space. Assume that ω satisfies the Fatou property. For any $x \in X_\omega$ and $r \geq 0$, define the modular ball

$$B_\omega(x, r) = \{y \in X_\omega : \omega_1(x, y) \leq r\}.$$

Note that since ω satisfies the Fatou property, then modular balls are ω-closed. An admissible subset of X_ω is defined as an intersection of modular balls. Denote by $\mathscr{A}_\omega(X_\omega)$ the family of admissible subsets of X_ω. Observe that $\mathscr{A}_\omega(X_\omega)$ is stable under intersection. For a subset A of a modular metric space X_ω, set:

$$\text{cov}_\omega(A) = \bigcap \Big\{ B : B \text{ is a modular ball and } A \subset B \Big\}.$$

Note that if A is ω-bounded, i.e., $\delta_\omega(A) = \sup\{\omega_1(x, y) : x, y \in A\} < +\infty$, then $\text{cov}_\omega(A)$ is well defined and $\text{cov}_\omega(A) \in \mathscr{A}_\omega(X_\omega)$.

Definition 8.7. Let (X, ω) be a modular metric space. We will say that $\mathscr{A}_\omega(X_\omega)$ is compact if any family $\{A_\alpha\}_{\alpha \in \Gamma}$ of elements of $\mathscr{A}_\omega(X_\omega)$ has a nonempty intersection provided $\bigcap_{\alpha \in F} A_\alpha \neq \emptyset$ for any finite subset $F \subset \Gamma$.

Remark 8.5. Very often we only need a weaker version of compactness. Indeed, let (X, ω) be a modular metric space. We will say that $\mathscr{A}_\omega(X_\omega)$ satisfies the property (R) if any decreasing sequence of nonempty elements $\{A_n\}_{n \geq 1}$ in $\mathscr{A}_\omega(X_\omega)$ has a nonempty intersection.

As for the classical metric spaces, we have the following technical lemma.

Lemma 8.1. *Let (X, ω) be a modular metric space. Assume that $\mathscr{A}_\omega(X_\omega)$ is compact. Then X_ω is ω-complete.*

Proof. Let $\{x_n\}$ be an ω-Cauchy sequence in X_ω. Set

$$r_n = \sup_{m,s \geq n} \omega_1(x_m, x_s),$$

for any $n \geq 1$. Since $\{x_n\}$ is an ω-Cauchy sequence, then $\lim_{n\to\infty} r_n = 0$. By definition of $\{r_n\}$, we get $x_m \in B_\omega(x_n, r_n)$, for any $n \geq 1$ and $m \geq n$. Hence for any
$n_1, n_2, \cdots, n_p \geq 1$, we have

$$x_m \in \bigcap_{1 \leq i \leq p} B_\omega(x_{n_i}, r_{n_i}),$$

for any $m \geq \max\{n_1, n_2, \cdots, n_p\}$. Since $\mathscr{A}_\omega(X_\omega)$ is compact, then

$$\Omega = \bigcap_{n \geq 1} B_\omega(x_n, r_n) \neq \emptyset.$$

If $z \in \Omega$, then we have $\omega_1(x_n, z) \leq r_n$, for any $n \geq 1$. Hence $\{x_n\}$ ω-converges to z,
which completes the proof of Lemma 8.1.
$\square$

The following result is the modular metric analogue to the main result on pointwise
contraction mappings of [125, 127].

Theorem 8.2. *Let* (X, ω) *be a modular metric space. Let* C *be a nonempty* ω-closed
ω-bounded subset of X_ω. Assume that the family $\mathscr{A}_\omega(C)$ is compact. Let $T : C \to C$
be a pointwise ω-contraction. Then T has a unique fixed point $x_0 \in C$. Moreover,
the orbit $\{T^n(x)\}$ ω-converges to x_0, for each $x \in C$.*

Proof. Since $\mathscr{A}_\omega(C)$ is compact, therefore there exists a minimal nonempty
$K \in \mathscr{A}_\omega(C)$ such that $T(K) \subset K$. It is easy to check that $\text{cov}_\omega(T(K)) = K$. Let
us prove that $\delta_\omega(K) = \sup\{\omega(x,y) : x, y \in K\} = 0$, i.e., K is reduced to one point.
Indeed, since C is ω-bounded, then $\delta_\omega(K) < +\infty$, i.e. K is also ω-bounded. As T is
a pointwise ω-contraction, so there exists a mapping $\alpha : C \to [0, 1)$ such that

$$\omega_1(T(x), T(y)) \leq \alpha(x)\, \omega_1(x, y), \quad \text{for any } x, y \in C.$$

For any $x \in K$, set $r_x(K) = \sup\{\omega_1(x, y) : y \in K\}$. We have $r_x(K) \leq \delta_\omega(K)$, for
any $x \in K$. Moreover, it is easy to check that $T(K) \subset B_\omega(T(x), \alpha(x)r_x(K))$, for any
$x \in K$. Hence $\text{cov}_\omega(T(K)) \subset B_\omega(T(x), \alpha(x)r_x(K))$. So

$$r_{T(x)}(K) \leq \alpha(x)\, r_x(K),$$

for any $x \in K$. Next we fix $a \in K$ and define

$$K_a = \{x \in K : r_x(K) \leq r_a(K)\}.$$

Clearly, K_a is not empty since $a \in K_a$. Moreover, we have

$$K_a = \bigcap_{x \in K} B_\omega(x, r_a(K)) \cap K \in \mathscr{A}_\omega(C).$$

And, since $r_{T(x)}(K) \le \alpha(x)\, r_x(K)$, for any $x \in K$, we get $T(K_a) \subset K_a$. The minimality behavior of K implies $K_a = K$. In particular, we have $r_x(K) = r_a(K)$, for any $x \in K$. Hence $\delta_\omega(K) = \sup\limits_{x \in K} r_x(K) = r_a(K)$, for any $a \in K$. Hence $\delta_\omega(K) \le \alpha(a)\, \delta_\omega(K)$.
And since $\alpha(a) < 1$, we get $\delta_\omega(K) = 0$, i.e., K is reduced to one point which is fixed by T. Hence the fixed point set of T is not empty. The remaining conclusion of Theorem 8.2 follows from the general properties of pointwise ω-contractions.
□

Next, we extend the previous theorems to the case of asymptotically pointwise contractions in modular metric spaces.

Definition 8.8. Let (X, ω) be a modular metric space. Let C be a nonempty subset of X_ω. A mapping $T : C \to C$ is called an *asymptotic pointwise ω-Lipschitzian mapping* if there exists a sequence of mappings $\alpha_n : C \to [0, +\infty)$ such that

$$\omega_1(T^n(x), T^n(y)) \le \alpha_n(x)\, \omega_1(x, y), \quad \text{for any } x, y \in C.$$

(i) if $\{\alpha_n\}$ converges pointwise to $\alpha : C \to [0, 1)$, then T is called an *asymptotic pointwise ω-contraction* ;
(ii) if there exists $k \in [0, 1)$ such that $\limsup\limits_{n \to +\infty} \alpha_n(x) \le k$, for any $x \in C$, then T is called a *strongly asymptotic pointwise ω-contraction* .

An extension of Theorem 8.2 to the case of asymptotic pointwise ω-Lipschitzian mappings may be stated as:

Theorem 8.3. *Let (X, ω) be a modular metric space. Assume ω is regular. Let C be a nonempty ω-closed ω-bounded subset of X_ω. Assume that the family $\mathscr{A}_\omega(C)$ is compact and $T : C \to C$ is a strongly asymptotic pointwise ω-contraction. Then T has a unique fixed point z. Moreover any orbit $\{T^n(x)\}$ converges to z, for each $x \in C$.*

Proof. Let $x \in C$ and define the function

$$\Phi(a) = \limsup_{n \to \infty} \omega_1(T^n(x), a),$$

for any $a \in C$. Since $\mathscr{A}_\omega(C)$ is compact, then

$$\Omega(x) = \bigcap_{n \ge 1} \mathrm{cov}_\omega\Big(\{T^k(x) : k \ge n\}\Big) \ne \emptyset.$$

For any $n, m, h \ge 1$, we have

$$\omega_1\Big(T^{m+n+h}(x), T^{m+h}(x)\Big) \le \alpha_h(T^m(x))\, \omega_1\Big(T^n(x), T^m(x)\Big).$$

If we let n go to infinity, we get

$$\Phi(T^{m+h}(x)) \leq \alpha_h(T^m(x)) \, \Phi(T^m(x)).$$

Next, we let h go to infinity to get

$$\limsup_{n\to\infty} \Phi(T^n(x)) \leq k \, \Phi(T^m(x)),$$

for some $k \in [0,1)$, which easily implies that $\limsup_{n\to\infty} \Phi(T^n(x)) = 0$. Fix $z \in \Omega(x)$ and notice that

$$\Phi(z) \leq \limsup_{n\to\infty} \Phi(T^n(x)).$$

Indeed let $a \in C$, then for any $\varepsilon > 0$, there exists $n_0 \geq 1$ such that for any $n \geq n_0$, we have

$$\omega_1(T^n(x),a) \leq \Phi(a) + \varepsilon.$$

In particular, we have $T^n(x) \in B_\omega(a, \Phi(a)+\varepsilon)$, for any $n \geq n_0$. So

$$\Omega(x) \subset \mathrm{cov}_\omega\Big(\{T^n(x); n \geq n_0\}\Big) \subset B_\omega(a, \Phi(a)+\varepsilon),$$

which implies $z \in B_\omega(a, \Phi(a)+\varepsilon)$. This is true for any $\varepsilon > 0$. Hence for any $a \in C$, we have $\omega_1(z,a) \leq \Phi(a)$. Therefore,

$$\Phi(z) = \limsup_{n\to\infty} \omega_1(T^n(x),z) \leq \limsup_{n\to\infty} \Phi(T^n(x)).$$

Therefore, we have $\Phi(z) = 0$, i.e., $\{T^n(x)\}$ ω-converges to z. This will force z to be a fixed point of T. Indeed, we have

$$\omega_2(z,T(z)) \leq \omega_1(z,T^n(x)) + \omega_1(T(z),T^n(x)),$$

which implies

$$\omega_2(z,T(z)) \leq \omega_1(z,T^n(x)) + \alpha_1(z) \, \omega_1(z,T^{n-1}(x)),$$

for any $n \geq 1$. Since $\{T^n(x)\}$ ω-converges to z, we get $\omega_2(z,T(z)) = 0$, i.e., $T(z) = z$, since ω is regular. Let $c \in C$ be another fixed point of T. Since C is ω-bounded, then we have $\omega_1(c,z) < +\infty$. Since

$$\omega_1(c,z) = \omega_1(T^n(c),T^n(z)) \leq \alpha_n(z) \, \omega_1(c,z),$$

for any $n \geq 1$. Hence

$$\omega_1(c,z) \leq \limsup_{n\to+\infty} \alpha_n(z) \, \omega_1(c,z) \leq k \, \omega_1(c,z).$$

Since $k < 1$, we get $\omega_1(c,z) = 0$. Because ω is regular, we get $z = c$. Therefore T has one fixed point z in C and any orbit of T ω-converges to z.
$\square$

Now we will relax the strong behavior of T. Before we state the next result, we need the definition of convex functionals defined on modular metric spaces.

Definition 8.9. Let (X, ω) be a modular metric space. Let $C \subset X_\omega$ be nonempty. A function $\Phi : C \to [0, +\infty]$ is said to be ω-convex if

$$\{x \in C : \ \Phi(x) \leq r\} \in \mathscr{A}_\omega(C), \quad for \ any \ \ r \in [0, +\infty).$$

We will say that C is *type-stable* if for any $\{x_n\}$ ω-bounded sequence in C, the function $\Phi : C \to [0, +\infty]$ defined by

$$\Phi(u) = \limsup_{n \to \infty} \ \omega_1(x_n, u),$$

is ω-convex.

Theorem 8.4. *Let (X, ω) be a modular metric space. Let C be a nonempty ω-closed ω-bounded subset of X_ω. Assume that the family $\mathscr{A}_\omega(C)$ is compact. Assume that $\mathscr{A}_\omega(C)$ is type-stable. Let $T : C \to C$ be an asymptotic pointwise ω-contraction. Then T has a unique fixed point $z \in C$. Moreover the orbit $\{T^n(x)\}$ converges to z, for each $x \in C$.*

Proof. As we did in the proof of Theorem 8.3, let $x \in C$ and define the function

$$\Phi(a) = \limsup_{n \to \infty} \omega_1(T^n(x), a),$$

for each $a \in C$. Since $\mathscr{A}_\omega(C)$ is compact and type stable, then there exists $z \in C$ such that

$$\Phi(z) = \inf\{\Phi(a) : a \in C\}.$$

Let us show that $\Phi(z) = 0$. Indeed, we have

$$\omega_1(T^{n+m}(x), T^m(z)) \leq \alpha_m(z) \ \omega_1(T^n(x), z),$$

for any $n, m \geq 1$. If we let n goes to infinity, we get

$$\Phi(T^m(z)) \leq \alpha_m(z) \ \Phi(z),$$

which implies

$$\Phi(z) = \inf\{\Phi(a) : a \in M\} \leq \Phi(T^m(z)) \ \leq \alpha_m(z) \ \Phi(z).$$

If we let m go to infinity, we get $\Phi(z) \leq \alpha(z) \ \Phi(z)$. Since $\alpha(z) < 1$, we get $\Phi(z) = 0$, which implies that $\{T^n(x)\}$ ω-converges to z. This will force z to be a fixed point of T. The end of the proof of Theorem 8.4 follows the same steps as the end of the proof of Theorem 8.3.

$\square$

8.3 Nonexpansive Mappings in Modular Metric Spaces

In this section, we study the fixed point theory of mappings which are nonexpansive in the modular sense. The main result one will try to prove in this setting is the Kirk's fixed point theorem [120]. A key property used in Kirk's theorem is the normal structure property. The following notations will be used throughout this section. Let (X, ω) be a modular metric space. Let C be a nonempty subset of X_ω. For any $A \in \mathscr{A}_\omega(C)$, not reduced to one point, ω-bounded, write

$$R_\omega(A) = \inf_{x \in A} r_x(A),$$

where $r_x(A) = \sup_{y \in A} \omega_1(x,y)$, and $\delta_\omega(A) = \sup_{x,y \in A} \omega_1(x,y)$.

Definition 8.10. Let (X, ω) be a modular metric space. Let C be a nonempty subset of X_ω. We will say that $\mathscr{A}_\omega(C)$ is *normal* if for any $A \in \mathscr{A}_\omega(C)$, not reduced to one point, ω-bounded, we have

$$R_\omega(A) < \delta_\omega(A).$$

We will say that $\mathscr{A}_\omega(C)$ is *uniformly normal* if there exists $c \in (0,1)$ such that for any $A \in \mathscr{A}_\omega(C)$, not reduced to one point, ω-bounded, we have

$$R_\omega(A) \leq c\, \delta_\omega(A).$$

Theorem 8.5. *Let (X, ω) be a modular metric space. Let C be a nonempty ω-closed ω-bounded subset of X_ω. Assume that the family $\mathscr{A}_\omega(C)$ is normal and compact. Let $T : C \to C$ be ω-nonexpansive. Then T has a fixed point.*

Proof. Since $\mathscr{A}_\omega(C)$ is compact there exists a minimal nonempty $A \in \mathscr{A}_\omega(C)$ such that $T(A) \subset A$. It is easy to check that $\mathrm{cov}_\omega(T(A)) = A$. Let us prove that $\delta_\omega(A) = 0$, i.e., A is reduced to one point. Suppose that $\delta_\omega(A) \neq 0$. For any $x \in A$, set

$$r_x(A) = \sup\{\omega_1(x,y) : y \in A\} \leq \delta_\omega(A) \leq \delta_\omega(C) < +\infty.$$

Since T is ω-nonexpansive, we have $T(A) \subset B_\omega\Big(T(x), r_x(A)\Big)$, for any $x \in A$. Hence

$$\mathrm{cov}_\omega(T(A)) \subset B_\omega\Big(T(x), r_x(A)\Big).$$

So $r_{T(x)}(A) \leq r_x(A)$, for any $x \in A$. Next we fix $a \in A$ and define

$$A_a = \{x \in A : r_x(A) \leq r_a(A)\}.$$

Clearly A_a is not empty since $a \in A_a$. Moreover, we have

$$A_a = \bigcap_{x \in A} B_\omega(x, r_a(A)) \cap A \in \mathscr{A}_\omega(C).$$

And since $r_{T(x)}(A) \leq r_x(A)$, for any $x \in A$, we get $T(A_a) \subset A_a$. The minimality behavior of A implies $A_a = A$. In particular, we have $r_x(A) = r_a(A)$ for any $x \in A$. Hence $\delta_\omega(A) = \sup_{x \in A} r_x(A) = r_a(A)$, for any $a \in A$. Since $\mathscr{A}_\omega(C)$ is normal, we get $\delta_\omega(A) < \delta_\omega(A)$, which is a contradiction. Thus we must have $\delta_\omega(A) = 0$, i.e., A is reduced to one point which is fixed by T.
$\square$

Next we give a constructive result discovered by Kirk [123] which relaxes the compactness assumption in the above theorem. The main ingredient in Kirk's constructive proof is a technical lemma due to Gillespie and Williams [77].

Lemma 8.2. *Let* (X, ω) *be a modular metric space and* C *be a nonempty* ω-*bounded subset of* X_ω. *Let* $T : C \to C$ *be a* ω-*nonexpansive mapping. Assume that* $\mathscr{A}_\omega(C)$ *is normal. Let* $A \in \mathscr{A}_\omega(C)$ *be nonempty and* T-*invariant, i.e.,* $T(A) \subset A$. *Then there exists a nonempty* $A_0 \in \mathscr{A}_\omega(C)$, $A_0 \subset A$, *which is* T-*invariant such that*

$$\delta_\omega(A_0) \leq \frac{\delta_\omega(A) + R_\omega(A)}{2}.$$

Proof. Set $r = \frac{1}{2}(\delta_\omega(A) + R_\omega(A))$. We assume that $\delta_\omega(A) > 0$, otherwise we can take the set $A = A_0$. Since $\mathscr{A}_\omega(C)$ is normal, we have $R_\omega(A) < \delta_\omega(A)$. Hence $R_\omega(A) < r$, which implies the existence of $a \in A$ such that $r_a(A) < r$. Therefore, the set

$$D = \{a \in A : A \subset B_\omega(a, r)\} = \bigcap_{x \in A} B_\omega(x, r) \cap A$$

is a nonempty admissible subset of C. Note that there is no reason for D to be T-invariant. Consider the family

$$\mathscr{F} = \{M \in \mathscr{A}_\omega(C) : D \subset M \ and \ T(M) \subset M\}.$$

Note that $\mathscr{F}$ is nonempty, since $A \in \mathscr{F}$. Set $L = \bigcap_{M \in \mathscr{F}} M$. The set L is an admissible subset of C which contains D. Note that $L \subset A$. Using the definition of $\mathscr{F}$, we deduce that L is T-invariant. Consider $B = D \cup T(L)$, and observe that $cov_\omega(B) = L$. Indeed, since $B \subset T(L) \subset L$ and $L \in \mathscr{A}_\omega(C)$, we have $cov_\omega(B) \subset L$. From this we obtain

$$T(cov_\omega(B)) \subset T(L) \subset B \subset cov_\omega(B).$$

Hence $cov_\omega(B) \in \mathscr{F}$, and $L \subset cov_\omega(B)$. This gives the desired equality. Define

$$A_0 = \{x \in L : L \subset B_\omega(x, r)\}.$$

We claim that A_0 is the desired set. Observe that A_0 is nonempty since it contains D (by definition of D). Using the same argument we can prove that A_0 is an admissible subset of C. On the other hand, it is clear that $\delta_\omega(A_0) \leq r$. To complete the proof, we have to show that A_0 is T-invariant. Let $x \in A_0$. By definition of A_0, we have $L \subset B_\omega(x,r)$. Since T is ω-nonexpansive, we have

$$T(L) \subset B_\omega(T(x),r).$$

For any $y \in D$, there holds $L \subset B_\omega(y,r)$. But $T(x) \in L$, so $T(x) \in B_\omega(y,r)$, which implies $y \in B_\omega(T(x),r)$. Hence $D \subset B_\omega(T(x),r)$ holds. Since $B = D \cup T(L)$, we get $B \subset B_\omega(T(x),r)$. Therefore, we must have

$$cov_\omega(B) = L \subset B_\omega(T(x),r).$$

By the definition of A_0, it follows that $T(x) \in A_0$. In other words, A_0 is T-invariant. Let $x,y \in A_0$, then $x,y \in L$, which implies $\omega_1(x,y) \leq r_x(L) \leq r$, i.e.,

$$\delta_\omega(A_0) \leq r = \frac{\delta_\omega(A) + R_\omega(D)}{2}.$$

$\square$

Next we give the analogue of the main fixed point result in [123] and of Theorem 5.8 in this book.

Theorem 8.6. *Let (X,ω) be an ω-complete modular metric space and C be a nonempty ω-closed ω-bounded subset of X_ω. Assume that the family $\mathscr{A}_\omega(C)$ is uniformly normal and $T : C \to C$ is ω-nonexpansive. Then T has a fixed point.*

Proof. Since C is ω-bounded, then we have $C \in \mathscr{A}_\omega(C)$ because

$$C = B_\omega(x,\delta_\omega(C)) \cap C, \; for \; any \; x \in C.$$

Now, let us take $C = A_0$ and $T(C) \subset C$. Since $\mathscr{A}_\omega(C)$ is uniformly normal, there exists $c \in (0,1)$ such that

$$R_\omega(A) \leq c\,\delta_\omega(A), \; for \; any \; A \in \mathscr{A}_\omega(C).$$

By Lemma 8.2, there exists $A_1 \in \mathscr{A}_\omega(C)$ such that $A_1 \subset A_0$, $T(A_1) \subset A_1$ and it satisfies

$$\delta_\omega(A_1) \leq \frac{R(A_0) + \delta_\omega(A_0)}{2}.$$

Using the induction argument, we build a sequence $\{A_n\} \subset \mathscr{A}_\omega(C)$ such that $A_{n+1} \subset A_n$, $T(A_{n+1}) \subset A_{n+1}$ and

$$\delta_\omega(A_{n+1}) \leq \frac{R(A_n) + \delta_\omega(A_n)}{2}.$$

Since $R_\omega(A_n) \leq c\, \delta_\omega(A_n)$, we get

$$\delta_\omega(A_{n+1}) \leq \left(\frac{1+c}{2}\right) \delta_\omega(A_n),$$

which implies that

$$\delta_\omega(A_n) \leq \left(\frac{1+c}{2}\right)^n \delta_\omega(C).$$

Since $\delta_\omega(C) < +\infty$, we get

$$\lim_{n\to\infty} \delta_\omega(A_n) = 0.$$

Let $x_n \in A_n$, for any $n \geq 1$. Then $\{x_n\}$ is ω-Cauchy sequence. Since X_ω is ω-complete and C is ω-closed subset of X_ω, then C is ω-complete. Thus $\{x_n\}$ ω-converges to $x \in C$. Since $A_{n+1} \subset A_n$, and A_n is ω-closed, for any $n \geq 1$, then $x \in A_n$, for any $n \geq 1$. Thus $A_\infty = \bigcap\{A_n : n \geq 1\}$ is not empty and clearly is reduced to the single point x. Indeed, let $y \in A_\infty$, then $y \in A_n$, for any $n \geq 1$. Hence

$$\omega_1(x,y) \leq \delta_\omega(A_n).$$

Since $\lim_{n\to\infty} \delta_\omega(A_n) = 0$, we get $\omega_1(x,y) = 0$. Since ω is regular, we get $y = x$, i.e., $A_\infty = \{x\}$. Since $T(A_n) \subset A_n$, for any $n \geq 1$, we get $T(x) = x$.
$\square$

We need the following technical proposition in order to show an analogue to the main result in [105].

Proposition 8.1. *Let (X,ω) be an ω-complete modular metric space and C be a nonempty ω-closed ω-bounded subset of X_ω. Assume that the family $\mathscr{A}_\omega(C)$ is uniformly normal. Consider the cartesian product $C_\infty = \prod_{n\geq 1} C$. Define*

$$\Omega : (0,\infty) \times C \times C \to [0,\infty]$$

by

$$\Omega_\lambda\Big((x_n),(y_n)\Big) = \sup_{n\geq 1} \omega_\lambda(x_n,y_n).$$

Then

(i) (C_∞,Ω) is an Ω-complete modular metric space.
(ii) C_∞ is Ω-bounded with $\delta_\Omega(C_\infty) = \delta_\omega(C)$.
(iii) For any $(x_n) \in C_\infty$ and $r \in [0,\infty]$, we have

$$B_\Omega\Big((x_n),r\Big) = \prod_{n\geq 1} B_\omega(x_n,r),$$

where

$$B_\Omega\Big((x_n),r\Big) = \Big\{(y_n) \in C_\infty : \Omega_1\Big((x_n),(y_n)\Big) \leq r\Big\}.$$

Then, for any $A \in \mathscr{A}_\Omega(C_\infty)$, we have $A = \prod_{n \geq 1} A_n$, where $A_n \in \mathscr{A}_\Omega(C)$.

(iv) $\mathscr{A}_\Omega(C_\infty)$ is uniformly normal.

Proof. The proofs of (i), (ii), (iii) are easy. Let us prove (iv). Let $A \in \mathscr{A}_\Omega(C_\infty)$ be nonempty and not reduced to one point. Then (iii) implies that $A = \prod_{n \geq 1} A_n$, where $A_n \in \mathscr{A}_\omega(C)$. Let $\varepsilon > 0$. Since $\mathscr{A}_\omega(C)$ is uniformly normal, there exists $c \in [0,1)$ such that for any $n \geq 1$ for which A_n is not reduced to one point, there exists $x_n \in A_n$ such that

$$r_{x_n}(A_n) \leq (c+\varepsilon)\, \delta_\omega(A_n).$$

Hence

$$r_{(x_n)}(A) = \sup_{(y_n) \in A} \Omega\Big((x_n), (y_n)\Big) = \sup_{(y_n) \in A}\Big(\sup_{n \geq 1} \omega_1(x_n, y_n)\Big),$$

which implies

$$r_{(x_n)}(A) \leq (c+\varepsilon)\, \sup_{n \geq 1} \delta_\omega(A_n) = (c+\varepsilon)\, \delta_\Omega(A).$$

So $R_\Omega(A) \leq (c+\varepsilon)\, \delta_\Omega(A)$. Since ε was arbitrary, we get $R_\Omega(A) \leq c\, \delta_\Omega(A)$. This completes the proof of (iv). $\qquad\square$

The following theorem shows that although we do not need compactness of the family of admissible sets in Theorem 8.6, its assumptions imply a weaker form of compactness, mainly countable compactness or the property (R).

Theorem 8.7. *Let (X, ω) be an ω-complete modular metric space and C be a nonempty ω-closed ω-bounded subset of X_ω. Assume that the family $\mathscr{A}_\omega(C)$ is uniformly normal. Then $\mathscr{A}_\omega(C)$ has the property (R).*

Proof. Let $\{A_n\}$ be a decreasing sequence of a nonempty subsets of C, with $A_n \in \mathscr{A}_\omega(C)$. Consider the modular metric space (C_∞, Ω) defined in Proposition 8.1. Set $A = \prod_n A_n$. Then $A \in \mathscr{A}_\Omega(C_\infty)$. Since $\mathscr{A}_\Omega(C_\infty)$ is uniformly normal, then $\mathscr{A}_\Omega(A)$ is uniformly normal. Consider the shift $T : A \to A$ defined by

$$T((x_n)) = (x_{n+1}).$$

Obviously T is ω-nonexpansive. Theorem 8.6 implies that T has a fixed point, i.e., there exists $(x_n) \in A$ such that $T((x_n)) = (x_n)$. The definition of T will force $\{x_n\}$ to be a constant sequence, i.e., $x_n = x$, for any $n \geq 1$. Obviously, we have $x \in A_n$, for any $n \geq 1$, which implies $\bigcap_{n \geq 1} A_n \neq \emptyset$. $\qquad\square$

References

1. Abdou, A.A.N.: On asymptotic pointwise contractions in modular metric spaces. Abstr. Appl. Anal. **2013**, Article ID 501631 (2013). http://dx.doi.org/10.1155/2013/501631
2. Abdou, A.A.N., Khamsi, M.A.: Fixed point results of pointwise contractions in modular metric spaces. Fixed Point Theory Appl. **2013**, 163 (2013)
3. Abdou, A.A.N., Khamsi, M.A.: On the fixed points of nonexpansive maps in modular metric spaces. Fixed Point Theory Appl. **2013**, 229 (2013)
4. Akimovic, B.A.: On uniformly convex and uniformly smooth Orlicz spaces. Teor. Funkc. Funkcional. Anal. i Prilozen. **15**, 114–220 (1972)
5. Aksoy, A.G., Khamsi, M.A.: Nonstandard Methods in Fixed Point Theory, 139 pp. Springer, New York (1990)
6. Alaoui, M.K.: On elliptic equations in Orlicz spaces involving natural growth term and measure data. Abstr. Appl. Anal. **2012**, 615816 (2012)
7. Al-Mezel, S.A., Al-Roqi, A., Khamsi, M.A.: One-local retract and common fixed point in modular function spaces. Fixed Point Theory Appl. **2012**, 109 (2012)
8. Alspach, D.: A fixed point free nonexpansive mapping. Proc. Am. Math. Soc. **82**, 423–424 (1981)
9. Alsulami, S.M., Kozlowski, W.M.: On the set of common fixed points of semigroups of nonlinear mappings in modular function spaces. Fixed Point Theory Appl. **2014**, 4 (2014)
10. Aronszajn, N., Panitchpakdi, P.: Extensions of uniformly continuous transformations and hyperconvex metric spaces. Pac. J. Math. **6**, 405–439 (1956)
11. Ayerbe, J.M., Dominguez-Benavides, T., Lopez-Acedo, G.: Measures of Noncompactness in Metric Fixed Point Theory. Birkhauser, Basel (1997)
12. Bae, J.S., Park, M.S.: Fixed point theorems in metric spaces with uniform normal structure. J. Korean Math. Soc. **30**, 51–62 (1993)
13. Baillon, J.B.: Nonexpansive mappings and hyperconvex spaces. Contemp. Math. **72**, 11–19 (1988)
14. Banach, S.: Sur les opérations dans les ensembles abstraits et leurs applications. Fund. Math. **3**, 133–181 (1922)
15. Batt, J.: Nonlinear integral operators on C(S, E). Studia Math. **48**(2), 145–177 (1973)
16. Beauzamy, B.: Introduction to Banach Spaces and Their Geometry. North-Holland, Amsterdam (1985)
17. Banas, J., Goebel, K.: Measures of noncompactness in Banach spaces. Lecture Notes in Pure and Applied Mathematics, vol. 60. Marcel Dekker, New York (1980)
18. Beg, I.: Inequalities in metric spaces with applications, Top. Meth. Nonlinear Anal. **17**, 183–190 (2001)
19. Belluce, L.P., Kirk, W.A.: Fixed-point theorems for families of contraction mappings. Pac. J. Math. **18**, 213–217 (1966)

20. Belluce, L.P., Kirk, W.A.: Nonexpansive mappings and fixed-points in Banach spaces. Ill. J. Math. **11**, 474–479 (1967)
21. Belluce, L.P., Kirk, W.A., Steiner, E.F.: Normal structure in Banach spaces. Pac. J. Math. **26**, 433–440 (1968)
22. Benkirane, A., Sidi El Vally, M.: An existence result for nonlinear elliptic equations in Musielak-Orlicz-Sobolev spaces. Bull. Belg. Math. Soc. Simon Stevin **20**(1), 57–75 (2013)
23. Berggren, J.L.: Episodes in the Mathematics of Medieval Islam, 197 pp. Springer, New York (1986)
24. Berinde, V.: Iterative Approximation of Fixed Points, 2nd edn. Springer, Berlin (2007)
25. Besbes, M.: Points fixes des contractions définies sur un convexe L^0-fermé de L^1. C. R. A. Sc. de Paris, Serie I **311**, 243–246 (1990)
26. Bielecki, A.: Une remarque sur l'application de la méthode de Banach-Cacciopoli-Tikhonov dans la theorie de l'équation s = f(x, y, z, p, q). Bull. Acad. Polon. Sci. Sér. Sci. Math. Phys. Astr. **4**, 265–268 (1956)
27. Brezis, H., Lieb, E.: A relation between pointwise convergence of functions and convergence of functionals. Proc. Am. Math. Soc. **88-3**, 486–490 (1983)
28. Bridson M., Haefliger, A.: Metric Spaces of Non-positive Curvature. Springer, Berlin (1999)
29. Birnbaum, Z., Orlicz, W.: Uber die Verallgemeinerung des Begriffes der zueinander konjugierten Potenzen. Studia Math. **3**, 1–67 (1931)
30. Brodskii, M.S., Milman, D.P.: On the center of a convex set. Dokl. Akad. Nauk SSSR **59**, 837–840 (1948) (Russian)
31. Brouwer, L.E.J.: Uber Abbildungen von Mannigfaltigkeiten. Math. Ann. **71**, 97–115 (1912)
32. Browder, F.E.: Nonexpansive nonlinear operators in a Banach space. Proc. Natl. Acad. Sci. U. S. A. **54**, 1041–1044 (1965)
33. Browder, F.E., Petryshyn, W.V.: The solution by iteration of nonlinear functional equations in Banach spaces. Bull. Am. Math. Soc. **72**, 571–575 (1966)
34. Bruck, R.E.: Properties of fixed-point sets of nonexpansive mappings in Banach spaces. Trans. Am. Math. Soc. **179**, 251–262 (1973)
35. Bruck, R.E.: A common fixed point theorem for a commuting family of nonexpansive mappings, Pac. J. Math. **53**, 59–71 (1974)
36. Bruck, R., Kuczumow, T., Reich, S.: Convergence of iterates of asymptotically nonexpansive mappings in Banach spaces with the uniform Opial property. Coll. Math. **65**, 169–179 (1993)
37. Bruhat, F., Tits, J.: Groupes réductifs sur un corps local, I. Données radicielles valuées. Inst. Hautes Études Sci. Publ. Math. **41**, 5–251 (1972)
38. Busemann, H.: Spaces with non-positive curvature. Acta Math. **80**, 259–310 (1948)
39. Bynum, W.L.: Normal structure coefficients for normal structure for Banach spaces. Pac. J. Math. **86**, 427–436 (1980)
40. Caristi, J.: The Fixed Point Theory for Mappings Satisfying Inwardness Conditions. Ph. D. Dissertation, Univ. of Iowa (May 1975)
41. Caristi, J.: Fixed point theorems for mappings satisfying inwardness conditions. Trans. Am. Math. Soc. **215**, 241–251 (1976)
42. Casini, E., Maluta, E.: Fixed points of uniformly Lipchitzian mappins in spaces with uniformly normal structure. Nonlinear Anal. **9**, 103–108 (1985)
43. Cerda, J., Hudzik, H., Mastylo, M.: On the geometry of some Calderon-Lozanovskii interpolation spaces. Indag. Math. N. S. **6**(1), 35–49 (1995)
44. Chen, S.: Geometry of Orlicz spaces. Dissertat. Math. **356**, 4–205 (1996)
45. Chidume, C.E., Mutangadura, S.A.: An example on the Mann iteration method for Lipschitz pseudocontractions. Proc. Am. Math. Soc. **129**, 2359–2363 (2001)
46. Chistyakov, V.V.: Modular metric spaces, I: basic concepts. Nonlinear Anal. **72**(1), 1–14 (2010)
47. Chistyakov, V.V.: Modular metric spaces, II: application to superposition operators. Nonlinear Anal. **72**(1), 15–30 (2010)
48. Cianchi, A.: Optimal Orlicz-Sobolev embeddings. Rev. Math. Iberoam. **20**(2), 427–474 (2004)

49. Ćirić, Lj.B.: A generalization of Banach's contraction principle. Proc. Am. Math. Soc. **45**, 267–273 (1974)
50. Clarkson, J.A.: Uniformly convex spaces. Trans. Am. Math. Soc. **40**(3), 396–414 (1936)
51. Crandall, M.C., Pazy, A.: Semigroups of nonlinear contractions and dissipative sets. J. Funct. Anal. **3**, 376–418 (1963)
52. DeGroot, M.H., Rao, M.M.: Multidimentional information inequalities. In: Krishnaiah, P.P. (ed.) Proceedings of an International Symposium on Multivariate Analysis, pp. 287-313. Held in Dayton, Ohio, June 14–19, 1965. Academic, New York (1967)
53. Diening, L.: Theoretical and numerical results for electrorheological fluids. Ph. D. Thesis, University of Freiburg, Germany (2002)
54. Dehaish, B.A., Kozlowski, W.M.: Fixed point iterations processes for asymptotic pointwise nonexpansive mappings in modular function spaces. Fixed Point Theory Appl. **2012**, 118 (2012)
55. Dehaish, B.A., Khamsi, M.A., Kozlowski, W.M.: Common fixed points for asymptotic pointwise Lipschitzian semigroups in modular function spaces. Fixed Point Theory Appl. **2013**, 214 (2013)
56. Dehaish, B.A., Khamsi, M.A., Kozlowski, W.M.: Fixed point processes for pointwise Lipschitzian semigroups in modular function spaces. Fixed Point Theory Appl. **2013**, 214 (2013)
57. DeMarr, R.E.: Common fixed-points for commuting contraction mappings. Pac. J. Math. **13**, 1139–1141 (1963)
58. Dhompongsa, S., Kirk, W.A., Sims, B.: Fixed points of uniformly lipschitzian mappings. Nonlinear Anal. **65**, 762–772 (2006)
59. Diestel, J.: Geometry of Banach Spaces - Selected Topics. Springer Lecture Notes in Mathematics, vol. 485. Springer, Berlin (1975)
60. Dobrakov, I.: On integration in Banach spaces I. Czech. Math. J. **20**, 511–536 (1970)
61. Dobrakov, I.: On integration in Banach spaces II. Czech. Math. J. **20**, 680–695 (1970)
62. Dominguez-Benavides, T.: Fixed point theorems for uniformly Lipschitzian mappings and asymptotically regular mappings. Nonlinear Anal. **32**(1), 15–27 (1998)
63. Dominguez-Benavides, T., Khamsi, M.A., Samadi, S.: Uniformly Lipschitzian mappings in modular function spaces. Nonlinear Anal. **46**, 267–278 (2001)
64. Dominguez-Benavides, T., Khamsi, M.A., Samadi, S.: Asymptotically regular mappings in modular function spaces. Sci. Math. Jpn. **53**, 295–304 (2001)
65. Dominguez-Benavides, T., Khamsi, M.A., Samadi, S.: Asymptotically nonexpansive mappings in modular function spaces, J. Math. Anal. Appl. **265**(2), 249–263 (2002)
66. Dorroch, J.R., Neuberger, J.W.: Linear extensions of nonlinear semigroups. In: Balakrishnan, A.V. (ed.) Semigroups and Operators: Theory and Applications, pp. 96–102. Birkhauser, Basel (2000)
67. Dugundji, J., Granas, A.: Fixed Point Theory. Polska Akademia Nauk, Instytut Matematyczny. PWN-Polish Scientific Publ., Warszawa (1982)
68. van Dulst, D., Sims, B.: Fixed points of nonexpansive mappings and Chebyshev centers in Banach spaces with norms of type (KK), Banach space theory and its applications. Proc. Bucharest 1981. Lecture Notes in Mathematics, vol. 991, pp. 35–43. Springer, Berlin (1983)
69. van Dulst, D., de Valk, V.: (KK)-properties, normal structure and fixed points of nonexpansive mappings in Orlicz spaces. Can. J. Math. **38**, 728–750 (1986)
70. Dunford, N., Schwartz, J.: Linear Operators, Part I. Interscience, New York (1958)
71. Edelstein, M.: The construction of an asymptotic center with a fixed-point property. Bull. Am. Math. Soc. **78**(2), 206–208 (1972)
72. Edelstein, M. and O'Brien, R.C.: Nonexpansive mappings, asymptotic, regularity and successive approximations. J. Lond. Math. Soc. (2) **17**(3), 547–554 (1978)
73. Ekeland, I.: On the variational principle. J. Math. Anal. Appl. **47**, 324–353 (1974)
74. Fréchet, M.: Sur Quelques Points Du Calcul Fonctionnel. Rend. Circ. Mat. Palermo **22**, 1–74 (1906)
75. Friedman, N., Tong, A.E.: On additive operators. Can. J. Math. **23**, 468–480 (1971)
76. Garcia-Falset, J.: The fixed point property in Banach spaces whose characteristic of uniform convexity is less than 2. J. Austral. Math. Soc. **54**, 169–173 (1993)

77. Gillespie, A., Williams, B.: Fixed point theorem for nonexpansive mappings on Banach spaces. Appl. Anal. **9**, 121–124 (1979)
78. Goebel, K.: On the structure of minimal invariant sets for nonexpansive mappings. Ann. Univ. Mariae Curie-Skłodowski **29**, 73–77 (1975)
79. Goebel, K., Kirk, W.A.: A fixed points theorem for asymptotically nonexpansive mappings. Proc. Am. Math. Soc. **35**, 171–174 (1972)
80. Goebel, K., Kirk, W.A.: A fixed point theorem for transformations whose iterates have uniform Lipschitz constant. Studia. Math. **XLVII**, 135–140 (1973)
81. Goebel, K., Kirk, W.A.: Topics in Metric Fixed Point Theory, 244 pp. Cambridge University Press, Cambridge (1990)
82. Goebel, K., Reich, S.: Uniform Convexity, Hyperbolic Geometry, and Nonexpansive Mappings. Series of Monographs and Textbooks in Pure and Applied Mathematics, vol. 83. Dekker, New York (1984)
83. Goebel, K., Sekowski, T.: The modulus of noncompact convexity. Ann. Univ. Mariae Curie-Sklodowska Sect. A **38**, 41–48 (1984)
84. Goebel, K., Sekowski, T., Stachura, A.: Uniform convexity of the hyperbolic metric and fixed points of holomorphic mappings in the Hilbert ball. Nonlinear Anal.: Theory Methods Appl. **4**, 1011–1021 (1980)
85. Göhde, D.: Zum Prinzip der kontraktiven Abbildung. Math. Nachr. **30**, 251–258 (1965)
86. Gossez, J.-P.: Nonlinear elliptic boundary value problems for equations with rapidly (or slowly) increasing coefficients. Trans. Am. Math. Soc. **190**, 163–205 (1974)
87. Gossez, J.-P.: Some approximation properties in Orlicz-Sobolev spaces. Studia Math. **74**(1), 17–24 (1982)
88. Gwiazda, P., Minakowski, P., Wroblewska-Kaminska, A.: Elliptic problems in generalized Orlicz-Musielak spaces. Cent. Eur. J. Math. **10**(6), 2019–2032 (2012)
89. Gwiazda, P., Wittbold, P., Wroblewska, A., Zimmermann, A.: Renormalized solutions of nonlinear elliptic problems in generalized Orlicz spaces. J. Differ. Equ. **253**, 635–666 (2012)
90. Hajji, A., Hanebaly, E.: Perturbed integral equations in modular function spaces. Electron. J. Qual. Theory Diff. Equ. **20**, 1–7 (2003)
91. Halpern, B.: Fixed points of nonexpanding maps. Bull. Am. Math. Soc. **73**, 957–961 (1967)
92. Harjulehto, P., Hasto, P., Koskenoja, M., Varonen, S.: The Dirichlet energy integral and variable exponent Sobolev spaces with zero boundary values. Potential Anal. **25**(3), 205–222 (2006)
93. Heinonen, J., Kilpelainen, T., Martio, O.: Nonlinear potential theory of degenerate elliptic equations. Oxford University Press, Oxford (1993)
94. Huff, R.: Banach spaces which are nearly uniformly convex. Rocky Mt. J. Math. **10**, 743–749 (1980)
95. Hudzik, H., Kaminska, A., Mastylo, M.: Geometric properties of some Calderon-Lozanovskii spaces and Orlicz-Lorentz spaces. Houst. Mt. J. Math. **22**(3), 639–663 (1996)
96. Hussain, N., Khamsi, M.A.: On asymptotic pointwise contractions in metric spaces. Nonlinear Anal. **71**, 4423–4429 (2009)
97. Ishikawa, S.: Fixed points by a new iteration method. Proc. Am. Math. Soc. **44**, 147–150 (1974)
98. Ishikawa, S.: Fixed points and iteration of nonexpansive mappings in Banach space. Proc. Am. Math. Soc. **59** (1976)
99. James, R.C.: Uniformly non-square Banach spaces. Ann. Math. **80**, 542–550 (1964)
100. Japon, M. A.: Some geometric properties in modular spaces and application to fixed point theory. J. Math. Anal. Appl. **295**, 576–594 (2004)
101. Kaminska, A.: On uniform convexity of Orlicz spaces. Indag. Math. **44**, 27–36 (1982)
102. Karlovitz, L.: Existence of fixed points for nonexpansive mappings in spaces without normal structure. Pac. J. Math. **66**, 153–156 (1976)
103. Kato, T.: Nonlinear semigroups and evolution equations. J. Math. Soc. Jpn. **19**, 508–520 (1967)
104. Khamsi, M.A.: Étude de la propriété du point fixe dans les espaces de Banach et les espaces métriques. Thése de Doctorat de l'Université, Paris 6 (1987)

105. Khamsi, M.A.: On metric spaces with uniform normal structure. Proc. Am. Math. Soc. **106**, 723–726 (1989)
106. Khamsi, M.A.: Nonlinear semigroups in modular function spaces. Math. Jpn. **37**, 1–9 (1992)
107. Khamsi, M.A.: Fixed point theory in modular function spaces. Proceedings of the Workshop on Recent Advances on Metric Fixed Point Theory held in Sevilla, September 1995, 31–35 (1995)
108. Khamsi, M.A.: One-local retract and common fixed point for commuting mappings in metric spaces. Nonlinear Anal.: Theory Methods Appl. **27**, 1307–1313 (1996)
109. Khamsi, M.A.: A convexity property in modular function spaces. Math. Jpn. **44**, 269–279 (1996)
110. Khamsi M.A.: On uniform opial condition and uniform Kadec-Klee property in Banach and metric spaces. Nonlinear Anal.: Theory Methods Appl. **26**(10), 1733–1748 (1996)
111. Khamsi, M.A., Kirk, W.A.: An Introduction to Metric Spaces and Fixed Point Theory. Wiley, New York (2001)
112. Khamsi, M.A.: On asymptotically nonexpansive mappings in hyperconvex metric spaces, Proc. Am. Math. Soc. **132**, 365–373 (2004)
113. Khamsi, M.A., Khan, A.R.: Inequalities in metric spaces with applications. Nonlinear Anal.: Theory Methods Appl. **74**, 4036–4045 (2011)
114. Khamsi, M.A., Kozlowski, W.M.: On asymptotic pointwise contractions in modular function spaces. Nonlinear Anal. **73**, 2957–2967 (2010)
115. Khamsi, M.A., Kozlowski, W.M.: On asymptotic pointwise nonexpansive mappings in modular function spaces. J. Math. Anal. Appl. **380**(2), 697–708 (2011)
116. Khamsi, M.A., Kozlowski, W.M., Reich, S.: Fixed point theory in modular function spaces. Nonlinear Anal. **14**, 935–953 (1990)
117. Khamsi, M.A., Kozlowski, W.M., Chen, S.: Some geometrical properties and fixed point theorems in Orlicz spaces. J. Math. Anal. Appl. **155**, 393–412 (1991)
118. Kilmer, S.J., Kozlowski, W.M., Lewicki, G.: Best approximants in modular function spaces. J. Approx. Theory, **63**, 338–367 (1990)
119. Kilmer, S.J., Kozlowski, W.M., Lewicki, G.: Sigma order continuity and best approximants in L_ρ spaces. Comment. Math. Univ. Carolin. **32**(2), 241–250 (1991)
120. Kirk, W.A.: A fixed point theorem for mappings which do not increase distances. Am. Math. Mon. **72**, 1004–1006 (1965)
121. Kirk, W.A.: Mappings of generalized contractive type. J. Math. Anal. Appl. **32**, 567–572 (1970)
122. Kirk, W.A.: Fixed point theory for nonexpansive mappings, I and II. In: Fadell, E., Fournier, G. (eds.) Fixed Point Theory. Lecture Notes in Mathematics, vol. 886, pp. 485–505. Springer, Berlin (1981)
123. Kirk, W.A.: An abstract fixed point theorem for nonexpansive mappings. Proc. Am. Math. Soc. **82**, 640–642 (1981)
124. Kirk, W.A.: Nonexpansive mappings in metric and Banach spaces. Rend. Semin. Mat. Milano, **LI**, 133–144 (1981)
125. Kirk, W.A.: Fixed points of asymptotic contractions. J. Math. Anal. Appl. **277**, 645–650 (2003)
126. Kirk, W.A.: A fixed point theorem in CAT(0) spaces and $\mathbb{R}$-trees. Fixed Point Theory Appl. **2004**(4), 309–316 (2004)
127. Kirk, W.A.: Asymptotic pointwise contractions. In: Plenary Lecture, the 8th International Conference on Fixed Point Theory and Its Applications, Chiang Mai University, Thailand, July 16–22, 2007.
128. Kirk, W.A., Xu, H-K.: Asymptotic pointwise contractions. Nonlinear Anal. **69**, 4706–4712 (2008)
129. Knopp, K.: Infinite Series, pp. 290–293 (1928).
130. Kozlowski, W.M.: Notes on modular function spaces I. Comment. Math. **28**, 91–104 (1988)
131. Kozlowski, W.M.: Notes on modular function spaces II. Comment. Math. **28**, 105–120 (1988)

132. Kozlowski, W.M.: Modular Function Spaces. Series of Monographs and Textbooks in Pure and Applied Mathematics, vol. 122. Marcel Dekker, New York (1988)

133. Kozlowski, W.M.: Fixed point iteration processes for asymptotic pointwise nonexpansive mappings in Banach spaces. J. Math. Anal. Appl. **377**, 43–52 (2011)

134. Kozlowski, W.M.: Common fixed points for semigroups of pointwise Lipschitzian mappings in Banach spaces. Bull. Austral. Math. Soc. **84**, 353–361 (2011)

135. Kozlowski, W.M.: On the existence of common fixed points for semigroups of nonlinear mappings in modular function spaces. Comment. Math. **51**, 81–98 (2011)

136. Kozlowski, W.M.: Advancements in fixed point theory in modular function spaces. Arab J. Math. (2012). doi:10.1007/s40065-012-0051-0

137. Kozlowski, W.M.: On the construction of common fixed points for semigroups of nonlinear mappings in uniformly convex and uniformly smooth Banach spaces. Comment. Math. **52**, 113–136 (2012)

138. Kozlowski, W.M.: Pointwise Lipschitzian mappings in uniformly convex and uniformly smooth Banach spaces. Nonlinear Anal. **84**, 50–60 (2013)

139. Kozlowski W.M.: An Introduction to fixed point theory in modular function spaces. In: Almezel, S., Ansari, Q.H., Khamsi, M.A. (eds.) Topics in Fixed Point Theory. Springer, New York (2014)

140. Kozlowski W.M.: On common fixed points of semigroups of mappings nonexpansive with respect to convex function modulars. J. Nonlinear Convex Anal. **15**(4), 437–449 (2014)

141. Kozlowski, W.M.: On nonlinear differential equations in generalized Musielak-Orlicz spaces. Comment. Math. **53**(2), 113–133 (2013)

142. Kozlowski, W.M., Lewicki, G.: On polynomial approximation in modular function spaces. In: Musielak J. (ed.) Function Spaces, pp. 63–68. Teubner-Texte zur Mathematik, Band 103 (1986)

143. Kozlowski, W.M., Lewicki, G.: Analyticity and polynomial approximation in modular function spaces. J. Approx. Theory **58**(1), 243–267 (1989)

144. Kozlowski, W.M., Sims, B.: On the convergence of iteration processes for semigroups of nonlinear mappings in Banach spaces, In: Bailey, D.H., Bauschke, H.H., Borwein, P., Garvan, F., Thera, M., Vanderwerff, J.D., Wolkowicz, H. (eds.) Computational and Analytical Mathematics. In Honor of Jonathan Borwein's 60th Birthday. Springer Proceedings in Mathematics and Statistics, vol. 50. Springer, New York (2013)

145. Kozlowski, W.M., Szczypinski, T.: Some remarks on on the nonlinear operator measures and integration. Coll. Math. Soc. Jan. Bolyai **35**, 751–756 (1980)

146. Kozlowski, W.M., Szczypinski, T.: Convergences theorems for integrals with respect to nonlinear operator measures. In: Constructive Function Theory 1981, Sofia 1983, 389–392

147. Krasnosel'skii, M.A., Rutickii, Y.B.: Convex Functions and Orlicz Spaces. P. Nordhoff, Groningen, 1961

148. Kuczumow, T.: Weak convergence theorems for nonexpansive mappings and semi-groups in Banach spaces with Opial' s property. Proc. Amer. Math. Soc. **93**, 430–432 (1985)

149. Kulesza, J., Lim, T.C.: On weak compactness and countable weak compactness in fixed point theory. Proc. Amer. Math. Soc. **124**, 3345–3349 (1996)

150. Kuratowski, K.: Sur les espaces complets, Fund. Math. **15** (1930), 301–309

151. Lami-Dozo, E., Turpin, Ph.: Nonexpansive maps in generalized Orlicz spaces. Studia Math. **86**(9), 155–188 (1987)

152. Lennard, C.: A new convexity property that implies a fixed point property for L_1. Studia Math. **100**(2), 95–108 (1991)

153. Lim, T.C.: A fixed point theorem for families of nonexpansive mappings. Pac. J. Math. **53**, 487–493 (1974)

154. Lim, T.C., Xu, H-K.: Uniformly Lipschitzian mappings in metric spaces with uniform normal structure. Nonlinear Anal. **25**(11), 1231–1235 (1995)

155. Lin, P-K., Tan, K-K., Xu, H-K.: Demiclosedness principle and asymptotic behavior for asymptotically nonexpansive mappings. Nonlinear Anal. **24**, 929–946 (1995)

156. Leustean, L.: A quadratic rate of asymptotic regularity for CAT(0)-spaces. J. Math. Anal. Appl. **32**(5), 386–399 (2007)

157. Luxemburg, W.A.J.: Banach Function Spaces, Thesis, Delft (1955)
158. Luxemburg, W.A.J., Zaanen, A.C.: Riesz Spaces I. North-Holland, Amsterdam (1971)
159. Luxemburg, W.A.J., Zaanen, A.C.: Notes on Banach function spaces I - XIII. Proc. Acad. Sci. Amst. **A-66 (1963)** , 135–153, 239–263, 496–504, 655–681; **A-64 (1964)** , 101–119; **A-67 (1964)**, 360–376, 493–543
160. Maluta, E.: Uniformly normal structure and related coefficients. Pac. J. Math. **111**, 357–369 (1984)
161. Mann, W.R.: Mean value methods in iteration. Proc. Am. Math. Soc. **4**, 506–510 (1953)
162. Menger, K.: Untersuchungen über allgemeine Metrik. Math. Ann. **100**, 75–163 (1928)
163. Milnes, H.W.: Convexity of Orlicz spaces. Pac. J. Math. **7**, 1451–1486 (1957)
164. Miyaderau, I.: Nonlinear ergodic theorems for semigroups of non-Lipschitzian mappings in Banach spaces. Nonlinear Anal. **50**, 27–39 (2002)
165. Mongkolkeha, C., Sintunavarat, W., Kumam, P.: Fixed point theorems for contraction mappings in modular metric spaces. Fixed Point Theory Appl. (2011). doi:10.1186/1687-1812-2011-93
166. Musielak, J., Orlicz, W.: On modular spaces. Studia Math. **18**, 49–65 (1959)
167. Musielak, J., Orlicz, W.: Some remarks on modular spaces. Bull. Acad. Polon. Sci. Ser. Sci. Math. Astronom. Phys. **7**, 661–668 (1959)
168. Musielak, J.: An application of modular spaces to approximation. Comment. Math. **I**, 251–259 (1979) (Tomus Specialis in Honorem Ladislai Orlicz)
169. Musielak, J.: Orlicz Spaces and Modular Spaces. Lecture Notes in Mathematics, vol. 1034, Springer, Berlin (1983)
170. Nakano, H.: Modulared Semi-ordered Linear Spaces. Maruzen, Tokyo (1950)
171. Oettli, W., Théra, M.: Equivalents of Ekeland's principle. Bull. Aust. Math. Soc. **48**, 385–392 (1993)
172. Oharu, S.: Note on the representation of semi-groups of non-linear operators. Proc. Jpn. Acad. **42**, 1149–1154 (1967)
173. Opial, Z.: Weak convergence of the sequence of successive approximations for nonexpansive mappings. Bull. Am. Math. Soc. **73**, 591–597 (1967)
174. Orlicz, W.: Uber eine gewisee klasse von Raumen vom Typus B. Bull. Acad. Polon. Sci. Ser. A, 207–220 (1932)
175. Orlicz, W.: Uber Raumen L^M. Bull. Acad. Polon. Sci. Ser. A, 93–107 (1936)
176. Penot, J.P.: Fixed point theorem without convexity, Analyse Non Convexe (1977, Pau). Bull. Soc. Math. Fr., Mem. **60**, 129–152 (1979)
177. Prus, S.: Banach spaces with the uniform Opial property. Nonlinear Anal. **18**, 697–704 (1992)
178. Peng, J., Chung, S-K.: Laplace transforms and generators of semigroups of operators. Proc. Am. Math. Soc. **126**, 2407–2416 (1998)
179. Peng, J., Xu, Z.: A novel approach to nonlinear semigroups of Lipschitz operators. Trans. Am. Math. Soc. **367**, 409–424 (2004)
180. Radon, J.: Theorie und Anwendungen der absolut additiven Mengenfunctionen. Sitz. Akad. Wiss. Wien **122**, 1295–1438 (1913)
181. Rao, M.M., Ren, Z.D.: Theory of Orlicz Spaces. Series of Monographs and Textbooks in Pure and Applied Mathematics, vol. 146, Marcel Dekker, New York (1991)
182. Reich, S.: Weak convergence theorems for nonexpansive mappings in Banach spaces. J. Math. Anal. Appl. **67**, 274–276 (1979)
183. Reich, S.: A note on the mean ergodic theorem for nonlinear semigroups. J. Math. Anal. Appl. **91**, 547–551 (1983)
184. Riesz, F.: Sur la convergence en moyenne I. Acta Sci. Math. **4**, 58–64 (1928/29)
185. Riesz, F.: Sur la convergence en moyenne II. Acta Sci. Math. **4**, 182–185 (1928/29)
186. Rus, I.A., Petrusel, A., Petrusel, G.: Fixed Point Theory, 531 pp. Cluj University Press, Cluj, (2008)
187. Ruzicka, M.: Electrorheological Fluids: Modeling and Mathematical Theory. Lecture Notes in Mathematics, vol. 1748. Springer, Berlin (2000)

188. Schu, J.: Weak and strong convergence to fixed points of asymptotically nonexpansive mappings. Bull. Austral. Math. Soc. **43**, 153–159 (1991)
189. Szczypinski, T.: Non-linear operator valued measures and integration. Comment. Math. **27**, 313–333 (1988)
190. Sine, R.: On nonlinear contractions in sup-norm spaces. Nonlinear Anal.: Theory Methods Appl. **3**, 885–890 (1979)
191. Soardi, P.: Existence of fixed points of nonexpansive mappings in certain Banach lattices. Proc. Am. Math. Soc. **73**, 25–29 (1979)
192. Suzuki, T: The set of common fixed points of one-parameter nonexpansive semigroup of mappings is $F(T(1)) \cap F(T(\sqrt{2}))$. Proc. Am. Math. Soc. **134**(3), 673–681 (2005)
193. Suzuki, T.: Common fixed points of one-parameter nonexpansive semigroup. Bull. Lond. Math. Soc. **38**, 1009–1018 (2006)
194. Suzuki, T., Takahashi, W.: Strong convergence of Mann's type sequences for one-parameter nonexpansive semigroups in general Banach spaces. J. Nonlinear Convex Anal. **5**, 209–216 (2004)
195. Taleb, A.A., Hanebaly, E.: A fixed point theorem and its application to integral equations in modular function spaces. Proc. Am. Math. Soc. **128**, 419–427 (1999)
196. Talenti, G.: Nonlinear elliptic equations, rearrangements of functions and Orlicz spaces. Ann. Mat. Pura Appl. **120**, 159–184 (1979)
197. Takahashi, W.: A convexity in metric space and nonexpansive mappings. I. Kodai Math. Sem. Rep. **22**(2), 142–149 (1970)
198. Tarski, A.: A lattice theoretical fixpoint theorem and its applications. Pac. J. Math. **5**, 285–309 (1955)
199. Tan, K-K., Xu, H-K.: An ergodic theorem for nonlinear semigroups of Lipschitzian mappings in Banach spaces. Nonlinear Anal. **19**, 805–813 (1992)
200. Wittmann, R.: Approximation of fixed points of nonexpansive mappings. Arch. Math. **58**, 486–491 (1992)
201. Xu, H-K.: Inequalities in Banach spaces with applications. Nonlinear Anal.: Theory Methods Appl. **16**, 1127–1138 (1991)
202. Xu, H-K.: Existence and convergence for fixed points of mappings of asymptotically nonexpansive type. Nonlinear Anal. **16**(12), 1139–1146 (1991)
203. Zeidler, E.: Nonlinear Functional Analysis and its Applications I: Fixed-Point Theorems, 897 pp. Springer, New York (1986)

Index

© Springer International Publishing Switzerland 2015
M. A. Khamsi, W. M. Kozlowski, *Fixed Point Theory in Modular Function Spaces*,
DOI 10.1007/978-3-319-14051-3